AF847839

A Risky Business

Catrin Townsend

A Risky Business

An Actuary's Guide to Quantifying and Managing Risk in Society

palgrave
macmillan

Catrin Townsend
Guildford, UK

ISBN 978-3-031-11672-8 ISBN 978-3-031-11673-5 (eBook)
https://doi.org/10.1007/978-3-031-11673-5

© The Editor(s) (if applicable) and The Author(s), under exclusive license to Springer Nature Switzerland AG 2022
This work is subject to copyright. All rights are solely and exclusively licensed by the Publisher, whether the whole or part of the material is concerned, specifically the rights of translation, reprinting, reuse of illustrations, recitation, broadcasting, reproduction on microfilms or in any other physical way, and transmission or information storage and retrieval, electronic adaptation, computer software, or by similar or dissimilar methodology now known or hereafter developed.
The use of general descriptive names, registered names, trademarks, service marks, etc. in this publication does not imply, even in the absence of a specific statement, that such names are exempt from the relevant protective laws and regulations and therefore free for general use.
The publisher, the authors and the editors are safe to assume that the advice and information in this book are believed to be true and accurate at the date of publication. Neither the publisher nor the authors or the editors give a warranty, expressed or implied, with respect to the material contained herein or for any errors or omissions that may have been made. The publisher remains neutral with regard to jurisdictional claims in published maps and institutional affiliations.

Cover illustration: eStudioCalamar

This Palgrave Macmillan imprint is published by the registered company Springer Nature Switzerland AG
The registered company address is: Gewerbestrasse 11, 6330 Cham, Switzerland

For my Mammy, who never understood what actuaries do but was proud of me anyway.

Preface

This book is not for actuaries.

It is for future actuaries, student actuaries, and people who work with (or live with!) actuaries. It is for those interested in practical applications of mathematics and statistics; for those who wish to understand their pension a little better. It is for the curious. It is for anyone even vaguely mathematically inclined who wishes to understand more about how society works, why it sometimes doesn't, and what the solutions could look like.

This book will give you an insight into what actuaries do all day, the sorts of problems they face, the kinds of methods they use, and what the hot topics of the industry are. Actuaries are a remarkably chameleonic group: they have found roles in a wide variety of industries and organizations, but often all but disappear into the background. This book highlights their work, and brings the mysterious art of actuarial science into the limelight. This book can't, and doesn't even try, to cover everything—so if you're a student actuary this book will not cover all the examinable content. Instead, see this as a travel guide to actuarial science; a smorgasbord of activities that actuaries are involved in for readers to explore. Use it to learn about what is most relevant to you, or explore what areas interest you most. Feel free to dip in and out according to your interests or, if you'd rather, read everything in order to build a holistic view of the actuarial profession and how it affects us all. Industry jargon is explained in the glossary (glossary terms are formatted **like this**).

More than anything, I hope this is a book that inspires you. Not necessarily to become an actuary (although the world needs more, and high-caliber, actuaries), but to give you confidence that despite the complicated problems the world faces, actuaries are quantifying the risks and creating workable solutions to make our lives a little more stable, certain, and equitable.

Interested in statistics, insurance fraud, data ethics, global warming, social justice, or societal change?

Then this book is for you.

Guildford, UK Catrin Townsend

Acknowledgments

A huge number of people, actuaries and non-actuaries alike, have contributed to the creation of this book. My heartfelt thanks to you all.

The Non-actuaries

I am grateful to Tula Weiss, Naveen Dass, and the team at Springer Palgrave-Macmillan whose expertise and guidance turned this book from an idea to a real-life book.

Many thanks also to all who provided images and photos included.

To Emily Griffiths, who gave me the numbers to prove that a lot of people google 'what is an actuary' and 'what do actuaries do all day'. Emily, you are a wonderful pal.

To excellent teachers: Neil Blackburn, Tony Reeves, and Dot Holloway, in particular, for encouraging my love of maths.

To Robert Flach, Andrew Colcomb, and Jeremy Bray, who introduced me to my first actuary.

To my wonderful pricing colleagues past and present: Rosie, Matt, Hannah, Kurtis, and Steve, thank you for keeping me sane. Particular thanks to Mark Josephs who gave me a valuable first review and a huge ego-boost.

I am grateful for educated and loving reviewers. Hilary, geologist in-chief; David, biochemist; and Yvonne, expert mathematician and self-proclaimed grammar pedant; I am thankful for your support all the time but particularly for reading this.

To all the children and pets whose names grace the examples in this book: you bring so much joy.

For my friends, who laugh with me, pray for me, and inspire me. I would be a much less interesting person without you in my life.

This book could not have been written without my dad, Dr. Rhobert Lewis. You have been my biggest supporter in my mathematical and scientific education, and you corralled my rambling explanations into a book. I hope you deem it educational and worthy of a place on your bookshelves.

The Actuaries

I am grateful to my fantastic actuarial colleagues for teaching me so much that I had enough knowledge to fill an entire book. I am grateful particularly for good advice (about actuarial matters and beyond!) from John Berry, Victoria MacDonald, Sophie Short, and Doug Warner.

To Anna Walklate, Darrell Chainey, Katherine Young, Ian Senator, and all at Acted who provided excellent tuition which helped me pass my actuarial exams. I can't recommend you highly enough.

Many thanks to Chris Rimmer and Michael Ward for reviewing chapters, and giving me encouragement and coffee throughout seemingly countless exams.

There are a few actuaries that have had a profound impact on me, professionally and personally. Firstly, I'd like to thank Alex Craig for sharing your pricing knowledge. Your sunny disposition belied the loss ratios and I learnt so much from you about Emblem, data ethics, and parenting. Claire Craig, thank you for sharing your vegan baking.

Thanks to Sarah Rushworth, who pushed me harder than I thought possible. You taught me to expect so much from myself, that I can achieve higher results than I even dreamed possible, and to be tenacious in challenging assumptions. You are an incredible role model for professional women everywhere.

I am grateful for the diagrams, humor, and encouragement from Antony Waring. Your tutorship got me through the bewildering first year of my career. I hope you appreciate the number of diagrams in this book. You are an excellent teacher.

To Rich Townsend: thank you for giving me the time and space required to write a book. Thank you for reading and reviewing it. Thank you for supporting everything I do with humor, patience, and love. For these reasons and thousands more, I think you are the best actuary and person in the world.

The Future Actuaries

To Rufus George, Flora Helen, and all future actuaries: I hope this book inspires you to become whatever you want to be.

Contents

List of Figures

List of Tables

1

Introduction

Chapter Objectives:

- Understand what defines a risk.
- Understand the stages of an insurance policy: underwriting, pricing, inception, provision of cover, insured loss event, claim notification, initial claim estimates, claim settlement, claim payment.

Where are you reading this?

Perhaps you are waiting for a train or a bus. You may be on a peaceful beach or in a busy city street. Or maybe you are curled up at home, with a cup of tea at your side. Wherever you are, you are exposed to risk. These might be physical risks—the risk of being assaulted, or hit by a car. They could be financial risks, such as losing your wallet and its contents, or perhaps losing all your money if a bank goes bust. It could even be a risk to your reputation, such as being spotted in a less than desirable location. Of course, you hope to mitigate the risks. You are well aware that you should stand back from the platform edge when a train is coming; you know not to leave candles unattended. You don't carry a wad of cash day-to-day, you hold onto the handrail, and you would check the weather forecast before hiking. But despite all of these precautions, risk remains. It is part and parcel of life.

© The Author(s), under exclusive license to Springer Nature Switzerland AG 2022
C. Townsend, *A Risky Business*,
https://doi.org/10.1007/978-3-031-11673-5_1

This book is about the ways that society deals with risk through the vehicle of organizations or companies. This could be by spreading the consequences of risk throughout the whole population, or by selling the risk on to willing parties, or by passing it to individuals to deal with alone. The commodification of risk, through insurance and pensions of various types, is just as much a characteristic of an advanced society as an effective and accessible legal system. But unlike the law, dealing with risk is rarely considered by most citizens. How do companies decide what your car insurance should cost? What is the balance between the risks we hold as individuals and the ways that societies or governments effectively take on those responsibilities?

The assessment of risk is the purview of many professions. Health and Safety Managers have a particular role in making assessments of the hazards posed by engineering foundations, slippery office floors, or the storage of lithium batteries. Crime commissioners and the government might examine the risks of violent crime, and a surgeon might weigh up the risks of proceeding or canceling an operation. An actuary, on the other hand, not only assesses but quantifies risk. It is relatively unknown profession, but actuarial work underpins much of modern life and is, in its own way, as important as a good legal system. But this book is not just for future actuaries: it intends to explain—through well-known examples—how risk is managed in and for society on large scales, and why it is that without this reassurance we would not dare step out of the door, never mind build skyscrapers, drive a car or move an oil rig.

1.1 Introduction to Risk

The concept of risk is not new: archaeological evidence suggests that dice-based gambling games were played in the Paleolithic period around 40,000 BC. The word itself has French origins, and was first recorded in 1621.[1] The Oxford English Dictionary defines risk as the possibility of loss, injury, or other adverse or unwelcome circumstance; a chance or situation involving such a possibility. This definition applies as much in actuarial work as it does in common parlance, although the actuarial focus is usually on the financial consequences of the event.

Actuaries quantify risk in order that the risk can be transferred, but how can a risk be transferred? Certainly, the event can't be transferred: a customer can't take out a policy and then expect a flood to divert and hit the insurer's

[1] The record shows that it was originally spelled 'risque', the same as the current French spelling. It was recorded in its current English spelling in 1655.

office instead of their home. But the financial consequences *can* be transferred. Then, if a customer gets flooded, there are no financial consequences for them. For the insurer, if their customer gets flooded, then they have to pay. By transferring the financial consequence, the risk—that is, the chance that something with negative consequences happens—is transferred.

Let's be clear, holding a risk doesn't mean that something will definitely happen—just that it might. The uncertainty is what makes it a risk, rather than an event that will happen. In fact, uncertainty in any variable can be viewed as a risk: whether an event will or will not happen, the timing of an event or its impact. In the life assurance business, it is certain that a customer will die, the uncertainty is when. In international mergers and acquisitions, the timing of a merger may be certain but due to currency exchange rates, the amount received is somewhat uncertain.

Actuaries talk about risk *a lot*. To avoid having to explain the event and the consequence of every risk each time, actuaries often just state the issue and then put 'risk' at the end. The common theme is that at any time in the future, something could change in a way that is unexpected, causing unexpected effects. Just a few examples are given below to give you an idea.

- **Inflation risk**: the chance that inflation might not be as expected, meaning future agreed prices may be worth more or less than expected
- **Operational risk**: the chance that a company's operations, such as the supply chain or processes, causes an event that has financial consequences
- **Investment risk**: the uncertainty around investment returns so that future returns are not as expected
- **Reputational risk**: the chance that a business might suffer an event that erodes customer confidence and therefore causes a financial impact, for example, loss of sales.
- **Credit risk**: the chance that a borrower won't pay back a loan owed as previously agreed
- **Regulatory risk**: the chance that laws or regulations change in a way that causes a material impact
- **Climate change risk**: the physical risks that a change in climate will bring about, as well as the risk of transitioning to a low-carbon economy
- **Security risk**: the chance that security measures fail and retrieval, recompense, or regulatory fines are levied

Exercise 1.1: For each risk, consider whether the uncertainty is due to uncertainty the event will happen, uncertainty around the timing of the event, or uncertainty of the scale or amount.

In the professional world, words may have specific meanings. We must therefore be unambiguous in defining terms, as well as demystifying acronyms and jargon. Industry-specific terms used in this book are highlighted in **bold**, and are included in the glossary.

In common parlance, risk is a scary and generally unwelcome word, usually associated with carelessness and stupidity. In actuarial work it is a necessity, to be measured, manipulated, analyzed, and passed around. And as a result of that process of discovery and quantification, risk becomes more manageable and doesn't seem so scary after all.

1.2 There's No Business Like...

Unsurprisingly, transferring risks is not straightforward: at least, not from a business point of view.

Put simply, in most businesses, you would:

1. Create and build a product
2. Incur some costs
3. Price your product
4. Sell the product
5. Generate revenue

This model applies to most industries because it allows for the price to be set given the costs incurred: the owner sets the price based on the cost of the product in question.

However, industries that transfer risk (which include insurance and pensions) necessarily work the other way around. They:

1. Create and build a product
2. Price your product
3. Sell your product
4. Generate revenue
5. Incur costs

This model cannot be avoided, as clearly, if the cost, timing, and occurrence were already certain then the nature of the eventuality would be known and there would no longer be any risk. But this means there is a big incentive for risk traders to accurately estimate their costs before they have been incurred. Insurance and pension schemes therefore rely on actuaries to quantify these risks quickly and accurately. Let's zoom in to one policy to look at its life cycle:

1. Underwriting/pricing

The first stage in an insurance policy is the underwriting stage. An intermediary called an insurance broker gathers all the details about the policyholder and the risk in question, and passes it to an **underwriter**. Underwriters then consider whether it is a risk they want to take on, at what price, and on what terms.

For personal lines of business (i.e., insurance for an individual rather than a company), this process is increasingly automated. To buy car insurance, you might put your details into a website that then retrieves quotes from multiple insurers. Here, the comparison website is acting as a broker. Alternatively, customers may go straight to an insurer to give their details, or choose to visit a broker in person, particularly if they have unusual circumstances, for example, health conditions, a heavily modified vehicle, or don't have a fixed address. These situations may require different terms of cover, for example, a smoker might find that lung cancer is specifically excluded from their health insurance policy.

Insurance brokers are more common for commercial business (insurance for a company rather than an individual), because they are likely to be purchasing more insurance and will likely have more complex or specific requirements. If you are spending a lot of money on an insurance purchase, you may want the extra expertise that a broker offers.

The underwriter is then presented with the risk details and decides whether this is a risk they wish to take on, decline, or whether they require further information to make a decision. This process is often automated now for standard products, although the algorithms can't generate responses for bespoke covers or unusual circumstances.

With this information, the underwriter determines a price for the insurance policy. This is increasingly calculated by statistical models, but for bespoke policies and unusual circumstances it will depend on the expertise of the underwriter to conjure a suitable premium. On top of the cost of covering the risk, other costs will have to be factored in: brokers' fees, office expenses for the insurer, insurance premium tax, plus a little extra to hopefully give some profit to the insurer. As in other businesses, this estimate of cost is called a quote.

The Cons of Comparison Sites

Comparison sites have been market disrupters, and largely in a good way. They help customers find the cheapest price for insurance, and also help smaller insurers become better known.

However, often focus is predominantly on getting the cheapest premium, in some cases only showing the cheapest three to the customer. This means that customers (and therefore insurers) can become very price-sensitive, and other metrics such as customer service and risk coverage become secondary to the level of premiums.

The other issue is that as they compare insurers, there is a risk that companies start to determine their prices by comparing themselves to others. This can lead to the whole industry underpricing a certain product, leaving them all at risk of failure. A high price to pay for chasing low prices!

The above process is the industry-standard way for 'writing' insurance and life assurance business in all locations other than Lloyds of London. Lloyds of London is a unique marketplace for insurance, and as such it has its own unique (and only slightly convoluted) way of placing business, which will be described in Chapter 2.

2. Policy Inception

If a customer accepts a quoted premium within a certain timeframe, then they pay that amount to the insurer (or to the insurer via a broker) and the policy is said to have been written. But unlike a physical product, insurance cover doesn't usually start right away—usually the next day as a minimum, but it can be months in the future. The point at which the risk starts to be covered by the policy is called the point of policy inception. This is when the transfer of risk actually occurs. There are lots of reasons why a policy might be written ahead of time. From an insurer's point of view, it helps to weed out anyone retrospectively taking out insurance cover, and it also helps administratively too. This is because a great number of companies want their insurance cover to run with the calendar year. This is also the most popular time for insurers to take out **reinsurance**, which is insurance to cover the risks that their risk costs will be higher than expected. As they wish for a lot of policies to begin on January 1st, underwriters would be very busy on December 31st! To avoid this post-holiday rush, many reinsurance policies and large commercial policies are written in November or December but incept in the new year.

Example 1.1: Dina has not bought health insurance. One Thursday morning, she starts to experience some pain in her abdomen and realizes she might require some urgent medical tests. She quickly applies for health insurance online, pays the cheapest quoted premium, and is covered within the hour. That evening she goes to the hospital for urgent tests, saying that the pain started suddenly on Thursday afternoon, conveniently only hours after she bought insurance. Appendicitis can come on very quickly, so her new insurance pays for the urgent surgery. Note that not declaring a medical condition or symptom that she goes on to claim for is likely against the terms and conditions of her policy, however it is exceedingly difficult to verify!

3. Risk is on cover

Once the policy has incepted, the risk is on cover. The policyholder can go about their business, feeling safe in the knowledge that they will be protected against the financial consequences of the risk.

Exercise 1.2: Insurance is often described as 'peace of mind'. However, if insurers are allowed to decline customers, then some people won't be able to access this feeling of security. With that in mind, do you think insurers should be allowed to decline individuals? Why/why not?

Top tip: Although this question has been framed from the point of view of the customer, to generate other ideas try to look at the situation from other points of view such as the insurance company, governments, the insurance company's shareholders and regulators.

i. A Loss Event

The customer has their peace of mind but probably has largely forgotten about their recent insurance purchase. But, in the course of daily life, the risk which they insured against has now happened. In fact, their first thought when they realize that something has happened might be to wonder if they do still have insurance! This book will refer to unfortunate situations like this as **loss events**—events where an insured loss occurs.

However, not all loss events are big and sudden catastrophes where someone knows immediately that they are experiencing a loss event. There are lots of reasons why a policyholder might not realize they've even had a loss. For example, they might be away from the insured item or location—perhaps they are on vacation for two weeks, and only realize their house has been burgled on their return. The damage might only be evident at certain times, like when roofs only leak when it rains, or if an annual maintenance check reveals that there is a small hole in a large boat. For liability claims, they might not realize the other party was injured at all until the injured person contacts the policyholder.

ii. Claim Notification

But just because a policyholder knows about a claim doesn't mean that the insurer, which is the party really holding the risk, does! The policyholder might need a few weeks to physically recover from the loss event before contacting their insurance company, or some time to find the right bit of paperwork with their policy number and the phone number of the insurer on it. It might not need urgent resolution, so calling the insurer goes on a 'to-do' list which never quite gets done. One example of a reporting delay that can be more than a year is if the policyholder might have forgotten they have relevant insurance, and so only claim once they are reminded of their cover. This might be the next year when they go to renew their cover, and notice the terms and conditions include a cover that they assumed was excluded, or it may be when they are reminded by another event. In any case, once the insurer is notified, the claim starts an entire claims journey of its own…

iii. Initial Claim Estimate

The customer will contact the insurer to notify them of a claim, and usually be transferred to the First Notification Of Loss team (the **FNOL**[2] team). The claims handler will input the claim details on the system, give the customer a claim number for their records, and make an initial estimate of how much the claim will cost.

Initial claim estimates used to be almost entirely experience-led: an experienced claims handler may roughly know the cost of a new door for a common make and model of car. Alternatively, in a bid for consistency, the FNOL team may agree to certain estimates: for example, a parking-based 'ding' will cost £400 to repair for a standard family car. Some teams may even have a policy of a standard initial estimate no matter what the claims details, so would always give the same estimate until it can be properly assessed by an expert.

[2] Pronounced Eff-Nol, not F-N-O-L.

Claims Insurtech

Insurtech—a portmanteau of insurance and technology—is any technology that could be used to benefit the insurance industry. One example is the use of smartphones or tablets to assess damage.

When a claim occurred, traditionally the policyholder would call the insurer to send a loss assessor out. In the case of widespread storms or flooding, the insurer would usually be so busy that this couldn't occur for a week or more. Sometime later, a repair company could be dispatched.

Nowadays, a builder can be sent immediately. They will take photos or videos of the damage on a smartphone or tablet, and sent to a loss assessor. The builder can start work straight away, as there is photographic or video evidence of the damaged state. Technology here allows the loss assessor to deal with more claims as they cut out travel time, and gives the customer a quicker resolution and continuity of service.

However, initial claims estimates are increasingly data-driven. This makes sense, as an FNOL handler will likely only see the initial estimates they give, and not the final and true cost of a claim. An algorithm can also collate all the thousands of similar claims and their costs, rather than just the ones that an individual handler can take. There are lots of possible ways of using data to produce initial estimates. A simple way would be to take tables of market values for different possible claim types, and use that to look up likely values. An extension of this approach would be to replace the market values with the historic claims costs of the insurer. This might be better than a market view, as an insurer might only insure specific segments or demographics of the market. It also allows for any agreements the insurer might have in place with repair networks. Data-driven initial claims estimates are sometimes classed as a form of '**Insurtech**'—technology to make insurance more efficient. This might be that an insurer is using machine learning to make better predictions about claims cost, or it might be that customers can use an app to upload a video of the damage to their insurer at the first notification of loss. While claims journeys have been streamlined and improved by technology, this is really an improvement in operational efficiency for businesses and not technological advancement in actuarial work.

Example 1.2: A large motor insurer receives a lot of chipped windscreen claims. They decide to partner with a national windscreen repair company. In future, the insurer won't pay any money to the customer for this type of claim but will instead arrange for the windscreen repair company to visit the customer at home and repair this damage.

This arrangement is beneficial to the insurer as they are likely to have negotiated a lower price per windscreen for providing the repairer with such a lot of business. Customers may find this arrangement more convenient as they don't have to find their own auto garage to carry out the fix. The FNOL team would certainly appreciate this partnership too, as now all windscreen repair estimates can be quickly and reliably set to the fixed amount, e.g., £78.

It should be noted that the claim process is different for a pension provider. If a pensions provider pays a set amount of money to a pensioner every month between their retirement and their death, the provider will require two notifications: one of retirement (that the claim is beginning and payments need to commence), and one that the customer has died (and so the payments can cease). In this case, after the notification of retirement, the handler will open a claim—but crucially will not give an initial estimate of the final claims cost to the customer themselves! They will however inform relevant administrative colleagues within the business that the pension needs to start, and inform the actuaries who set money aside for claims payment that the claim is open and requires estimation.

iv. Claim Movements

Initial estimates are unlikely to be the final cost of a claim, as FNOL handlers don't have all the information about the loss event. They are unlikely to be there in person to assess the damage, and can't be experts in how to repair or replace absolutely everything covered by the policy. After an initial estimate is made, the insurer sets about getting more information on how much this will cost. They may dispatch a surveyor to assess claims cost in person, or commission an independent expert to give an opinion of the cost. This gives another estimate of the claim cost and it could be higher or lower than the previous one. The expert could even revise the estimate to zero if they find that the insurer is not liable for this cost, perhaps if the loss event happened outside of the time the risk was on cover, or if the policyholder is found not liable for a liability claim.

It is during this time that actuaries will try to estimate the final claim amount in a process called **reserving** (or provisioning). They want to calculate the likely final cost of all claims in order to advise the business on how much money needs to be kept aside to pay claims. The aim of reserving is effectively to budget for future payments that will need to be made, much in the same way that individuals need to ringfence some of their paycheck to pay the bills at the end of the month. Analyzing the current claim estimates is analogous to keeping an eye on the electricity meter: if they are higher than expected, or are being revised upward unexpectedly, then that is a sign that more money should be kept aside and reserved for paying these claims. In a pension setting, claims movements will be based on the number of customers that have died or retired since the last time the estimates were carried out.

For some claims, there won't be any movements at all and the initial estimate will be correct. This is likely to be because the loss event leads to a fixed benefit of some kind, or because the **FNOL** team are incredibly accurate in their estimates. This makes the role of reserving actuaries much easier!

v. Claim Settlement

At some point, the claim will be settled. A settled claim is one where the policyholder and insurer are in agreement about the full amount of benefit that will be paid. This could be months or even years after the insurer was initially notified of the loss event! But why would this agreement take so long?

Before the claim amount can be agreed, the customer or insurer might want the damage to be repaired in full, perhaps if they are concerned that the quote given by the repairer will spiral upward, or if the insurer thinks the costs might be less than previously expected. For a rocket, satellite, or a large building, it could take years to finish the project entirely, delaying claims settlement.

For claims where the policyholder is being sued for injuring a third party, there may be other delays. Firstly, there may need to be litigation to determine liability. This decision could then be appealed, and settlement be delayed by the appeals process. The court may also need to pronounce a settlement amount (which will then be paid by the insurer), but waiting on legal decisions is rarely swift.

If the injured party is in a coma, or waiting on some other medical judgment about their recovery, then there may be further delays, even to producing an initial estimate. A consultant might suggest two claim costs: one for the case in which the patient wakes up and has some level of function, and another for the other outcome. The insurer must then wait until the outcome is more known.

In this period of waiting, actuaries will continue to reserve for the unsettled claim, and perhaps anticipating future claim movements. We'll dive into reserving in more detail in Chapter 10.

vi. Claim Payment

In the finale of this claim journey, full payment is made and the claim is closed. It is not immediately obvious why payment wouldn't occur at the moment of settlement, but there are some good reasons. Firstly, it should be noted that partial payment can in fact occur before settlement! If the insurer and customer are waiting for a rebuild to complete before a final claim amount can be settled, the insurer may agree to pay for the building work that has already been completed. This won't be the full claim amount, however that can only occur after settlement.

Equally, claim payment may not be instantaneous after settlement as it might be a very large sum of money. Sometimes, in these cases, the claim settlement agrees not only on the amount of money but the payment schedule. If the claim is for an injured third party that is claiming for a lifetime's salary, then a lump sum might not be the most appropriate way of paying the money. In these cases, a **Periodic Payment Order** (PPO) may be court-ordered, which is a legal agreement that the claim payment must be in the form of a regular payment instead of a lump sum. There is more information on PPOs and their consequences to insurers in Chapter 10.

From an actuary's point of view, after a claim is settled there are no longer any claim movements which makes reserving a great deal easier. However, the money still needs to be kept safe until the payment is made. But once a claim is paid in full, it no longer needs to be reserved for: once the money is paid, you no longer need to consider future payment! The final scene in the claim journey is the relief on a reserving actuary's face when a particularly large and difficult claim has been paid.

4. Cover ends

At the end of the cover period, regardless of whether a loss event occurred or not, the insurer will no longer provide **indemnity** against the risk. It is up to the policyholder what to do next: they may wish to allow the policy to lapse and seek cover elsewhere, or extend the cover with the same insurer by renewing the policy and the same terms and conditions. Some risks go away or lessen with time and so a policyholder may no longer wish to transfer these risks. For example, someone who previously had a life assurance policy in order to take care of their dependent children may allow such a policy to lapse once their youngest had turned 21.

It is important to remember that cover may end before the claim is settled, and that's ok. It should even be expected for some claim types—particularly claims where the policyholder has caused some kind of injury to a third party, or for things that take years to repair or rebuild. Policies will usually have some kind of terms or conditions about when the insurer must be notified of claims, but as long as those are complied with, then it is perfectly possible for a policy to end before a claim is presented to an insurer.

The life cycle of a policy has the standard stages—underwriting, risk on cover, cover ending, policy renewal/lapse—but within that timeframe, if a loss event occurs then dealing with the claim has a process of its own. Once triggered, the claims journey runs parallel to the stages of a policy. Types of business that have (relatively) short claims journeys are called **short-tailed**. This typically means 3–5 years from the end of cover until the overwhelming majority of claims are settled, and examples include motor, pet insurance, and mobile phone insurance. Lines of business that involve complex injuries or liability-based disputes are described as **long-tailed**.

Actuarial work touches all points of the policy and claims journeys. Actuaries who create models to set the price of policies (pricing actuaries) will need to understand the cost of claims that could arise, but will need a good understanding of underwriting and broker behaviors to be able to realistically model price elasticity or market reaction to price changes. They need to work closely with underwriters and product managers to understand if the mix of customers buying the product is consistent over time, or if any changes have been introduced to either the prices or the cover provided. Actuaries who work in reserving will be primarily concerned with the claims journey, and the length of time a claim spends at each stage in the process: how long they typically take to be notified, how long they take to settle, and common causes of movements. But pricing and reserving affect each other: if a different mix of policies are sold, a different mix of claims may be expected, perhaps with different timings to their notification and settlement. Actuaries involved in capital modeling may alert the business that this requires more money may be held to allow for this extra uncertainty. But then the business has more money tied up, requiring more input from investment actuaries, and might need to increase its premiums to give it more **cashflow**. But increasing premiums might attract a different mix of customers.... The challenge is that actions taken today will take years to flow through the whole policy and claims processes, so today's actuarial work will echo for years into the future.

1.3 Can I Insure That?

The previous section described how a policy works, without saying much at all about what a policy might do. Policies pay out in the case of a loss event, but what exactly could these events be?

There are thousands of different types of policies, which cover a huge range of events: broken phone screens to broken legs, people falling ill to rockets falling out the sky, getting hacked, getting sacked, and starting to frack—insurers see it all.

But not all events can be insured, not least because an insurer might not know how! Even an experienced underwriter or specialist insurer would struggle to find a mutually agreeable payout for the loss of a sentimental item, as the item's value to the customer is vastly different to the insurer's valuation. And that is even if a customer feels that the sentimental item is replaceable! If the purpose of having insurance is to restore the policyholder to their original (pre-loss event) position, then replacing something unique and sentimental with something as common as money is not in the spirit of insurance. Therefore, the first criterion for an insurable loss is that it must be something that is financially quantifiable. That's not to say you can't insure an heirloom necklace, but that the amount of money paid out in the case of its loss will only be the financial value of the necklace, usually the current market value.

Did you know?
The tiaras worn by Princess Eugenie, Princess Beatrice and Meghan, Duchess of Sussex on their wedding days are worth a combined £17 million, and would cost around £300,000 per year to insure! However, the British Crown Jewels are deemed uninsurable as the historic nature of them means their value far exceeds their financial worth.

An insurer needs to not only assess the value of an item, but also estimate the likelihood of a loss event. To do this with any kind of accuracy it will need relevant data, without which an insurer is unlikely to want to take on the risk. This can mean that new risks, perhaps related to new technology or new geographic locations, may be uninsurable initially. Lack of data is one of the reasons that people in developing countries may not be able to access insurance even if they had the money to do so: without reliable government or crime statistics, and no previous experience writing business in a country, insurers can be reluctant to begin writing business in new areas. Lack of insurance means that after large natural disasters, the affected area can struggle to rebuild, unless the government or charitable agencies step in with funds. In this sense, widening access and take up of insurance in developing nations is important for reducing global inequality—and it all starts with good data!

Of course, it is not just the accuracy with which loss event likelihoods can be estimated that is important to the insurer, it is the likelihood itself! Events that have a high chance of happening are unlikely to be insurable, as it would be expected that the loss would happen and the insurer would have to pay the claim. To give the policy a chance of being profitable, the premium would need to be very high, which the customer would likely not be interested in. Although events with a high likelihood of occurring theoretically could be insured, the practicalities around pricing mean that they tend to be considered uninsurable unless there is a material uncertainty with respect to timing, e.g., as there is with life insurance.

Insurers are not afraid of a loss—in fact, without loss events happening, people would no longer need to buy insurance! It is not about avoiding all losses, but rather about making sure the insurer wins more than they lose. So, it's okay if 10 policies have high claims costs, if the cumulative profit from the other 90 policies is higher—i.e., the set of policies in aggregate is still profitable, even if an individual one isn't. This is the idea of **pooling** risk: if you hold lots of similar risks, then you win some and you lose some, but you gain plenty of data and can apply the **Law of Large Numbers** to help you quantify the risks. If you only are insuring a single risk, then either you win big, or you lose big—a very uncertain and precarious position for a company to be in. Pooling risk is the foundation of success risk transfer, and is a similar idea to efficiencies of scale: in the same way an individual can't make their own clothes as cheaply and efficiently as a large-scale factory, it is cheaper and more efficient for a large insurer to hold risk than it is for lots of individuals to do so. Risks that can be pooled are risks that can be insured.

Even if a risk is financial in nature, has plenty of available data, is unlikely to happen and can be pooled, it still might not be insurable. At least, not by you.

For a risk to be insurable, the prospective policyholder must incur a financial loss in the case of a loss event. Insurance usually[3] works on the principle of **indemnity**—it will try to restore the policyholder to the position they

[3] There are some notable exceptions: if the loss event involves death or a health event, then the policyholder can't be fully indemnified. Insurers are good, but they can't bring people back from the dead! However, they will usually aim to financially indemnify the policyholder, for example by paying a lump sum based on the deceased's salary, or paying the cost of any treatment for an illness.

were in before the loss. If the policyholder does not experience the downside of a loss event, there is no restoring to do! If a person is not affected in any way by the outcome of an event, then to take out a contract which pays in the case of a specific outcome happening, is just betting on the outcome. Betting is not insurance! It is not just the event, but the relation of the policyholder to the loss that matters in the case of insurance. If there is such a relationship, the person is said to have an **insurable interest** in the loss event.

Examples:

- Harrison will receive £100 if a certain hockey player injures themselves in the next match. Harrison has no stake in this other than this bet. So, if the player is injured, Harrison gains £100. This is gambling.

  Conversely, Hetty will receive £100,000 if she is injured in her next hockey match, as she will lose an important sponsorship deal. If she is injured, she will lose £100,000 in lost sponsorship, but gain £100,000 from the insurer payout. This is insurance.

- Amy will receive $5000 if it rains heavily next Saturday. It doesn't matter at all to Amy what the weather is on that day (other than this policy payout), as she has no plans. So, if it rains on this day, Amy is $5000 better off (less the cost of the bet) than she is today.

  However, Amy's sister Carys will receive $5000 if it rains heavily next Saturday, flooding the ground for her outdoor wedding. If it rains, Carys will have to rebook the outdoor ceremony for another date, costing her around $5000. In this case, she would be no better off if it rains or if it doesn't—if it rains, she loses $5000 and gains $5000. If it doesn't, she doesn't lose anything, other than the policy premium.

For a risk to be insurable, there needs to be this interest, but a further condition applies as well: the amount paid in the case of a loss event must be related to the size of the loss. This is basically an extension of insurable interest, as if someone loses $200 in a loss event but receives $2000 from the insurer, they are in a much better position than before the loss. They would not only be restored to their original position, but actually better off if it happened! Most insurers wouldn't write a policy in this circumstance, and would consider it 'over-insurance': the risk is insured for well over the true cost of indemnity.

Did you know…?
In 1911, the United States Supreme Court ruled that although a policyholder must have insurable interest when buying a life insurance policy, the validly obtained policy could be legally sold to someone without insurable interest, for example to settle a debt.

However, selling a policy that pays out on your death to someone aggressively chasing a debt seems like the first episode of a crime drama, and so is not recommended.

The other problem with over-insurance may be spotted by devious (or dishonest!) readers immediately: if a profit can be made if a loss event occurs, then a certain profit can be made if the event can be engineered. This is another compelling reason why insurers won't knowingly over-insure a customer. But some insurers, in a bid to make life as easy as possible for a customer (and to avoid being targeted by money-laundering criminal gangs) will avoid giving a cash benefit and instead replace or repair the damaged item. Since finding an identical item in terms of age, condition, brand, and color would be time-consuming for the insurer, insurers work on a 'new-for-old' basis whereby the damaged old item is replaced with a new equivalent. A certain cash payout is very attractive, but as gameshows have proven, there is a certain allure of new kitchen appliances too. So, even if a customer is not over-insured, they still might see a benefit to claiming on their policy and receiving a new item. While intentionally engineering loss, damage, or death constitutes fraud, policyholders might just be a little less careful with their insured items than those uninsured.

More generally, if you are indemnified against a loss, there are no longer any great consequences of that event happening. A customer who has insurance might take on more risk by driving more quickly, or by playing a stupid game because they are insured.[4] This **moral hazard** is a downside that comes when risk-takers are no longer risk-holders. Put simply, it isn't their loss anymore, so they don't particularly care if it happens or not. In reality, any indemnity-based policy has some degree of moral hazard, but for most people the annoyance of having to deal with insurers means that they too want to avoid the loss event. Insurers have come up with other ways to reduce moral hazard as well, such as requiring customers to pay the first $200 of any claim (called an **excess**, as the insurer pays the claim in excess of this amount), a risk sharing situation such as a co-pay (where an insurer and a customer split

[4] *Friends* fans might recognize the concept of moral hazard from "*The One Where Joey Loses His Insurance*".

the costs in some proportional way, common in health insurance), and using bonus-malus (also called No-Claims-Discount) schemes, so the customer will get cheaper future premiums if they don't claim. Using one of these systems may be beneficial to the customer as well, as they are usually incentivized by a lower premium. Moral hazard is really difficult to prove so insurers tend to use strict terms and conditions, as well as the 'carrot' of bonus-malus schemes to avoid it. It's hard to tell if a customer took on additional risk because they were insured, or if they always engaged in this kind of behavior. This is one instance where the customer will have more information about the risk than their insurer.

1.4 Risk by Numbers

The ultimate aim of actuarial science is to quantify risk. To corral something as wild as uncertainty into a neat, concise framework, suitable for business lunches and boardrooms, cannot easily be done. However, this section will introduce some of the ways in which actuaries try. None of these metrics are perfect for capturing risk profiles (and that's partly why there are so many of them!), and most of the terms here have alternative definitions or calculation methods. However, this list includes the main terms that a publicly available financial results pack of an insurer might use in order to quantify and communicate the risk it holds.

1.4.1 Risk Metrics

- *Number of policies*: this does what it says on the tin, and simply counts the number of policies written in a given period. This might be used in life insurance or pensions, where all the policies will be of a similar magnitude of risk. It's also used as a measure of market share, or if a certain number of policies need to be written to make a product viable from an expenses standpoint. It's prone to underestimating the level of risk if policies can cover multiple items, for example a property policy covering multiple business locations or multiple vehicles. For heterogenous risks such as in commercial or bespoke insurance, where one policy could cover a small risk and another could cover a multimillion-dollar yacht, this metric is almost useless in conveying a quantity of risk, but is great as a fun fact on marketing material.
- ***Written premium***: this is another simple metric—the total premium for policies written in that period. This is very similar to how other industries

calculate sales: the money gets counted as soon as it hits the till. However, as per the life cycle of insurance, that's not very realistic view for insurance products, where the main costs aren't incurred for a while after the product has sold. If it is assumed that the premium is a good measure of the size of the risk, then this measure can be used as a proxy to risk. However, if market forces pushed prices down one year, then it may seem like the company has taken on less risk when actually it took on the same amount of risk for a lower price. It's not ideal, but it is very common, and specifically **Gross Written Premium** (abbreviated to GWP, meaning the written premium before the cost of reinsurance) is one of the main metrics for most insurers in their reporting and risk management.

- ***Earned Premium***: this is the total premium that relates to the risk exposure for that period. An insurer earns a premium by providing cover for a risk for a period of time, so if that cover has been provided already, the premium has been earned. As most policies don't run exactly to an insurer's accounting system, there are going to be some policies where some of the risk has already been covered, but some is still left remaining. Only the proportion of the premium that relates to the former counts as earned premium. They can be gross or net reinsurance, but most commonly seen as Gross Earned Premium (GEP).

  Of course, if there are earned premiums then there are also unearned premiums which will (hopefully) be earned in future time periods. However, it's not certain that they will be earned, as the customer could cancel the policy. Therefore, in order to avoid counting chickens before they hatch, unearned premiums usually aren't included in the headline financial statements. Despite not being included for historic reporting, it very important that they are not forgotten about. When deciding how much risk to take on in the next time period, the unearned premium plays a substantial part in the calculation. As there is a limit to how much risk a company can hold at a time, remembering that there is already a portfolio of risk that the company has to service is a key part of deciding how much can be written in the next few months. Earned premiums can be thought of as the written premiums on a time delay, so immediate changes to business strategy will take a little while to flow through to the earned premiums. It is important to remember that earned premiums can be higher than written premiums for a given period if business volumes decline.

Example 1.3: A company writes $5m of pet insurance in 2021. In 2022, the company directors decide that they want to write less of this business to allow them to write more rocket insurance instead, so pet insurance GWP drops to $1m. However, the previous business still needs to be covered, and some of that cover will have to be provided in 2022. So the 2022 earned premiums are actually a portion of the $5m plus a portion of the $1m—so GEP is higher than GWP in that year.

- *Exposure units*: if an insurer wanted to do away with the idea of premium entirely (for example, because of a volatile market meaning that **GWP** or GEP were not comparable year on year), then they could use an exposure unit, a unit of measurement to measure the amount of risk being held. Exposure units are a very good way to measure risk, as that is exactly what they are designed to do! But they aren't that useful in financial statements for a few reasons. Firstly, there are different exposure measures for each different line of business. This means that they can't be aggregated in a meaningful way, which is often what company financials try to do. So, in the same way that you can't add one centimeter and 80 decibels together, you can't add one vehicle year with $100,000 of Sum Insured value. Secondly, different insurers will use slightly different versions of exposure measure, even for the same line of business. This might be because of variations in the exact product they're selling, but is more likely to be for legacy reasons,[5] just as some countries use the metric system and some use imperial. Exposure units will be covered in more detail in Chapter 3.
- ***Expected Maximum Loss** (EML)*: insurance works by **pooling** lots of risks together, so that the whole is less risky than the sum of its parts. However, for commercial insurers, an individual risk could still be quite big! Therefore, they might want to convey a sense of how much their largest single risk contributes to the pool. The Expected Maximum Loss is the largest loss that is reasonably expected to arise from a single event in respect of a specific insured object, often a given property. It might be less than the true market value or replacement cost—it's not meant to convey the full extent of the policy coverage after all—, but gives a sense of how big a likely loss will be (if a loss occurs at all). The EML is also used as an exposure measure for some commercial property business.

[5] 'Legacy reasons' is a term that crops up a lot in insurance and pension companies, and can be applied to lots of different methods and systems. If something is done for legacy reasons it usually means that that is how it has always been done, and no one has had time or inclination to change it, usually due to cost or because of fear of inconsistency.

- Possible Maximum Loss: If an insurer wanted to be a little more prudent in its reporting, it might want something a little stronger than EML. After all, the EML only considers claims that are reasonably expected to occur. But the unreasonable may still happen, causing a larger loss than an EML suggests. The Possible Maximum Loss is similar to the EML but considers a wide range of less likely scenarios, so it is higher than the EML. It is therefore a good measure of individual risk, but it does run the risk of report readers thinking 'yeah, but that's never going to happen to us…'—so EML is more widely used.

An insurer taking on risk is not a wholly altruistic act; to make this a sustainable business model, insurers seek to make a profit. As such, metrics need to be devised to understand how profitable the policies written have been. Of course, as per the life cycle of a policy, the emergence of claims won't necessarily be immediate. Settling and paying a claim, and therefore understanding a policy's ultimate profitability (or lack thereof) is therefore not something a business can determine right away. The following metrics are designed to quantify the profitability of a pool of policies, and can either be given as actuals (i.e., the final, true value of the metric, although this is likely to only be available years after the set of policies were written), or as estimates, as calculated by actuaries.

1.4.2 Success Metrics

- ***Claim ratio*** *or* ***Loss Ratio*** *(abbreviated to LR)*: as the biggest cost of insurance are the claims, monitoring how much the claims are is very important to insurers. This can be done by summing up all the claims (both paid and estimates) relating to the period in question to give an Incurred Claims figure. However, this doesn't help the reader. Is \$10m of incurred claims good or bad? Of course, it depends on how much business was written in the first place. The Claims Ratio is therefore the cost of claims divided by the corresponding premiums. Ideally, earned premiums should be used, as a policy which has had very little time on cover is very unlikely to have had a claim. Therefore, it would have a low incurred claim cost, which may be very low compared to the written premium. Using the earned premium is a less volatile measure, although written premiums are sometimes used for ease. Like premiums, loss ratio can be calculated as either gross or net of reinsurance. It should be as low as possible (although if it is too low, there may be questions from the regulator about fair pricing). There are lots of different varieties of loss ratio, depending on how far through the claims journey the policies are. The main ones are the Reported Loss Ratio (RLR), which only includes payments and estimates for claims that have been reported to the insurer, and the **Ultimate Loss Ratio** (ULR) which is an actuary's view of where the loss ratio will end up after all claims (including ones that the insurer is not yet aware of) are taken into account. Exactly how the ULR is calculated by reserving actuaries will be explained in Chapter 11.

- ***Expense ratio***: the other big cost of policies are the expenses. For some lines of business, particularly those that use brokers and therefore have to pay hefty commission on sales, this can be very high. However, a high expense and commission bill might mean that the company is selling lots of policies, so it is more informative to show the ratio of expenses (including commission) to premium. This is usually to written premium, as that is when commission and administrative expenses are incurred, but it doesn't have to be—check the small print of the report to understand exactly what is being presented.
- ***Combined Operating Ratio***[6] *or Underwriting Ratio*: The Combined Operating Ratio typically adds up the Claim Ratio and Expense Ratio to give an overall view on profitability. The ultimate insurance misnomer, the COR is not actually a proper ratio! Remember, the LR is best shown as a proportion of earned premium but the expense ratio is usually given as a proportion of written premium. As they have different denominators, the sum of them is itself not a proper ratio. This can sometimes make the COR behave erratically, but if business volumes are stable then this inconvenient technicality can be ignored without too much harm. In practice, for the purposes of the COR, insurers express their expense ratio as a proportion of earned premium.

  The COR is also called the Underwriting Ratio as it is the profitability of the underwriting activity. Ideally, it would be as low as possible, but anything less than 100% means the insurer is making a profit on these policies. Most large companies target a loss ratio of 80–95% in order to give shareholders the required profit, but individual lines of business may run significantly higher or lower. Note that the underwriting profit (i.e., the profit implied by the COR) is not the final profitability of the company, as insurers invest huge amounts of capital due to all the premiums and claim reserves they are holding. Because of this, investment profit is a substantial (and welcome) addition to an insurer's profit and loss statement. However, an insurer can't control its investment return any more than any investor

[6] Abbreviated to COR—pronounced either C-O-R or 'core'.

can, and investing is really a side-hustle for insurance: it's not its main business activity and it comes with its own risks and uncertainties. As such, insurers tend to report on their underwriting profit and underwriting ratios, with the investment return confined to a line or two at the end.

- ***Solvency Ratio****:* shareholders need insurers to be profitable, but customers need insurers to be solvent. A customer needs to have faith that their insurer will still be able to pay their claim in a year or two, even if a policy is written today. Because of this, insurers need to set aside money in order to pay future claims, called their claim reserves. But that is money that is already spoken for—that is what the actuaries think the claims will cost. Any test of solvency can't include this money, and instead must only include the other assets a company has (that are not part of the claim reserves in any way). This is divided by the written premiums, usually net of reinsurance, to create a ratio. The higher the solvency ratio, the more money the company has for a rainy day and the better it will be able to withstand figurative and literal storms. A solvency ratio around 200% would indicate a financially strong company, whereas anything lower than 120% would raise eyebrows. For more solvency calculation fun, turn to Chapter 12.

These metrics give some indication of how actuaries quantify risk: the key features being how much risk is being held, how profitable holding the risk is, and how likely the bearer can withstand a catastrophic loss. Now the metrics are determined and the stage is set, let the art of actuarial science begin!

2

A Short History of Actuarial Work

What's past is prologue

—William Shakespeare, The Tempest

Like any industry, insurance has own idiosyncrasies, quirks, and jargon. Although this book is not a history book, to understand quite why it was set up in this way, we need to go back to the beginning…

The first evidence of insurance comes from London in 1426, however, the contract was written in Italian by a merchant from Florence. Merchants regularly sailed between Southampton and various Italian ports, and merchants would approach one another and effectively 'swap' risk. Shipping is a very risky business: it would cost 10,000 Dutch Guilders[1] to build a ship, or else charter one, and then the crew must be paid for a voyage that would last months. But if the ship made it to its destination, be it America or India, the vessel could be packed full of treasures which could be sold in London for a worthwhile profit. Unless the ship didn't come back. This wasn't common, but it was catastrophic for a merchant, who would usually have taken out a large amount of debt to pay for the ship and its crew. A ship devoid of cargo (for this was also the age of pirates!) meant the voyage made a loss, but at least the ship could be used again. A loss of both the ship and its contents would mean penury for the merchant and his family.

[1] It is notoriously difficult to translate the value of money from previous centuries into today's value, but this was likely more than a million dollars in today's money.

© The Author(s), under exclusive license to Springer Nature Switzerland AG 2022
C. Townsend, *A Risky Business*,
https://doi.org/10.1007/978-3-031-11673-5_2

The premiums charged by these entrepreneurial Italians were a percentage of the cost of the ship, and depended on the destination port: Florence to Southampton in 1442 would cost 12–15%[2]; from Genoa it would be 8–13% of the ship's cost. Written insurance policies were not uncommon, and there are over 100 policy documents preserved from 1500 to 1527. They are all written in Italian, but English merchants were quick to copy. The early influence of the Italians on insurance is evident today: in London many insurers still have their offices on or around Lombard Street.[3] Almost a century later, English merchants vastly out-numbered Italians, and by 1560 the default language for policies was English. This move proved to be prescient as Dutch and Italian relations deteriorated due to religious differences, and so increasingly London became the center for insurance.

Although the concept of insurance was there, this ad hoc system was in fact very unstable. It was not regulated, so anyone could be an **underwriter** if they chose—and lots did! But lots of individuals were merely betting: data from the late 1500s shows that 67% of underwriters only wrote one policy. This meant the supply of insurance was very patchy: after an underwriter made a profit on his one policy, they would often never write insurance again.

Not only was this a problem for merchants, who need a steady supply of insurance, but it also meant that some underwriters didn't pay up if there was a loss: they would take the premium and disappear into the winding streets of London, never to be seen again. They didn't care about their reputation among merchants as they weren't planning on participating in insurance in future. The law courts decided that they could tackle the outright fraud, but not the chaos of contractions to supply, the huge number of underwriters, and the variety in terms and conditions imposed. Buying insurance was no guarantee of a risk transfer, and the industry was a Wild West.

2.1 I'll Be There For You: The Beginning of Lloyd's

It's the early '90s, and five friends are sitting in a coffee shop in the capital. While they sip their coffee, they share their news and chat about nothing much. Suddenly, a stranger bursts in, looking a bit bedraggled and lost. One of the friends recognizes them—it's an old acquaintance—and they all listen as they tell the tale of how they lost all their money. The five decide to support

---

[2] (Leonard, 2015, pp. 131–150).

[3] Lombardy is in northern Italy, and was used to mean Italians generally.

their new friend until they're back on their feet, and all six agree that whatever life throws at them, they will be there for each other.

But these friends might not be the ones you are thinking of, in fact, you might be in the wrong century entirely! This was the scene in 1690, in a London coffee shop on Tower Street, and there were certainly no women in the group—women weren't even allowed in coffeehouses.[4] Tower Street was the place where the financial and shipping worlds collided: bankers brushed past merchants on the way to one of the many coffee houses in the area to get a drink and hear news from around the globe.

This setup, where several people share risks between themselves, solved quite a few issues for these merchants. They were all in the same position of needing insurance, so knowing that the group would insure them without having to find a new underwriter each year was likely a relief. If you hate renewing your insurance online, imagine if you had to approach strangers in a London pub in order to insure your house! It also meant that one member was unlikely to cut and run, or he would lose his insurance—and his favorite coffeeshop!

The significance of the coffeeshop itself—this particular one was called Lloyd's—is that this group had a base where they could often be found: they gained a reputation there as regulars, and were unlikely to run off with a premium as their business model depended on trust. Coffeehouses were also where shipping news was spread first, so this was a sign that they were engaged in the industry and experts in shipping. It wasn't long before other Lloyd's customers wanted to get their insurance from this group rather take their chances elsewhere.

Nowadays, it seems obvious that insurance is the solution to many individuals holding huge amounts of risk, but doing so in an organized way that doesn't create unnecessary risk was novel. The original group of friends, who eventually became known as the founders of Lloyd's of London, are still much quoted today in policy wording and insurance terminology.

While risk transfer has certainly been around since the early 1400s, the Lloyd's merchants had hit upon the idea of pooling risk.

[4] It was the discovery of tea, and the subsequent creation of much more respectable 'tea-houses' that were deemed suitable for all that allowed women to participate in café culture.

2.2 Pooling

The concept of **pooling** risk is a key part of many actuarial activities. Pooling risk means accepting multiple similar risks, but dealing with them as one, larger risk. This is what the merchants at Lloyd's were doing: they each had their own ship(s) that they wholly owned, but they acted as if they all owned a small portion of everyone's, including their own. Instead of one 'all-or-nothing' risk, they were holding a bundle of little risks.

It may not be immediate apparent that this is advantageous, but the following example illustrates one benefit.

Example 2.1: Imagine I am insuring a single car. If it crashes, it will cost £1,000 to repair. The probability of a crash is 10% over one year. How much money should I keep aside to pay for any repairs this year?

It is tempting to whip your calculator out in response, but the answer here requires no calculation: I need to hold £1,000 for this purpose. Otherwise, if my car crashes, I won't have enough money to pay the full cost of the repair. Even if I were an insurer holding just this one risk, I would still need to hold £1,000.

That's because an insurance contract is basically just a promise that the money will be there when you need it. Not being able to provide the promised money would be very damaging to the reputation of the company—no one would ever want to buy insurance from them again! It is therefore very important that an insurer holds enough to pay its claims, at all times.

Of course, it's not reasonable to have to hold enough money to cover every eventuality all of the time, even though extreme and unexpected events do happen. The 2008 banking crisis showed that institutions can and very occasionally do fail and insurers are no exception, but customers need to feel assured that a failure of an insurer is a very unlikely and extreme situation. For this reason, insurers tend to hold enough money to pay all its obligations 199 times out of 200—that is to say in any given year, the chance the insurer will not be holding enough money is 0.5%. We'll learn more about holding money in Chapter 12.

In the example given above, there is a one in ten chance of needing to pay £1,000. By holding any less than that, the insurer would fail 10% of the time. So, an insurer would need to hold £1,000 in this circumstance.

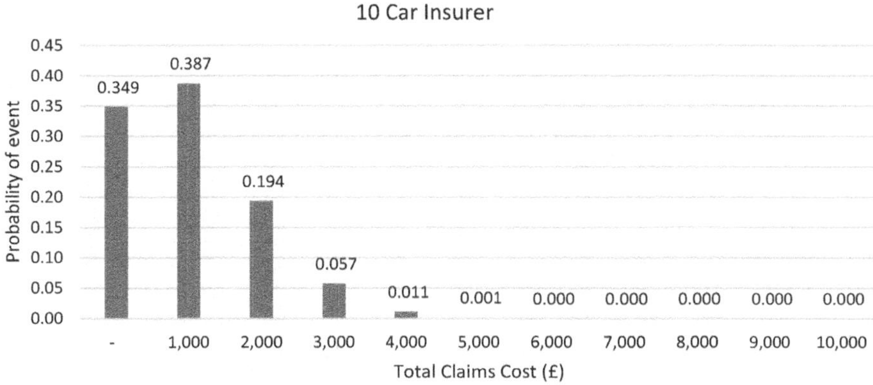

Fig. 2.1 Outcomes for an insurer of 10 cars

To explore the benefits of holding multiple, similar risks, let's imagine this insurer is now insuring two cars,[5] each with the same cost to repair and chance of a crash as the first example. Now, the possible outcomes are:

Outcome 1: Neither car crashes.

No money is needed to pay claims. This will happen with probability $0.9 \times 0.9 = 0.81$.

Outcome 2: Exactly one car crashes and £1,000 is needed.

This will happen with probability $0.1 \times 0.9 + 0.9 \times 0.1 = 0.18$.

Outcome 3: Both cars crash, costing £2,000. This is likely to occur $0.1 \times 0.1 = 0.01$ (i.e., 1% of the time).

How much money should be held now? Using the 1 in 200 level of failure, an insurer should hold £2,000, i.e., £1,000 per car. Otherwise, there would be a 1% chance of an insurer failing to pay its obligations, which is more than the desired 0.5% level. This doesn't seem like much of a benefit—the insurer is still holding £1,000 per car! But extending this example to 10 cars gives the following possible outcomes (Fig. 2.1).

The graph shows that for these 10 cars, the most likely case is that the insurer experiences a claim, whereas previously the most likely scenario was to experience no claims. Notice too that the chance that every insured car crashing in the insured year is very, very low; a one in a billion event, whereas for two cars it has a 1% chance of happening.

[5] Which we assume to be independent—that is, one having a crash has no bearing on whether the other will.

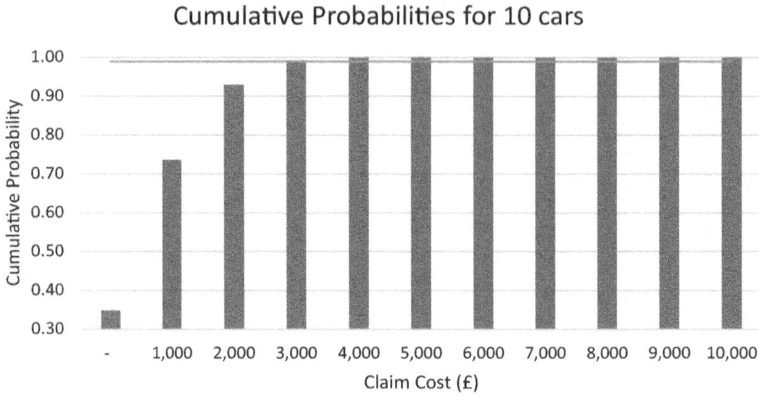

Fig. 2.2 Cumulative probabilities of outcomes for a 10-car insurer. The orange line marks 99.5%

The graph above shows the probability of each claim cost, so summing these gives a cumulative probability distribution. This is what we'll use to find the amount of money that should be held (Fig. 2.2).

The first point where the cumulative probabilities cross the 99.5% line is £4,000, so the insurer should hold £4,000 in total to have less than a 0.5% chance of failure. Observe how now, the insurer doesn't need to hold £1,000 per car—it only needs to hold £400! The money it doesn't need to keep immediately available (in this case the £600) can be invested in the hope of further profit.

As seen in this example, pooling risks together means that less money is required to be held to service claim costs per risk. This means that commercial insurance is a more efficient use of money than thousands of customers keeping £1,000 aside.

But beware, as in every calculation there is usually an assumption lurking. In this example, the probability of ten cars crashing was calculated as the probability of them each crashing, multiplied together. The implicit assumption here is that each car crash is an independent event. This is likely to be a reasonable assumption for motor insurance, although it is possible to think up convoluted examples of motorway pileups or similar.

However, the assumption of independence shouldn't always be relied upon. In the car example, one car crashing didn't increase the probability of another car crashing, but for a home insurance policy which covers against fire damage, a next-door neighbor's home being on fire dramatically increases the chance of neighboring buildings catching fire. This can be a big problem on industrial estates particularly, where one workshop can start a fire that spreads to several nearby buildings (which themselves are usually large and expensive).[6] This is one disadvantage of pooling: if the risks are not sufficiently independent, pooling risk escalates the level of risk to the insurer, and therefore makes total claims costs more unpredictable, not less! In this instance, it is usually called an **accumulation** of risk. Risk accumulations most commonly arise due to premises being geographically near each other, but could also be relevant in cyber risk, if policyholders use the same technology that is remotely hacked.

What else can be pooled?
Any type of policy where there are lots of similar risks can be pooled, including health and life insurance.

Pooling leads to more efficient use of money and decreases the likelihood of extreme events, so planning is easier. Therefore, administrative practicalities aside, the bigger the pool, the better.

Universal healthcare systems, such as the UK's National Health Service, and universal state pensions are therefore better placed than multiple private companies to benefit from the efficiencies of pooling, as their pool is larger.

Unfortunately, some pools had to learn the danger of risk accumulation the hard way. The *Hamburger Feuerkasse* (The Hamburg Fire Office) was one of the earlier fire insurance schemes and was officially set up in 1543. Residents would be put into groups called guilds, and if one resident had a fire, all the members of the guild would pay a certain sum. However, a series of large fires in Hamburg in the 1600s exposed the fatal weakness: if one home has a fire, they often all do. Not all members could be compensated, so the small guilds were broken up.

The Great Fire of London in 1666 sparked the demand for insurance for private homes. Over 13,000 homes were burnt down due to one errant flame (although there were only six deaths attributed to the fire), so Londoners clamored for some kind of risk transfer system to financially take the burden should such an event ever happen again. The architect in charge of rebuilding London after this tragedy, Sir Christopher Wren, even created space in his blueprints for an Insurance Office, such was his belief that commercial insurance was the solution. The *Hamburger Feuerkasse* was initially the blueprint, but one Nicholas Barbon realized that avoiding the geographical proximity of risks, while still insuring multiple properties was the key. As a property

[6] There is an unfortunate tale of how a large insurer declined cover for a welding workshop as it was deemed too high a fire risk. When a fire started there, the insurer felt vindicated—before realizing it insured (and had to pay a large claim for) the papermill next-door!

developer and economist, he understood that brick and frame homes would be lower risk than all-timber, so with some associates, set up the "Insurance Office for Houses" to insure 5,000 (presumably brick and frame!) London homes.

Others soon followed suit, and decided to not only provide financial assistance after the event, but also a service to prevent and minimize fire damage to buildings in the first place. These fire departments would only help their own policyholders. While paper contracts were used, 'fire insurance marks' were also issued. These metal plaques were displayed above the front door to indicate which insurer—and therefore which fire service—was liable for the building. Many of these fire insurance companies still exist and provide fire cover and other property insurance, such as the Sun Fire Office which was founded in 1710.

Did you know?
As many as 90% of wildfires in the US are caused by people.
From the Insurance Information Institute, 2021.
https://www.iii.org/fact-statistic/facts-statistics-wildfires.

This system was still flawed—uninsured homes, or homes insured by competitors, were often left to burn. Eventually, municipal fire departments were created, funded in part by a levy on private insurers, with the explicit aim of reducing fires in the city. Insurers of properties in London are still required to pay a levy to the London Fire Brigade today (Fig. 2.3).

Merchants continued to meet at Edward Lloyds's coffee house to conduct insurance business until 1774, when they moved to the Royal Exchange building instead to accommodate the growing number of underwriters and customers. They called themselves 'The Society of Lloyd's'. During the eighteenth and nineteenth centuries, cover expanded from marine and fire to include all kinds of property damage. The newly discovered tools of calculus[7] and statistics[8] were put to work by underwriters to determine premiums, and start to put numbers to the industry. Some tried to reconcile traditional

[7] Leibniz and Newton independently discovered the rules of calculus in 1675 and 1666, respectively.

[8] A proof of the **law of large numbers**, completed by Jacob Bernoulli, was published in 1713 (8 years after his death)—this is the mathematics behind pooling. The Bernoulli distribution is also named after him. Jacob's brother Johann Bernoulli was also a statistician, but he was pivotal in the creation of L'Hôpital's rule. The beginnings of probability theory were rounded out by Abraham de Moivre, who published his *Doctrine of Chances* in 1718 (while very much alive).

Fig. 2.3 Firemarks were plaques on buildings to show that the premises were insured. This is a replica of a Sun Fire firemark with the policy number 1710. It was issued between 1707 and 1714, during the reign of Queen Anne. (Photograph by Andrew Lines)

'pensions' to the new risk transfer business known as insurance, the full story of which is detailed in Chapter 7. The Royal Exchange building itself burnt down in January 1838 and was rebuilt in its current form. The **general insurance** industry—that is, insurance that isn't life, health, and pensions—settled into its new home in London.

2.3 San Francisco and Uberrima Fides

Across the pond, the newly formed America had its own insurers from 1732. Even Benjamin Franklin and associates got in on the act, founding the *Philadelphia Contributionship for the Insurance of Houses from Loss by Fire* in 1752. While this company was initially innovative both in the way that it encouraged all property owners to have insurance and in its efforts toward fire prevention generally in South Carolina, insurance developments were largely imported from Europe. It is ironic then, that an American city would be pivotal to the success of the Society of Lloyd's.

At the turn of the twentieth century, there were insurers in all major cities across Europe, as well as around the globe. Lloyd's of London (as it had become known) was still a major player, but non-marine customers increasingly chose a more local company for their insurance needs.

On the 18th of April 1906, as the London underwriters of Lloyd's came back from their lunch, they had no idea what was happening on the other side of the globe, and certainly no idea that the next 60 seconds would make history. But in the US, California was hit by a huge earthquake. At 8.25 on the Richter Scale,[9] the damage was catastrophic. 3,000 people died in the earthquake and resulting fires, which continued to burn for three days despite the best efforts of the horse-drawn fire engines. San Francisco was the worst affected due to the amount of brick buildings and tower blocks in the city. Half the population of the city was made homeless in a week, after 25,000 homes, 30 schools, and 80 churches were destroyed. The damage caused was estimated to be $400 million[10] in 1906 terms.

The extent of the destruction was not solely due to the earthquake, but also that the city was not well prepared for large-scale fires. The water tanks used by the fire department quickly ran dry as the pipes were broken by the quake, and the city's fire chief died in earthquake, leaving the team without clear leadership. Both the fire department and the military recalled that in the Great Fire of London, gunpowder was used to create firebreaks and so set about with dynamite to replicate this. With no experience in this technique, it is highly likely that some fires were directly caused by the fire department initially (Fig. 2.4).

But the US government didn't spring into action as they might today. There was no precedent and no expectation that the government would supply relief funds or aid, so although the military helped to maintain law and order in the city after the event, there wasn't a state-funded response. It fell to insurers to rebuild San Francisco.

It was in some sense fortunate that central Baltimore had been a victim of a fire in 1904, as it meant that 90% of homeowners had fire insurance—a much higher rate than today! However, the terms and conditions of this cover was generally limited to fire. In fact, it was common for fire insurance to contain a 'fallen building' clause, which specifically excluded any events where a part of the building had fallen because of anything other than a fire. It is no surprise then that of the first 2,000 claims received by insurers, not a single policyholder mentioned anything about the huge earthquake, and instead attributed all the damage to the resultant fires. As this was clearly

[9] The Richter scale is logarithmic, so that an earthquake assigned 8.0 is 10 times more powerful than one assigned 7.0. The largest earthquake ever recorded was a 9.5, recorded in 1960 in Bio-Bio, Chile.

[10] According to National Oceanic and Atmospheric Administration estimates.

Fig. 2.4 Fires rage in San Francisco after the 1906 earthquake (Library of Congress Public Domain Archive)

nonsense, many insurers denied all liability and refused to pay. Political pressure meant that some insurers did row back on this by the end of the summer and expanded the coverage, but certainly not all insurers did so. Some insurers agreed to partially pay the claim, admitting some liability for the fire component but reducing the paid amount to account for the damage caused by the earthquake.

Local insurers (some of whom were themselves victims of the earthquake, losing offices and employees) were inundated with claims, and were unable to send employees to verify the extent of every loss. Undoubtedly, some claims were inflated, that is, the policyholder claimed for a higher amount than the damage actually was. Upon discovering a few instances of this, some insurers reduced all claim payments accordingly.

But Lloyd's of London was somewhat of an outlier. While other insurance companies were vehemently denying their liability, the leading earthquake underwriter, Cuthbert Heath, told the agent in San Francisco to "*pay all of our policyholders in full, irrespective of the terms of their policies*". It was an unusual and decisive act which cost Lloyd's $50 million.

Eventually, about 100,000 claims were settled by insurers, at approximately 80% of the full value of the claim. The payouts—if they occurred—were remarkably swift, and San Francisco was rebuilt rapidly. 59 insurance companies refused to pay all their claims, or refused to pay in full. 12 insurers were bankrupt because of this event. Lloyd's of London shouldered almost a quarter of the claims paid, but was rewarded with an international reputation for paying claims swiftly and fairly.

A good reputation is vital for insurers. Insurance policies are essentially just promises that the insurer will do something in the future, contingent on other events. But clearly, there are lots of possible events and not all can be explicitly listed in the policy document. Also, like in San Francisco, insurers may dispute liability, or the value of the claim. They may not pay out at all. Because they are just promises, insurance contracts are bought in *uberrima fides*—the utmost good faith: the customer must trust that the insurer will fulfill its promises. More recently, this trust has been tested by business interruption claims due to Covid-19: some businesses had insurance to cover the loss of income if they had to shut down for whatever reason, but some of those policies didn't pay out. This was because some insurers had policy wording that excluded certain categories of respiratory viruses. In some cases, claims were declined because the business didn't technically have to shut down, it was just that customers were ordered by the government to stay at home. Of course, insurers shouldn't pay invalid claims, but the lessons from 1906 and the concept of *uberrima fides* hangs over the decision: you might save a few bucks, but lose your good name.

Lloyd's went on to facilitate the insurance for many globally significant catastrophes, some of which are listed here. The quantity of their losses does not reflect poor underwriting judgment but rather the degree to which Lloyd's has been at the center of world history:

- The HMS Lutine, 1799. Carrying gold and silver bullion from London to Hamburg with the intent of saving the German economy from collapse, this enormous 32-gun frigate was wrecked off the Dutch coast. In addition to the bullion retrieved mostly by fishing vessels, the ship's bell was found in 1858 and is now hung in the Lloyd's building, where it is rung once to announce bad news, and twice to announce good.
- The HMS Titanic in 1912 was insured at a premium of only £7,500 for this 'unsinkable' ship. The claims were all paid in full within 30 days (Fig. 2.5).
- The first case of a 'smash-and-grab' jewelry theft at Harman's on Bond Street, where an £8,000 emerald necklace was stolen.

Fig. 2.5 The Lutine Bell in Lloyd's of London (Copyright free image from Lloyd's Photo Gallery)

- The Hindenburg Zeppelin, 1937: Lloyd's insured a series of airships from World War I onward, but when the German Hindenburg airship was destroyed, killing 35 passengers, zeppelins (and zeppelin insurance) were finished.
- The Exxon Valdez, 1989. The oil tanker ran aground in the Gulf of Alaska, drowning the area in crude oil. It cost Lloyd's nearly $6m—almost half the total insured cost. In 2010, Lloyd's paid out $600m after the Deepwater Horizon oil spill in the Gulf of Mexico.
- Hurricane Katrine, 2005. The most costly US disaster to date, Katrina caused $150 billion of damage. The effects of this one event will be considered in more detail in Chapter 14.
- 9/11. Lloyd's (and the world) watched in horror as events unfolded in New York. It was Lloyd's largest ever single loss, paying out billions of dollars to businesses, airlines and airports, injured individuals, and the families of those killed. In the face of an unthinkable tragedy, Lloyd's reasserted itself as a global insurance leader and essential part of a modern society. The Secretary to the US Treasury at the time, John Snow, said about Lloyd's support in the aftermath: "*We are indebted to you*".

2.4 Lloyd's as a Marketplace

The San Francisco earthquake cemented Lloyd's of London as being the center of insurance once again, but it was vastly changed from the coffeehouse system set up in the 1690s. The insurance written now extends to virtually anything (including virtual any-things), and although marine still accounts for about 8% of Lloyd's business, property and casualty is the largest type of business. In total, £30 billion of insurance business is insured at Lloyd's. As it was in 1690, Lloyd's of London is the venue where insurance is written, but not an insurer in its own right.

While the building is remarkably modern,[11] some things haven't changed. Anyone with sufficient money can still participate in insurance. But instead of loitering in a coffeehouse, the process of participating at Lloyd's is now a little different (Fig. 2.6).

A wealthy person wishing to make money from insurance would first need to become a member of Lloyd's. There are some administrative hurdles to doing this, but once those checks have been performed, the member decides the maximum amount of insurance business they would want to participate in.

One way that Lloyd's adds value is by providing the market framework for insurance; it makes the rules that everyone in the building must abide by. These rules promote the stability and trustworthiness of insurance written. To increase trustworthiness, and to avoid the age-old problem of someone disappearing when a large claim emerges, Lloyd's now requires members to deposit some money with Lloyd's itself. Lloyd's has two pots for this: one is the money that is linked to the member. This is like setting up your own safe deposit box at Lloyd's: it will be used in the case of a very large claim, but otherwise it will be returned to you when you leave. To give you some sense of scale, there is about £25 billion in these accounts. These are called **Members' Funds at Lloyd's**.

On top of that, members also need to make a contribution to the Lloyd's Central Fund, which won't be returned to them. This savings pot is much

[11] The new Lloyd's Building on Lime Street was built in 1986 by architect Richard Rogers. It is interesting architecturally as the building services such as elevators, stairs, and pipework are all on the outside of the building. This leaves the inside as a vast space which is filled daily with underwriters and brokers. It is the youngest building in the UK to be Grade I listed. On top of this architectural honor, the building is immortalized in Lego at Legoland Windsor, complete with working elevators. It is fitting that the external pipes and elevators mean the building has a slight resemblance to a coffee machine. After Lord Rogers' death in December 2021, a service of remembrance was held in the Lloyd's building, and the Lutine Bell was rung once in his memory.

Fig. 2.6 The underwriting floor at Lloyd's of London (Copyright free image from Lloyd's Photo Gallery)

smaller—only £4 billion—and is the last to be raided in the case of a catastrophically expensive year. The Central Fund is not entirely cash, but they are assets that could be liquidated in case of catastrophe. By managing this fund itself, Lloyd's has control and oversight of how much money is in the system in case of emergency. This transparency increases the stability of Lloyd's: the Members' Funds at Lloyd's are a member's back-up, and then there is a back-up in the case of the back-up being exhausted in the form of the Central Fund.

Let's return to our wealthy individual, who has decided the amount of insurance they wish to participate in, and they have deposited money into their Members' Funds at Lloyd's and into the Central Fund. So far, so expensive! Now the member needs to find some business. They need to become part of a **syndicate**.

In some ways, a Lloyd's syndicate is very similar to being part of a lottery syndicate. In a lottery syndicate, a group of people instruct one person to buy a ticket, splitting the cost of doing so. If that ticket wins the lottery, the group will split the winnings in the same proportions. A syndicate will usually agree upfront which lottery game and which numbers it plays. Of course, a player can be involved in many syndicates: it increases their chance of winning (and losing!).

A Lloyd's syndicate works in a very similar way: members join a syndicate and collectively tell an underwriter how much business what type of business and how much to write. In practice there may be hundreds of members, so members interact with the managing agent of the syndicate, who operates the syndicate and instructs the underwriter on behalf of all members. Any claims that arise are split between the members involved, and when everything is totted up, any profit is split too. This is the principle of co-insurance: multiple parties sharing or splitting the transferred risk. For practical reasons, a managing agent will usually require members to deposit some money with the syndicate to pay claims and expenses. This means that there are three pots of money which with to pay claims if required: the money lodged with the syndicate, the Members' Funds at Lloyd's, and the Central Fund. Claims will be paid using these assets in the order given, so that only when the first two pots are exhausted is the Lloyd's Central Fund used to make up any shortfall. Individuals may need to top up their syndicate level assets or Members' Funds at Lloyd's if there are more claims than expected. This three-tier system is called the **Lloyd's Chain of Security**, and is one reason why policyholders want their business to be insured at Lloyd's.

In a lottery syndicate, every week the draw occurs and the syndicate wins or loses. It effectively disbands, then comes together again (sometimes with new players or fewer players) to buy another ticket the following week. The same is true at Lloyd's. At the end of the period (which usually runs to a calendar year), the syndicate disbands and reforms, perhaps with some different members. All the business written in the year is known as the **Annual Venture**. However, the claims journey for those policies may not have come to a close yet, and so it is too early to count the winnings. Instead, a fund is formed relating to that Annual Venture. As claims continue to emerge, they will be paid from this fund. Three years after the Annual Venture began, the syndicate will tot up the profit[12] and allocate it to the members in their pre-agreed share.

This may seem like a hugely convoluted way to get involved in insurance, but it does generally make a reasonable profit, with a return on capital of between 8 and 15%. Of course, some years make a loss, notably 2011 and 2017 due to natural catastrophes, but historically the profit made is worth the initial administrative effort. It is also a good way to participate in insurance if you want to limit your loss: all new members have limited liability.

[12] In fact, this 'totting up the profit' is a complicated process that requires actuaries to estimate the cost of any claims not yet settled. Then, the syndicate will buy insurance so that the liability of the Annual Venture is limited. The cost of this is deducted from the fund relating to the Annual Venture. This insurance is called 'Reinsurance To Close', and the complexities will be discussed in Chapter 13.

Notice how co-insurance works in the case of losses here: a member is not required to pay a higher share of claims just because another member can't. Each individual member is only and solely responsible for their proportion of claims, up to the limit of their liability (but also solely in receipt of profit). The other benefit is that once they are a member of Lloyd's, an individual can participate in lots of syndicates. This is an easy way for them to diversify and write a little bit of lots of risks, helping smaller investors get the benefit of pooling.

But none of this happens in the Lloyd's building itself. The Lloyd's building, particularly the underwriting floor, is dedicated to risk transfer. An insurance broker strides onto the floor with a risk they need insurance for. For each risk, they will have a placing slip, a document summarizing the proposed contract and premium, and the details of the risk. Now they need to find an **underwriter** willing to take the risk on. The Lloyd's market is based on face-to-face negotiation of complex risks, and is driven by relationships between brokers and underwriters. Underwriters quickly develop reputations for specializing in certain risks, so a broker will usually head straight for an underwriter they know in that field. The same underwriter may even have written this risk previously. The broker will explain the risk and the underwriter will scrutinize the placing slip. If the underwriter has been tasked by the managing agent of the syndicate to write this kind of business, they may want to take the risk on using the terms set out on the placing slip. If so, they sign or *write* their name *under* the risk details.

An underwriter may not want or be able to write the whole risk. Risks taken to Lloyd's are often large or particularly specialized, so an underwriter may not have authority from the managing agent to write the whole risk—or perhaps they just don't want to. However, Lloyd's works on a co-insurance basis! As such, the underwriter can decide to only write a certain proportion of the risk. Then the broker needs to find other underwriters to take on the rest of the risk, so will visit the desks of other underwriters to sign more up.

Sometimes, the first underwriter won't agree to the terms set out on the placing slip. In this case, they may cross out the disagreeable part and write something new. Once one underwriter has signed, the other underwriters can't change the placing slip. For this reason, the first signatory is called the lead underwriter. Having a lead underwriter with a strong reputation can help a broker find more underwriters willing to take on the risk. Occasionally a broker may not be able to find underwriters to write all of the risk. In this case, they must tear up the placing slip, change the terms and price to something more favorable for the underwriters, and start again.

Although the building and the regulatory structure around it has changed, the process of writing business at Lloyd's—a broker finding an underwriter to physically sign their slip—remained largely unchanged for over 300 years. In 2018 however, Lloyd's introduced a mandate that syndicates must write at least 10% of risks electronically. This quickly ratcheted up, and by the end of 2020, syndicates were required to write at least 80% of business in this way. This not only allows better data collection and more efficient processes, but also opens up the possibility that brokers could place business at Lloyd's of London from anywhere in the world. Lloyd's became a global leader in insurance based out of just one square mile: by using electronic systems and opening it up to the world, it renews its intention to innovate and lead the insurance industry for another 300 years.

2.5 The Role of Actuarial Institutes Around the World

Lloyd's provides a market framework for insurance to take place. It creates the rules, checks the credentials of members to ensure its reputation, and to encourage innovation and market-leading practices in insurance.

In the same way, actuarial institutes provide a framework in which actuaries can work. They create the conditions of membership (usually a grueling set of technical exams, along with skills based tests and workplace assessment), they guard against members behaving in ways that would harm the reputation of the profession, and they encourage innovation and best practice by insisting on continuous development for qualified actuaries, by running discussions and seminars, and by bringing actuaries together to draw on their combined knowledge to suggest solutions to the big problems in society.

An actuary is someone who is numerically-minded, creative, inquisitive, trained to quantify risks, and is a member of a relevant actuarial society. The benefits of the first qualities may be obvious, but being a member of an actuarial society is more than just a badge or a set of letters after your name.

Actuarial institutes are the professional body for actuaries. Actuarial work is found in many industries, the most common being pensions, life insurance, **general insurance**, and finance—the traditional stomping grounds for actuaries due to the risks caused by their inverted production cycle. Some societies, such as the American Casualty Actuarial Society (CAS) are only for actuaries in certain industries, whereas others span a range of topics and industries. Actuarial institutes also write papers that are sometimes used to inform government policy, or to suggest regulatory changes to the insurance or pensions industry. More practically, they are often an invaluable source of information for working actuaries, particularly the working parties about hot topics, or industry data that is used to create benchmarks.

Example 2.2: The Institute of Actuaries' Third Party Working Party compiles data from large UK-based motor insurers about the cost of third-party claims (for example, if a motorist injures another road user), to inform understanding of trends and inflation in the broader market that had driven up insurance prices in the past decade. This industry data is used by actuaries to provide a comprehensive, stable data to analyze the inflation for these claims, as an individual insurer may only experience a handful of these large claims.

Your local institute will have more information about the relevant training, exams, continuing professional development, and code of conduct for your area. Some institutes are listed below:

- UK: Institute and Faculty of Actuaries (IFoA)
- US: Society of Actuaries (SOA) or the Casualty Actuarial Society (CAS)
- Canada: Canadian Institute of Actuaries
- Austria: Aktuarvereinigung Osterreichs
- Australia: Actuaries Institute
- China: China Association of Actuaries
- Germany: Deutsche Gesellschaft für Versicherungsmathematik e.V
- Iceland: Felag Islenskra Tryggingastaerofraeoinga
- Ireland: Society of Actuaries in Ireland
- Kenya: The Actuarial Society of Kenya (TASK)
- New Zealand: New Zealand Society of Actuaries
- Poland: Polskie Stowarzyszenie Aktuariuszy
- South Africa: Actuarial Society of South Africa (ASSA)

No matter where they are in the world or which society they are subscribed to, an actuary's responsibility is to the public. Society needs competent, trustworthy individuals to help companies and governments manage risk. And with that in mind, let's begin…

3

Products and Perils

Chapter Objectives:

- Identify perils such as fire, flood, accidental damage, and personal injury and estimate their impact in terms of frequency and severity.
- Understand how cover for perils is commonly 'packaged' up to create insurance products.
- Be familiar with the names, covers, and challenges of different types of general insurance.
- Understand common health insurance products and fixed benefit covers.
- Consider how and why new products are created, and what the challenges of creating new products are.

3.1 What Could Possibly Go Wrong?

> *She was caught in a mudslide, eaten by a lion, got run over by a crappy purple Scion…*
>
> —Train, 50 Ways to Say Goodbye

The previous chapter covered the life cycle of an insurance policy. But which policy should a customer take out? Given that the customer (and remember, the customer could be a business) feels the need to have insurance to transfer

© The Author(s), under exclusive license to Springer Nature Switzerland AG 2022
C. Townsend, *A Risky Business*,
https://doi.org/10.1007/978-3-031-11673-5_3

some risk, presumably they can specify at least one risk they wish for the insurance to cover. Specific things that can go wrong are called **perils**.

Example 3.1: A customer inadvertently built their business premises on a floodplain. They now feel very vulnerable to flooding, and so buy flood cover. This policy would pay for any flood-related damage, but not for any damage due to fire.

The easiest way to identify which perils might be relevant for a customer, or which perils an insurer might be exposed to, is to ask yourself: what could possibly go wrong?! Depending on the circumstance, this could be a very long list! And don't confuse possible with probable here: it is important not to forget the risk of unlikely-but-expensive events. You might find that the answer to 'what could possibly go wrong?' is usually 'everything'!

Top Tip!
Identifying perils for an unusual circumstance or made-up scenario can even be an exam question, so it's a useful skill.

When given a strange circumstance, work through the event chronologically to ensure you don't miss any perils. We've used this technique in the example here.

Example 3.2: A telecoms company is planning a satellite launch. What could possibly go wrong pre-launch?

While building the satellite: Accidental damage to the satellite (delaying the project and causing financial loss); Fire cover in case of spontaneous combustion; Flood cover in case the workshop floods.

When towing the machine to the required position: Accidental damage to the satellite; tow truck could hit a pedestrian and cause injury (Third-Party Injury) or cause damage to their vehicle (Third-Party Damage).

Exercise 3.1: Identify the perils for the satellite launch.

Each peril gives rise to claims with certain characteristics. There are lots of different points of interest, such as,

- How often claims occur (the **claim frequency**)
- The size of claims (the **claim severity**)
- Whether the claims are impacted by seasonal trends
- The impact of court awards or litigation
- What kind of inflation affects the claim cost (for example, does it follow a Retail Price Index, or an index relating to wages)
- The currency that the claims will be paid in
- The length of the average notification and settlement delays.

Now we are familiar with some claims characteristics, let's match these with some perils. To introduce all perils and their claim characteristics would be very repetitive, but by grouping perils into one of four categories that tend to have similar claim characteristics, we can get a sense of the broad themes and considerations for most types of **general insurance**. There are four broad categories:

- Property damage
- Liability
- Financial loss
- Fixed benefit

3.1.1 Property Damage

This is the form of insurance that most people will be most familiar with: the policyholder's property has been lost or damaged and the insurer replaces it, or gives the policyholder the money to indemnify[1] against the loss. The property in question can be almost anything from a diamond ring to a diamond mine, a bottle of fine wine, or an aircraft carrier. The perils here are all the ways that the item can be damaged or lost. For cargo ships, this list might include piracy or explosion, but for a more mundane item such as a mobile phone, accidental damage or theft will be the main concerns.

Accidental Damage: this is usually any event caused by the policyholder, where damage to their property was unintentional. A motor example would be that the insured accidently reversed into a wall (causing no damage to the wall, but smashing their own headlights), or a household example might be that a toddler threw a wooden block and broke the television screen. These events happen with relatively high frequency, but tend to be low cost.

[1] **Indemnity** means that the insurer restores the policyholder to the position they were in before the loss.

Fire damage: as the name suggests, this covers any damage caused by a fire. Although claims are relatively infrequent, they are a significant source of claim costs—not least because fires frequently cause a total loss or severe damage, and can spread to nearby properties too. As fire claims often require a rebuild, inflation applied to these should be measured with a cost of building materials index, rather than a retail price index. Because of the total loss nature of the event, sometimes fires are caused by unscrupulous individuals in order to receive a large insurance payout, particularly if the business is unprofitable anyway. It's estimated that a third of fire claims for UK-based commercial properties are deliberate! This behavior means that fire claims occur more often in economic recessions, so fire claim frequency depends on the wider economic conditions.

Storm damage: this covers any damage to the insured's property that was caused by a storm. Note that insurers will usually set their own criteria as to what constitutes a storm claim, for example, perhaps it needs to be named as a storm by relevant meteorological agencies. The frequency of these events depends on the geographic location of the policyholder's property, but are also highly seasonal. The cost of storm damage claims depends on the size of the storm, but also the extent of secondary damage such as storm surge. This peril might cover other costs such as hotel costs if the property is uninhabitable, so for big events this can be expensive. Finally, the cost of individual storm claims can vary hugely depending on the number of claims: if every house in an area has been damaged, there will be high demand on local builders and building materials, increasing the cost of them. This demand surge is a source of considerable uncertainty in estimate future storm claim costs, and this can be a significant loss for insurers if they have unwittingly aggregated risk by writing a lot of policies in an area. Storms and other natural catastrophes are covered in more detail in Chapter 14.

3.1.2 Liability

Liability insurance is where the policyholder must legally pay compensation to someone else, because of an accident or through negligence. There is often a maximum cover level for liability insurance (although there are some exceptions, for example in some jurisdictions the default is to provide unlimited cover for third-party injury as part of motor insurance).

Third-Party Injury: this is when the insured injures someone else,[2] for example, in motor insurance, the policyholder hitting a pedestrian. The cost

[2] The first and second parties being the insured and the insurer.

of the claim depends on the damage done: soft tissue injuries such as whiplash are relatively common but low cost, whereas a severe injury would thankfully occur infrequently but could cost millions. The cost of these claims is quite sensitive to the cost of medical treatment and care, so using a retail price index is not appropriate. Instead, a medical-specific index such as the (US-based) Milliman Medical Index should be used. The occurrence of third-party injury claims also depend on society: some cultures and countries are more litigious and therefore more likely to claim. Landmark rulings or new regulation can therefore create step changes in the frequency or the severity of injury claims. One example is whiplash reform in the UK, where the government introduced fixed payments for soft tissue injuries, in order to reduce the occurrence of whiplash claims. This creates a sudden change in customer claiming behavior so historic data on the number of claims per year may be obsolete. To complicate injury claims further, severe injury claims are sometimes paid as **Periodic Payment Orders** (PPOs), where a partial claim payment is paid at regular intervals, rather than a lump sum on settlement. These will be discussed in more detail in Chapter 10, but is important to consider as PPOs create payment obligations that can extend many years into the future.

Third-Party Damage: A policyholder can also be liable for damage to a third party's property. This cover is technically a liability cover, but because the damage is to physical property, this peril has claim characteristics that are similar to property damage perils, such as short reporting delays and the relevant type of inflation.

Product Liability: this is injury to someone else caused by a faulty or harmful product made or designed by the policyholder. This could be a manufacturing oversight (such as missing or loose screws) or a design flaw meaning safety is compromised. Big pharmaceutical companies are also purchasers of product liability cover, as new drugs can have unexpected adverse side effects that may give rise to large numbers of claims.

Other liability perils include trips in a workplace, liability insurance for professionals (such as medical negligence liability for doctors), or harmful substance exposure. Some examples of harmful substance exposure are detailed in Chapter 16.

3.1.3 Financial Loss

Unlike property damage or liability, there may be no physical damage; rather, the perils are monetary losses.

Business Interruption: this covers the insured if they can't trade as normal. It can either be sold alongside a damage peril (and the business interruption must be due to physical damage) or less commonly, on its own (and interruption could be due to anything out of the owner's control). For example, if a bakery has a fire and needs new ovens, it won't be able to make any baked goods until the ovens are fixed, causing a loss of profit. Business interruption cover would pay the baker for the profit lost, usually based on a recorded, historic profit figure, or a certain amount per week. Business interruption doesn't always have to link to physical damage: it could cover supplier failure (for example, if the mill had a fire and couldn't provide flour to the bakery), or infectious disease in the area meaning that a local government lockdown is in force. The frequency of these claims therefore depends on whether claims are attached to physical damage. The severity depends largely on the profitability of the business. However, business interruption claims are not always independent of each other: if an insurer insures a large number of suppliers and their customers, it may spark a cluster of business interruption claims from one physical event. Due to the increasing amount of business being carried out virtually, these risk **accumulations** are getting harder to detect. For example, because of Hurricane Sandy, which hit New York in 2012, a large number of business interruption claims were paid in the Gulf of Mexico as oil producers were unable to trade as normal with their Manhattan partners.

Other perils would be anything that would cause a financial loss, such as fraud by employees, or if a debtor went bankrupt and their debt canceled, to the loss of the lender. An example perhaps more familiar to readers is travel insurance, which compensates a person for the money they lose due to travel arrangements changing. Like liability insurance, financial loss insurance usually has a maximum cover amount. To reduce **moral hazard**, which would otherwise be high for this type of cover, **excesses** are usually in place. This means that the policyholder will incur a (small) financial loss if there is a claim, so risky behavior such as lending to those likely to default still holds some risk for the policyholder.

3.1.4 Fixed Benefit

Fixed Benefit insurance is an unimaginatively titled form of insurance where the benefit (the amount the insurer pays to the policyholder) is fixed in advance.

Loss of Limb: this cover pays a fixed amount to a policyholder if they lose a limb for any reason. It's not quite like other forms of insurance in that it doesn't work on the principle of indemnity, for what amount of money would restore you to your former position after losing an arm? This type of insurance usually comes with a long list of perils and how much the policyholder receives in each case. For example, for the loss of a limb, there may be a table giving the claim payment depending on whether the limb is on the dominant side, whether it is a leg or arm, and the degree of loss (loss from the shoulder vs elbow vs wrist for example).

It is important to remember that the benefit need not be fixed in absolute terms either: it may be a percentage of the cost for something. Health insurance is a form of fixed benefit insurance where the benefit is often paid as a proportion of the hospital bill, and the perils here would be things such as hospital treatment or routine joint replacement surgery. Fixed benefit insurance has very different claim characteristics to other forms of insurance as the claim cost is to some extent known as soon as the insurer is notified. However, the benefit may be fixed in relative terms and therefore sensitive to the relevant measure of inflation.

3.2 Boats, Buildings, and Beckham: An Introduction to Types of Insurance

Understanding perils is a valuable skill for an actuary, but it is very rare for a customer to request an insurance cover for just one peril. More commonly, a customer (whether an individual or a business) has a circumstance where many things could go wrong, and so they require cover for several perils. To save the customer listing every situation they want covered, insurers package up cover for multiple perils and sell them together as one product. As with any product you buy, different companies will sell slightly different versions: perhaps they will include slightly different covers, different limits to the cover, different terms and conditions and a different price, but there are some common combinations.

3.2.1 Motor Insurance

Lots of things can go wrong when driving a vehicle. Motor insurance is therefore a bundle of covers designed to cover these eventualities, covering both property damage and liability perils. Third-party liability cover is mandatory in many countries, as without this, road traffic accident victims (and perhaps also the government) would be forced to pay for others' bad driving. Therefore, all motor policies will cover third-party injury and third-party damage.

It is also common (but optional) to cover damage to the policyholder's vehicle and the contents of the vehicle, whether through accidental damage, windscreen damage, fire, or theft: this is called comprehensive insurance. Damage due to natural catastrophes such as flooding or storms is often included as part of this comprehensive property damage component.

Did you know?

One in five road accidents is caused by an uninsured driver.

In the UK, the cost of an uninsured driver collision is paid by the Motor Insurance Bureau. This is funded by a levy on all UK motor policies. According to the MIB, between £15 and 30 of a driver's premium goes to this fund.

Exercise 3.2: Estimate the frequency of motor claims based on your own experience. Start by recalling how long you have (or someone you know has) been a driver. Compare this with the number of accidents you've had.

Commercial vehicles may also want cover for any goods or cargo they are carrying. This is called Goods in Transit cover, and a premium for this is usually determined by the quantity that can be carried by the vehicle and the type of good it is, plus cover for the cost of any clean-up required. Premiums for milk floats are high for this reason: environmental agencies have discovered that dairy products are 400 times polluting as sewage because of the way bacteria break it down, so clean-up needs to be quick and thorough. Insurers definitely do have a reason to cry over spilt milk it seems!

The combination of property damage with some liability cover extends to all vehicles: ships, aircraft, cars, bikes, cranes, and even drones. There are some quirks of terminology which are worth noting:

- Marine insurance can be categorized as ocean marine or inland marine. While inland marine seems like an oxymoron, it covers the goods or equipment that support shipping that are frequently relocated by land, for example between or within a port.
- For marine and aviation insurance, physical damage cover is described as 'hull insurance'.

3.2.2 Property Insurance

Property insurance follows a similar model: a mixture of property damage covers with some liability covers. Specifically, home insurance will often cover accidental damage, escape of water (that is, water where water shouldn't be, perhaps due to a burst pipe, leak, or other plumbing issue), theft, fire, and subsidence (when the foundations start to sink or slip). Natural catastrophes are also commonly covered, however in areas where these events are likely, insurers may explicitly exclude these from cover. This creates a difficult situation: homeowners who most need coverage are the most likely to be declined. However, from an insurer's point of view, these events may be so likely that they don't meet the criteria for insurable events (one solution to this problem is discussed in Chapter 13). There is also a liability element to home insurance, for any injury to members of the public that occurs on your property, and this often includes the legal fees to defend such a claim. Homeowners insurance is often a condition for getting a mortgage, as the mortgage provider would hate for the secured asset to be rendered worthless due to a **loss event**. Therefore, the market for this insurance is second only in size to motor insurance, with $147,775 million of gross premiums written in the US alone in 2018.

For commercial properties, all of the above perils still apply, although public liability insurance is much more likely to be used than in domestic properties due to the number of people entering the building. Business interruption cover can be added onto a property product to provide financial **indemnity** of profit after a physical loss stops a business trading. Depending on the business, there might also be cover for loss or damage to stock, as well as loss or damage to the contents or equipment in the building. The perils covered in commercial and domestic property insurance are largely the same, but the differences in the use can affect claim characteristics: large commercial properties are likely to be clustered on industrial estates, whereas domestic properties are often more evenly spread throughout a locality. This can affect how claims are modeled, as most models assume that claims are independent. Remember, if one fire on an industrial estate is able to quickly set several other commercial premises alight and cause further claims, assuming independence is not valid.

3.2.3 Business Insurance

As well as the risks to their property, businesses are exposed to a host of other liability and financial perils. The largest products to cover these are public liability, employer's liability, and workers' compensation. Workers' compensation pays for any injury incurred by an employee of a company in the course of their work. This cover pays the benefit regardless of whether it was the employer's fault (although they don't pay if the employee was negligent), and payment is made straight to the employee. This cover is mandatory in Canada and all US states apart from Texas, although large companies can self-insure if they meet certain criteria (including sufficient capitalization). In the UK, employer's liability is more common. This is similar to workers' compensation, but is taken out by the employer and indemnifies them in the case of an employee injury. The key difference is that employer's liability cover only pays out if the employer is found to have been negligent, whereas worker's compensation pays irrespective of fault. In the UK, employer's liability cover is mandatory for up to £5m for businesses, in order to ensure that employees will have some access to funds in case of an accident. As well as covering accidents such as slips and trips or injury due to faulty machinery, this insurance also pays for injury due to exposure to hazardous substances, although it is harder to prove a direct link. Asbestos, a material now known to cause lung problems, is covered in more detail in Chapter 16. However, products that are predominantly exposed to liability perils have a higher probability of **latent claims**, and are therefore often written on a **claims-made basis**. This means that the policyholder claims on their insurance for any losses reported to them during the cover period (contrast this with a **losses-occurring basis**, where the loss event must occur during the cover period).

Depending on the nature of the business, a policyholder may require additional insurance to cover any accidental damage to or theft from the item they are working on, or cover for their own tools. A financial service company owner might require Directors' and Officers' insurance (abbreviated to D&O), which pays legal fees and for any personal losses if an individual is sued because they hold a senior role in a business. Senior actuaries might wish to have this cover if they are signing off reserves or disclosures and are at risk of being sued for inaccuracy. An add-on to this cover is entity insurance, which is similar to D&O but covers the company, rather than an individual. Professions who give advice or a service, such as solicitors and medical professionals may require specific insurance, indemnifying them if they are sued for accidentally giving bad advice. These policies are often carefully worded to reduce **moral hazard**: the bad advice needs to have been accidental and not due to negligence on the professional's fault.

A business policy might therefore bundle property insurance, employer's liability, public liability, business interruption, and D&O, as well as any trade-specific covers such as tool theft, as a product for small to medium size businesses. Large businesses often prefer to negotiate their own insurance to cover the specific perils they are exposed to.

3.2.4 Bespoke Financial Loss Policies

Insurance products covering multiple perils are popular because they are convenient for the customer. However, large companies and very wealthy individuals are less concerned with convenience because they probably have a person to procure insurance for them. They may also have unusual requirements not met by a standard product. As such, for these customers, insurers will create bespoke packages of cover to address all the insurance needs.

Just as an electrician might insure their tools, singers, actors, and athletes might want to insure their body against anything that damages their unique tool and therefore might reduce their earning power. Because of the erratic nature of celebrity earnings, a traditional business interruption policy that pays out based on an average of previous months' revenues is unlikely to be suitable or acceptable to the client. As such, these tend to be fixed benefit policies. The amount of benefit can be related to a promotional campaign with a company, like when Colgate took out a policy to cover the teeth and smile of their new spokesperson, America Ferrara, for $10 million. Alternatively, the amount could be based on future earnings, or how much they would need to effectively retire from their industry, although this is fairly subjective. Non-celebrities can also buy this kind of cover under the name of 'personal accident', however the benefit paid out for each body part is usually an industry-standard amount, e.g., $50,000 for loss of a leg.

However, the most bizarre claims come from insuring body parts that are not essential to the celebrity's talent, but are perceived to be part of their brand nevertheless. This can be anything from legs (Mariah Carey's were insured for $1 billion), chest hair (Tom Jones insured his for $7 million), or breasts. There are an alarming number of examples here, but Holly Madison, Madonna, and Dolly Parton are insured for $1 million, $2 million, and $3.8 million, respectively.

Did you know?
A record for the largest insurance policy for a sportsperson was reportedly set in 2006 for David Beckham, who insured his entire body for $195 million.

Alongside this, a celebrity might also want cover for financial loss due to reputation loss. This is unlikely to be their own reputational loss (as that may lead to **moral hazard**), but that of a sponsor or co-star. For example, a TV show would be likely to be canceled if the lead actor was found guilty of multiple historic sexual assaults, and a co-star would lose their income if the show was canceled. An insurer might write a policy that includes this as an add-on to personal accident.

Perils could be combined in a multitude of ways to create an endless list of products, but in reality, there are a few common products (although the exact terms and conditions and the levels of cover vary). However, in **general insurance** there are some key features to policies:

- Policies are usually for a short time period, usually a year or 3 years, but sometimes the period of a project. For example, an aviation policy to cover a rocket launch may be valid until the rocket is launched, or a construction policy lasting until the project is complete.
- Claims are usually settled with respect to current prices, so there is an element of inflation baked into most claims. The notable exception here are fixed benefit policies, where the claim cost is fixed in advance. The policy wording may allow for the benefit to be indexed to some predefined level of inflation, but for annual policies this is unlikely.
- There can be multiple claims. This is in contrast to life assurance policies, where you can only die (and therefore claim) once.

General insurance policies are usually for short time periods, unlike life insurance, so it is tempting to think that they are less risky: after all, there is less risk of inflation, big changes in regulation or the market, and less risk that the model is parameterized incorrectly. This is all true, but there are still quite considerable challenges for general insurance.

Firstly, the number of claims, claim amounts and the delays to reporting and settlement are big areas of uncertainty. In contrast to life insurance/assurance or pensions where there are few delays to claiming the benefit, the reporting and settlement lag in general insurance means that **reserving** has to be ongoing for at least a few years after the cover period has ended. This not only requires actuaries' time, but means that some capital will need to be held for more than a couple of years, although recall that Lloyd's of London has a special (and only slightly convoluted!) method of dealing with this issue in order to release capital after 3 years called **Reinsurance To Close**.

The short period on cover can also lead to challenges. Of course, it's an administrative pain having to check policyholder details and issue a new policy every year, but it is more of a pain when a customer doesn't renew their policy and goes elsewhere, which they could do annually. A high level of churn (also called a low persistency rate) means that much more effort has to be made to attract new customers.

Another risk is the nature of the business itself. There may be significant potential for **accumulations** of risk, i.e., lots of risks geographically close to each other, or that will somehow all be exposed to one event. The obvious examples are natural catastrophes which damage properties in one area, but another example could be an insurer who writes a lot of one type of business and is therefore hit harder by an event in that sector than diversified insurers. Insurers who wrote a lot of travel insurance were hit harder by a global cancelation of flights due to the unforeseen pandemic than those who wrote multiple lines of business. But the risk of writing many different products in different countries in order to diversify is also great: the **underwriters** may not be able to become experts on all risks, and so the selection of risks may be not quite as good as if the business had become very niche, so claims experience may generally be slightly worse.

Trading in risks is always going to leave you exposed to some risks and general insurance is no exception: the key is deciding which risks to take on and how to mitigate them.

3.3 New Products, New Possibilities, New Problems

The previously mentioned products are now common offerings in general insurance, but of course it wasn't always that way. With new technology and enterprise comes new risks, and therefore demands new insurance products and covers. But how do insurers create new products?

The steps in creating a new insurance product start off similarly to those of any new product: identify demand, decide exactly what is being offered, market it, and sell the new item. For insurance, identifying a demand can be done by considering what cover is being demanded by brokers or clients. For example, if an insurer currently gets 50 brokers asking for cover of a specific item, say, musical instruments, per year, then it may already expend a considerable amount of resource assessing and writing these bespoke risks. It therefore might decide that creating a generic musical instrument product (which would cover multiple damage perils) would be more efficient, as well as attracting customers who wouldn't necessarily go to the effort of involving a broker but would like to have cover if it were more convenient. An insurer might also consider what others in the market are offering, or by anticipating future technology and customer behavior that might drive an insurance gap.

The more difficult part of the process is to decide exactly what perils and level of cover should be offered, so that customers are not underinsured and the product meets their needs, but the premium is not too high in order to provide cover that the customer does not value. Here, actuaries and underwriters also need to decide the acceptable level of risk to the business: should there be any exclusions to cover? Should there be a limit on how many policies (or how much **gross written premium**) can be written per year? The answers to these questions will be determined by many factors, including capital requirements of the new business. You can look forward to the calculation of these requirements in Chapter 12, but it is sufficient here to say that all insurance products need a certain amount of assets to call upon just in case of large losses.

Who's at fault?

As insurers scramble to design products for driverless cars, one question keeping cropping up: who is at fault if an autonomous vehicle crashes?

Should the driver always have responsibility? If so, then what is the benefit of an autonomous vehicle?

Should it be the car manufacturer? If the car is on autopilot and it crashes, surely that is a fault of the car?

Should it be the computer programmer, who may have made an incorrect assumption in their code? If so, there may be a sudden shortage of coders for those roles!

Should it be the other car on the road or the pedestrian that was injured, as they may have acted in an unpredictable way that the car failed to anticipate?

Until this question can be resolved, fully autonomous cars are unlikely to get insurance—and without scope for transfer, the risk of this tech is likely to be too high for most people. In the meantime, the assistance systems are advertised as ways of making driving easier, but always with the driver in constant charge of the vehicle.

All businesses need some level of capital to remain solvent in case of bad times, but for insurance this is required by regulation in order to protect customers and society. The design of the product impacts heavily on what the level of assets required might be: a new product that diversifies an insurer won't require as much extra capital as one which embeds a company further into a niche, and a product with lots of natural catastrophe exclusions is likely to have much lower capital requirements than one without.

Even after a new product design has been agreed upon, the difficulties continue. Like all products, there is the risk that the product doesn't sell well or that it is mispriced. There are also the initial expenses of designing, marketing, and launching the product. For most new products, this outlay can be recouped during the first few sales of the product. However, for insurance products, there is an additional challenge as profit is not earned at the point of sale. Instead, it is earned over the course of the policy, as time elapses and cover is provided. Unlike a loaf of bread where the profit can be banked immediately, this means that the premium for a new insurance product can't go toward paying back the initial outlay until it is earned. Remember too that the profitability of a policy can't be determined until all claims are settled. So even if the cover period has ended, for policies on a **losses-occurring basis**, it may still be possible to report a claim a year or two after the cover period has ended. Then the claim needs to be settled, and only then is the actual profit known. This all means that there could be substantial period of time—likely to be at least 3 years for an annual policy of a **short-tailed** product—between incurring product design costs and making any actual profit. This impacts the accounts, and therefore the apparent profitability of the entire company. This is called new business strain, as the new product puts a financial strain on the company for at least one accounting period.

Unlike tangible products, it would be a risk for new insurance products to increase the price for the first few years in order to earn profit more quickly and recoup initial outlays, particularly in an established market. This is because of **anti-selection**, where the high price point turns off good customers and so only the bad risks remain with the company. This vicious circle is covered in more detail in Chapter 8, and is unique to insurance and pension industries: another reason why new business causes new problems for actuaries. Conversely, volume and premium growth is no guarantee of long-term profit and success.

Designing a new insurance product needs to take into account first and foremost what the needs of the customer are and then consider the profitability of that. Like all new products, consideration should be given to market acceptability and what other competitors are doing, and how the product fits in with others that the company offers. Insurance providers need to think carefully about what regulation and capital requirements would apply to the new product, and as part of that, make sure that the terms and conditions are fair to customers. The business needs to be sure it can shoulder the new business strain until the first few policies' profits are earned, and of course it also needs to be easy for customers to understand what the new product covers. The result is that coming up with successful new insurance products is less like Dragons' Den[3] and more like a labyrinth of ethical, regulatory, legal, and financial issues.

3.3.1 Environmental, Social, and Governance Sustainability in Insurance and Pensions

When designing a product, regulatory restrictions of all types (both capital and non-capital based) need to be complied with, including fair-pricing considerations and equality laws. The business itself also may have its own rules and guidelines too around what it will and will not write. Questions of product sustainability and ethics now come under the business term **ESG**: environmental, social, and governance. This is a catch-all for deciding and discussing product impact on the wider world, evaluating the extent to which a corporation works on behalf of social good beyond purely commercially driven goals. Taking on risks for a customer is often crucial for them to function, so by doing you are enabling that activity. Great if they are making cupcakes to raise money for a new school, but not so much if they are engaging in morally dubious or downright illegal activities.

Example 3.3: A new product is being designed to insure factories. During the design phase, some underwriting rules are suggested by the ESG committee.

- Environmental impact: Factories that are 100% coal powered will be declined, as the insurer has an aspiration to be carbon neutral in the next 5 years. An insurer with a different path to net-zero may not decline this business, but only allow a certain proportion of its gross written premium to be for coal powered properties.
- Social impact: Factories that rely on child labor will be declined, as that is a human rights violation (and a huge reputational risk to the insurer).
- Governance: Companies which engage in certain tax strategies may need to provide proof of legitimacy, as the insurer does not wish to be associated with any illegal activity.

An ESG committee may impose rules around which risks must be declined, or rules around appetite for certain risks (in terms of gross written premium), or require that some companies provide additional information before the policy is written. For example, companies that have not published a Modern Slavery Statement (which is a logo to confirm that the business and supply chain are not reliant on modern slavery) made be asked additional

[3] Also known as *Shark Tank* in the US, *Money Tigers* in Japan, and *Die Höhle der Löwen* (Lions' Cave) in Germany.

questions about their supply chain, or may need to explain why they don't have a statement. This type of underwriting rule is more time-intensive than an outright decline, but can mean an insurer writes business that is valid but doesn't meet the tick-box exercise. For example, someone who rears sheep to spin their own wool may not have such a statement as they are a small business with no supply chain.

One insurer declining to write certain policies may not seem like it would have a big impact, but by decreasing the supply of insurance for harmful activity the insurance premium may rise. Eventually, as most insurers announce that after a certain time, they too will decline the business, the industry will have no choice but to reform or to take on the risk itself. It also sends a very powerful message: insurers are putting the wider good ahead of profit. Actuarial input into ESG committees is one way that actuaries can make tangible positive change on these topics. The insurance industry is moving very quickly with surprising unity: in July 2021, eight of the largest global insurers committed to writing portfolios with net-zero emissions by 2050. While that is quite a while away yet, the intent itself is powerful. This does not come at the expense of profit for insurers however. In fact, insurers stand to gain a considerable amount (in the form of lower claim payments) by averting the worst effects of climate change: more on this in Chapter 15. Insurance that makes no profit will soon disappear, and with it the social good and protection provided by insurance itself.

ESG considerations are becoming more prominent and are by no means limited to general insurance; pension providers may have an ESG committee to decide upon the criteria for ethical investing too. While ESG is a bit of a buzzword at the moment, it is an old concept that has origins in religious beliefs. For Muslims, Sharia law gives guidance on investing and prohibits doing so in certain industries such as weapon-making. There is also a Sharia-compliant insurance, called takaful,[4] which is a form of mutual insurance. In the UK and US, Quakers and Methodists used to create investing pools that were based upon their own religious beliefs. However, with complex international supply chains, there has been renewed interest from modern companies and investors to ensure that their money is supporting positive practices and discouraging harmful activity.

The most obvious facet of ESG is environmental concerns. This area is traditionally thought of in terms of carbon emissions, waste dumping, and polluting behavior, but should also include deforestation and biodiversity loss, water and energy efficiency, and resource depletion. There have

[4] The Arabic word *takaful* heartwarmingly translates to 'solidarity'.

been several high-profile examples of ESG-related policy declines, including Apollo Lloyd's announcement to not renew the policy for Adani's Carmichael coalmine when the policy expired in 2021.

Social concerns might include things such as working conditions, employee diversity, child labor, and human rights. It also encompasses philanthropy and community engagement, including the topic of CSR—corporate social responsibility. Governance can be anything relating to the ethics of the business, encompassing executive pay, encompassing tax strategy, the company's role in corruption, the diversity of the board, and any political lobbying. Governance is probably the least talked-about topic in ESG, probably because a discussion about tax strategy is somewhat less appealing than a picture of cute wildlife. Yet a company's top level of management and strategy influence almost everything about the way they work.

The prevalence of ESG in financial discussions may give the impression that insurers and pension providers have suddenly become wholly altruistic entities. While most employees certainly care about people and the planet, it would naïve to think that profit is no longer important to shareholders. In fact, there are lots of financial reasons for insurers to focus on ESG in their product design and investments. Firstly, there is the reputational risk to the insurer. With the help of social media, public opinion can seemingly go from unaware to outraged in a matter of days, and that poses a risk that a company can become reputationally tainted in days. Public opinion and public action have an impact on stock price, so investors want to feel secure that there are no skeletons in a company's closet ready to spark a backlash. This is especially important in businesses reliant on trust such as insurance.

Secondly, insurers have a good reason to want to slow climate change: claims for events attributed to or worsened by climate change cost a lot of money! In the US alone, property claims from catastrophes have increased from $16.7 billion in 2010 to over $50 billion in 2018. By encouraging renewable energy use, the worst effects of climate change may be averted or lessened, reducing the uncertainty, volatility, and potential scale of claims. Even health and life insurers have a financial incentive to slow climate change: it has a negative impact on health, particularly in respect of heat-related illness, respiratory conditions, mental health, and zoonotic diseases.

Even for insurers without catastrophe exposure, studies have shown that companies who score well against ESG metrics and act in a responsible manner have lower claim costs than those who don't. For example, companies with strong health and safety initiatives and good employee engagement are less likely to claim for employee accidents and fatalities. In fact, research by Allianz (a top 10 global insurer) showed that poor scores in ESG-based metrics are highly correlated with higher claim payments, more workforce accidents, higher regulatory fine payments, more future controversies, and a poorer financial outlook. This analysis, quantifying that being ESG conscious is financially beneficial, backs up many years of anecdotal industry experience. One example in December 2020 was the protests of the Las Bambas mine in Peru, one of the world's largest copper mines. Disputes over paying tax to the local community and building roads on community land led to local protesters blocking the roads out of the mine. This prevented trucks from transporting copper concentrate over disputes. The strikes reduced the mine's annual output by over 10%. This kind of business interruption and any damage to the property caused by protesters may be avoided if companies acted in a way that is more socially conscious, so insurers may want to penalize companies for not mitigating that risk themselves.

3.4 Case Study: Drone Insurance

Flying robots have been a staple in sci-fi novels for decades. We may not have hoverboards (more's the pity) but drones—more formally called radio-controlled aircraft—are an example of a science fiction prediction that is not only possible but now commonplace. Drones come with a variety of purposes, sizes, and prices. At the time of writing, you can buy a hobby drone for $800, or for a pricier option, the Ehang 184 is a manned drone without a pilot that costs about $300,000. The unmanned ScanEagle has a price tag of $800,000!

Drones can be used as delivery systems, and have been used to deliver aid parcels in disaster. They are also a convenient alternative to manned aircraft for making small deliveries to remote locations. For example, drones have been used for delivering textbooks and other supplies to homes in the Australian outback. Drones are incredibly versatile, and can be used both to target chemical sprays on affected crops without wasting it on healthy plants and also for spraying large areas of crops quickly. As delivery systems alone, there is a large potential market for drones, particularly if ubiquitous online retailers show interest.

Another use for drones was predicted by the 1989 film, *Back to the Future Part II*, where a camera drone was first at the scene of an arrest, reporting the breaking news.[5] There are lots of reasons why a drone is a good choice

[5] *Back to the Future Part II* also predicted that by 2015, drones would be used by families to walk the dog. While this hasn't come to pass, the film did correctly predict the widespread use of videocalls, biometric devices, contactless payment, hands-free gaming consoles, and wearable tech.

for this type of news reporting: it might be quicker than getting a person on location, and certainly easier for the camera operator to be sitting comfortably at home operating a drone than carrying a heavy camera on their shoulder! Drones can also catch perspectives that humans physically can't. Drones are used to film sports games in order to provide the bird's eye view, and have been used in the filming of blockbusters such as *Wolf of Wall Street*.

The camera functionality of a drone can also be used for more serious activities. They have been used by the UK Fire Service to survey the extent of large fires and identify access points to the blaze. The camera needn't be a regular one either—heat sensitive or radio-frequency cameras can be used to quickly identify humans on a landscape. This has boosted the surveillance powers to users, from the US border security force monitoring the US-Mexico border, to park rangers in Kenya spotting poachers. A trial in one Kenyan national park in 2014 showed that using drones reduced poaching by 96%, saving the lives of hundreds of elephants and rhinos. In fact, almost any scanning technology can be attached to a drone to scan large areas at a time. This is useful for marine biologists looking for whales in thousands of kilometers of open ocean, archaeologists trying to map buried ancient settlements, and more uses are being tested every year.

Did you know…?
40% of Japan's rice crop is sprayed using drones. This requires a swarm of over 2,400 of them!

This rapid expansion of drone technology created a new market for specific insurance, so the product design process began by trying to understand the market. However, as we've seen, the uses for drones are almost endless, and increasing all the time! If your customer base is changing and increasing every year, it becomes difficult to do relevant market research. The customer base is also heterogenous: how could one product ever cover the needs of researching grad students, big news corporations, farmers, government agencies, and Sunday afternoon hobbyists? The trick here is to work out what is scalable as a product. Big corporations will probably only buy a single policy to cover all their drone equipment, so although the premiums will be huge, the policy is likely to be a bespoke cover written through a broker. Some Lloyd's **syndicates** will undoubtedly get a drone expert to help set up a basic tariff with the help of pricing actuaries, but this is unlikely to ever be an off-the-shelf product. Instead, the product design team might focus on customers that will only have one or two drones, but sell hundreds or thousands of policies. This strategy also allows the insurer to create a pool of risk, rather than have fewer large risks, so they are less exposed to one very large claim.

With the target market identified, the covers need to be decided. Some perils will obviously be covered, such as accidental damage to the drone (drones are still rather expensive and have the unfortunate habit of falling from height if operated incorrectly) and theft of the drone from a locked storage facility. As with any vehicle that is operated in public, liability covers should also be included: coverage for members of the public who are injured or sustain property damage due to the drone would be important for public safety. There is precedence for these types of liability claims: in April 2014 a triathlete was injured by a drone in Western Australia. The drone was filming the race at a height of 10m when it fell from the sky. Drones are otherwise ideal for triathlons as they can move between water and land smoothly. In this case, the controller of the drone said the drone had been attacked, perhaps virtually via a smartphone, which may suggest there is a need for terrorism or cyber-attack cover. This is even more important for drones that are doing dangerous tasks such as spraying crops with concentrated chemical pesticides. If these were hacked, they could be directed to spray people and cause very large liability claims.

Another cover that might be useful is that for legal fees. Drones are still a relatively new technology, and the legal system hasn't had time to create laws and precedents around their usage. This means that any litigation due to drone-inflicted injury might be time-consuming and therefore expensive. Legal cover clearly is important for such cases, but could be expensive for insurers to provide. This therefore might be provided as an 'add-on'; customers could choose whether to include this cover on their policy for a higher premium.

Did you know…?

According to a report by Kennedys and the Insurance Post, the top three insurable risks associated with drones are:

1. Bodily injury/death to those on the ground
2. Mid-air collision
3. Property damage on the ground

Illegal use (including terrorism) was listed fifth, and cyber-attack seventh. https://kennedyslaw.com/media/2102/kennedys_droneswhitepaper.pdf.

Now it is decided what the policy should cover, thoughts should be turned to the exclusions of cover, as well as the terms and conditions. Some of these terms could be copy and pasted from car or gadget insurance. For example, introducing **excesses** (so perhaps the customer pays the first $100 of any claim), and the conditions that the drone must be locked securely overnight. Looking at the applicable regulation for the area this policy will be sold in, the public liability cover will need to be at least £700,000[6] in the EU. However, there have been some drone-related lawsuits regarding claims for higher amounts than that, so to avoid a difficult conversation with a policyholder about claim limits, it might be better to set a higher limit of claim amounts—perhaps £1m instead.

In the terms and conditions there could also be some exclusions around who uses the drone: that it must be operated by an adult, or only by named people. Unfortunately, another issue with new technology is that it can take a long time for competency checks or licenses to be created and mandated. Currently there is no mandatory license for private drone operation, although the UK's Civil Aviation Authority has created certification for using them for commercial purposes, which include the requirement of maintaining the drone and keeping a logbook. This is particularly difficult to enforce until licenses are required. These qualifications are recognized in other countries, and so having a drone license could either be a condition of cover, or a factor used for setting the premium.

The price of drone insurance very much depends on the structure of the cover. Some insurers have opted for a traditional annual policy, like car insurance or gadget insurance, and that tends to suit commercial operators who are using the drone regularly. However, hobbyists may fly less frequently and so insurers have created a 'pay-as-you-go' policy, whereby the customer uses an app to buy cover for the next 24 hours. This is obviously much more flexible for the customer and may work out cheaper if they don't fly much. However, unless the insurer can request the flight logs for the drone, it may be difficult to ensure that the damage did actually occur in the policy window.

The factors that determine the riskiness of the flight include the flying ability of the customer, the area of use, and what it is being used for. However, as you can't directly ask how good someone's ability is (not only can it seem rude, but they also have a financial incentive to lie), the factors used to determine a premium for drone insurance (either annual or ad hoc) should include:

---

[6] According to EU Regulation 785/2004 on Insurance Requirements for Air Carriers and Air Operators. Who said insurance wasn't exciting?!

- Whether the customer has a drone license/training and whether the training agency is reputable
- How long they have owned this drone
- The customer's postcode or likely flight area (this could be augmented with telematics data from the drone log to ascertain if the drone really did stay in this area)
- Whether flights will be in rural or urban areas
- Whether it is being used for commercial or personal use
- What exactly the use is
- If the drone is customized or has any specialist equipment or hazardous chemicals on it
- If the policy is only for a single flight or short time period, the likely flight path and weather conditions could be used.

To collect data about drone claims in order to set a price, the insurer will look at market data, perhaps trawling through legal records or newspapers to find examples of possible costs. It would also want to price up the different parts of the drone to ascertain the cost of repairing or replacing different models. Another way would be to use the premiums in the market levels as a guide and then revise them up or downward depending on the differences in cover. Once the product starts to sell, the insurer will want to carefully monitor the cost of any claims that come in. With a dataset of their own drone claims, they can then revise their premiums, then monitor how that goes, then revise again, etc. An actuary's job is never done!

4

Data

Chapter Objectives

- Understand how data and technology has improved the actuarial toolkit
- Identify good exposure measures and, for different products, suggest suitable exposure measures
- Consider the conflict between granular data and customer privacy in actuarial work
- Learn the six behaviors in the (UK-based) Actuaries' Code and understand their importance in actuarial work
- Appreciate the data that new technology could yield, and its benefits and challenges

In God we trust, all others bring data.

—W. Edwards Deming

4.1 More Is More: Data for Beginners

You have probably used data before in some form or another. For example, you may have looked up the average temperature in Florida before packing for holidays, or quickly considered the odds of drawing a certain card in a game before making a bet. You may have also calculated summary statistics (mean,

© The Author(s), under exclusive license to Springer Nature Switzerland AG 2022
C. Townsend, *A Risky Business*,
https://doi.org/10.1007/978-3-031-11673-5_4

mode, median, variance, skew) for a set of data in a preliminary statistics or math course. For most people, summary statistics are more than adequate for their purpose. When deciding what to pack for a holiday, receiving time series of temperature and rainfall for the area, along with recent charts of air pressure, humidity, and a year's worth of odometer readings would be overwhelming, and totally disproportionate for your purpose—less is more, right? But when actuaries use data, there is only one rule: more is more.

4.1.1 More data gives more certainty

Like other people, actuaries use data to make predictions and decisions: but for an actuary the decisions often have a material impact on other stakeholders, so actuaries need to be confident in their assertions.

Try this for yourself

What comes next in the series? 1, 2......
What comes next in this series? 1, 2, 4, 8 ....
Which sequence do you think you are more likely to have guessed correctly?[1]

More data gives more information. Actuaries use data to make predictions and solve complex problems, and usually there are lots of different factors that have an effect on the outcome of a model. Using as many of these as possible means that your model will be more realistic (that is, it will make predictions that are more likely to be true) and this may help you make a better decision.

Try this for yourself

Imagine an animal with 4 legs and a tail. If you saw this creature in front of you, how would you react?
Imagine an animal with 4 legs, a tail, that weighs 250 kg and is carnivorous. If you saw this creature in front of you, how would you react?

[1] The second one is much easier to guess, because having double the number of points help you rule out other possibilities. For example, for sequence 1, the rule may be 'add 1', 'double each time', or 'only number that are factors of 2'. The extra data points in the second series rule out two of these possibilities.

More data gives more possibilities. A person on the street may not want a lot of data as it can be overwhelming. Understanding data takes time, and taking in the salient information can be tricky. However, actuaries have a toolkit of various statistical techniques in order for them to make sense of the data, spot trends, and make predictions. In the same way that an airplane needs more fuel than a car to get going, some of these advanced methods need enough data—usually thousands or millions of observations—to give good and stable results.

Try this for yourself

Using simple mathematical operators (+, −, × and ÷), what numbers can be made from the inputs in these lists?
List A: [2, 3]
List B: [2, 3, 5, 8, 9, 10, 13, 17, 18, 24]

Top Tip—Working with Data
Working with data is a core part of an actuary's skillset. Checking data thoroughly to spot errors and issues is an essential step before any analysis. In a small dataset you might be able to do this by eye, but with large datasets, you might use the following checks:

- *'5M' check* (mean, mode, maximum, minimum, missing): For every variable, what is the mean and mode? What is the smallest and largest value? Do these values make sense? What does that tell you about the skew of the data? How many values are missing?
- *Compare to last time*: If this dataset is produced regularly, compare to the last update. What has changed? Does that seem reasonable in the time?
- *Count the number of data entries*. If you have attempted to link or manipulate your data, you may have accidentally duplicated data entries. Doing a quick count of your rows and columns will help identify this: has your action impacted the size of your data as expected?
- *Spot checks*: check 5 or 10 data entries at random (try to check some from the top, middle and bottom of your data). Read through carefully. Is the story consistent? Is there anything strange here? Spot checks can give a deeper understanding of the data, but they are slow to do.
- *Reconcile to other sources*: use an audited or verified source that contains similar data to yours and compare summary statistics, or aggregate your data to a high level. Is the premium consistent? Is it in the right order of

magnitude? Data sources will rarely match perfectly, but this can flag any large errors.

- *Variable formats and lengths*: if you know something should be a number, check it is stored as a number and not a text string, as this can impact functionality of your dataset. Also check that each field is long enough to contain all you need so that digits or text is not truncated.
- *Common sense check*: an oldie but a goodie, common sense is not to be underrated. Check if the dates are what you expect (it is surprisingly common for computers to assume year 00 means 1900 not 2000, or if you have any dates input in the wrong format). For each variable, what are the possible values it can take and what do you have? This can help spot misalignment of columns and general nonsense.

Remember, checking data thoroughly initially can save time later on, so it is worth doing well.

4.2 Data Is (a Bit Like) the New Oil

Data is the new oil.

—*Clive Humby, British mathematician*

Comparisons between data and oil have been doing the rounds in business conferences for almost two decades, and was originally intended to mean that data is a valuable commodity with lots of possible uses, and needs a bit of refinement to reach its full potential. Whether you think this is a good simile or not, the comparison to oil highlights a few issues that data creates:

- *Like oil, data needs a lot of expensive infrastructure to get it and hold it securely.* Like an oil rig, data needs a virtual 'rig' of a warehouse to store it in, and a data pipeline for the data warehouse to be updated regularly. Big companies may hire a team of (data) architects and (data) engineers to set all of this up for them—not dissimilar to an oil rig. Of course, like a rig, this is all pretty expensive to build and maintain. This means that data is more easily accessed and leveraged by rich and powerful companies than by small businesses or individuals. Unsurprisingly, the data is often prized and protected!

- *Data is produced by the world around us—but it's not scarce.* As we are all acutely aware, fossil fuel reserves are being depleted. Conversely, the accumulation of data is just beginning. In some ways, the better analogy here is solar energy: the sun has been shining for millions of years, but solar energy has only just started to be captured. Likewise, data has always existed, in what we buy, what we watch, and where we go, but we rarely thought to capture the data of daily life before this century. Once you start capturing it, you realize quite how abundant it is. Even reading this book, your smart watch could be capturing your heart rate, your movements, and your temperature. Smart glasses could go further and capture eye movements, and so on. Of course, whether data truly exists if it isn't ever captured, is a question for philosophers, not actuaries!
- *Data needs a lot of expensive refinement.* It may not require a distillation column, but refining and cleaning data is still expensive. Other than the infrastructure cost of storing and analyzing it, it also needs skilled people to corral it into ordered datasets (data engineers), and then more skilled people to analyze it (sometimes actuaries, but more generally data scientists).
- *Technology will be created to use it.* Technology had been created to enable this data deluge, but technology is being created that uses this data as input as well. This creates a self-perpetuating cycle: new ideas arise from data currently available, which create new data sources or require a different type of data. That in turn creates a new data source and enables a new technology… So much data and innovation means that those who deal with data (including actuaries) need to constantly be developing their skills and be alert to the opportunities and challenges these will bring.
- *It is potentially harmful.* We are all painfully aware of the damage that oil can do, through extraction, political instability, spillage, and through climate change, but it took around 100 years for all of this to be understood. Large datasets are still relatively new, so who knows what damage they may wreak? This chapter will touch on some of the ethical and privacy issues later, but it is difficult to imagine the potential damage that might be uncovered in decades' time.
- *Data is a valuable commodity that can be used in lots of ways… but only used well by those who know how.* If you had a quantity of oil, no matter how big or small, you could sell it at the market price. Alternatively, you could use it yourself. Data is not like oil in this respect: it is valuable, but only really if you have lots of it. One individual's data is not particularly valuable at all, unless it can be combined with a million others'. Big tech companies such as Amazon, Google, and Facebook are the new oil barons: data has

concentrated power in the hands of the biggest and wealthiest companies in the world. Small personal data essentially funds these free online services, and the platforms collate and commoditize it.

- *You can't always opt out of giving data.* Oil products are everywhere, they are literally the fabric of modern life. Similarly, technology that depends and creates data is so ubiquitous now that it is incredibly difficult to opt out of creating or giving your data to big companies. When you buy a coffee, your credit card tracks the amount of the purchase and the location. The coffee chain scans your loyalty card and it tracks what drink you bought, and figures out what incentives would make you spend more. And you haven't even had breakfast yet! This kind of information can be packaged and sold on by data brokers. Data brokers can use cookies and location signals to categorize people into increasingly small interest-based groups. There is increasing concern about this practice and the lack of regulation.

Data is not of itself a bad thing. But like any new material, it needs to be used carefully and responsibly, sourced ethically, and the risks and challenges should be understood and well communicated. This chapter is all about the possibilities and pitfalls that actuaries may face when using this most valuable commodity.

4.3 Exposure Data

Given how much data is now available, what data should actuaries collect to help them to quantify risk?

In the same way that a shop owner needs to know how much stock they have in order to trade efficiently, an actuary needs to know how much uncertainty they are dealing with. They could count the number of policies written, much like a shopkeeper counts the number of items, but that doesn't allow for the fact that some policies are insuring against low cost and low **frequency** events, and some are against very expensive ones. Some policies open us up to the possibility of bigger **loss events**, and are therefore at bigger risks. How should we count those? This is made more complicated because when quantifying the risk of the policies we wrote previously, we are not saying that for any policy the loss event *did* happen, but that it *could* have happened.

The idea of exposure to risk is something that is used in common parlance as well. For example, the more sun exposure you have (that is, the more time you spend in the sun), the higher your risk of sunburn. This is a physical exposure and a physical risk, but in actuarial work, exposure is not usually so obvious. Of course, being actuaries, we are not content with the mere idea of being exposed to risk: naturally, we wish to quantify it. In the case of sunburn, we might choose to measure the time spent in the sun or the amount of skin left uncovered. This would allow us to collect some data on how badly sunburnt people got *per hour in the sun.*

This is useful because it allows us to quantify, then specify and control the amount of risk we want to take. I am only happy with a surface level of sunburn, so I must spend no more than 30 minutes in the sun. If I didn't mind turning lobster-red but am unhappy with heat exhaustion, then I might be able to spend up to 3.5 hours outside. Notice how the exposure measure can be quantified before the risk event happens—I can state how long I will spend in the sun, without knowing if I will or will not be sunburnt in that time until it occurs.

Some people will be more likely to get sunburnt than others, and some days have much higher UV levels. We can use our exposure measure with these risk factors to quantify the risk:

$$\text{Risk cost} = \text{Base cost} \times \text{Exposure} \times f(\beta)$$

where $f(\beta)$ denotes a function of the **risk factors**—here, that might be your tendency to tan, the UV levels, if you are in the shade or direct sun, etc.

It is not obvious why it is helpful to split the risk cost up in this way, or why the exposure measure was chosen to be time, instead of the intensity level of UV, or the amount of skin exposed. But the advantage of this formula for risk cost is that it allows a 'cost per unit exposure' to be calculated, which is a very powerful tool as it allows us to calculate a base cost—the cost for 1 unit of exposure, if the risk was 'average[2]' in terms of risk factors.

Example 4.1 The base risk of anyone getting sunburnt in an hour on a hot day is 0.1. Ella burns easily, so her risk factor is twice as high as the average person and she will only spend 1 hour in the sun. Hector doesn't burn easily, but often forgets his sunscreen and likes to sunbathe for 3 hours at a time. Given his forgetfulness, his risk factor is 20% higher than the average person. Which person is more likely to get sunburn on a hot day?

Ella's sunburn risk = Base cost × time × risk = $0.1 \times 1 \times 2 = 0.2$

Hector's sunburn risk = Base cost × time × risk = $0.1 \times 3 \times 1.2 = 0.36$.

---

[2] We don't have to use the average or the most common value here, but it is statistically advantageous to do so: we want to have a lot of confidence in the accuracy of our base price as it determines all other prices, and choosing a level with a lot of data is a good way to ensure this. Base levels and risk factors are discussed further in Chapter 8.

From this we can see that Hector has a higher sunburn risk that day than Ella; specifically, he is almost twice as likely to burn than she is.

Exposure measures are used in much more than pricing. They also help an organization quantify how much risk they are holding at any given time. This is useful for capital allocation, buying **reinsurance**, and operational logistics, such as organizing engineers to survey risks being covered. But before we get too deep into the risks, we must consider how rewards—the insurance profits—are earned.

If I purchase an item from a shop, I pay at the till, and take the item home. The shop makes a profit, and our deal is done. I received the item, so they have earned their profit. In insurance, the day the customer pays the premium is only the beginning. They have entered into a contract to provide insurance for a given time frame, and until that time frame is over, the premium isn't really earned. As insurance is the business of risk transfer, then the insurer has to be exposed to the risk in order to earn the premium. So, when are we exposed to risk?

The simplest assumption is that exposure to risk is uniform over time: the risk on the first day of cover is the same as the risk on the final day, and all days in between. This is a common assumption for most annual forms of insurance, mainly because of its simplicity.

Sometimes exposure is definitely not earned uniformly. Sometimes, more risks emerge the longer the cover continues. This is true in construction: at the beginning of the project, there is only a patch of land. Any floods, storms, or fires really won't make too much of a difference to the site, and it won't cost much to restore the patch of land to its pre-event state. As building gets underway, the risks increase: the same flood, storm, or fire will now mean the project needs to be restarted, or somehow salvaged. This would be a bit expensive, but it depends on how finished the building is. In the final months of construction, the risk is relatively high: an event now would not only cost money for a rebuild, but also the time delay may mean the insurer might have to pay for business interruption as well. For construction therefore, exposure and profit are usually earned in a predefined increasing pattern, depending on the building schedule. For example, in a 4-year project, 10% of the exposure might be earned in the first year, 20% in the second year, etc., with the assumption that within the year, exposure is earned uniformly. This is a still simplification of course, but it is commonly done.

Jan	Feb	Mar	Apr	May	June	July	Aug	Sep	Oct	Nov	Dec
5%	5%	5%	8%	12%	15%	15%	12%	8%	5%	5%	5%

Fig. 4.1 An example of a symmetrical seasonal exposure pattern. Note how the exposure over the year sums to 100%

In some lines of insurance, the risk is seasonal. In these lines of business, exposure might be earned seasonally, with higher proportions of the exposure earned throughout the year. Common examples are natural catastrophes, which tend to occur in either winter or summer months, but items that are only used seasonally are also interesting examples. Yachts and classic cars tend to be dusted down in April or May (or whenever the weather turns that year), used throughout the summer, and carefully tucked away in a garage or boathouse, safe from harm in the winter months. Of course, there is still a level of risk in winter, perhaps if there was a windstorm or similar, but the level of risk in winter is minimal compared to summer when the vehicle is being used. Exposure would be earned in a custom pattern, but most commonly in a symmetrical pattern (Fig. 4.1).

Seasonal exposure patterns may use a sine function instead of a stepped pattern. Calculating exposure for fractions of months can then be done by integration.

Because of the additional complexity of calculations with non-uniform exposure patterns, most lines of business use uniform exposure unless the earning pattern is materially different to uniform (e.g., in the case of construction or natural catastrophe), or if the additional complexity is reasonable given the objective of the work.

Sum Insured, or Sum Unsure?

Most homeowners are not very good at estimating what the sum insured (often known as loan to value ratio in the US) for their home should be. Most would know what they bought the property for, but not many would know the rebuild cost. The sum insured (SI) should include material costs, labor costs, building surveyors and architects' fees, and any site clearance.

But these poor estimates are a real problem for customers—experts believe that 40% of homeowners are underinsured. This could mean that in the event of a claim, insurers refuse to pay out as the customer has provided false information, which voids the policy. Consumer fairness has become a big topic in insurance: why and how should a homeowner be expected to correctly estimate such a cost? And is it fair to void a policy due to lack of financial understanding?

Two main ideas have sprung up: firstly, that insurers should ask the right questions to elicit a good estimate of sum insured, including help and online calculators as a guide if required. Secondly, that a policy should not be voided in the case of underinsurance: rather than canceling the insurance, an 'average clause' should be applied to provide some payout to the customer. An average clause takes the declared (wrong) sum insured over the correct sum insured, and the insurer pays that proportion of the claim.

There are lots of potential, reasonable choices for how exposure measures are earned. The main consideration should be materiality and proportionality—what level of accuracy does the work require, and how much of a simplification can be tolerated? There are also lots of potential choices for the exposure measure itself. However, an ideal exposure measure needs to not only define the risk, but have the following additional qualities:

- It needs to be objective and measurable, so that calculations can be performed.
- It needs to be something that either the policyholder or the insurer can understand and obtain easily.
- It needs to be something that can't be manipulated, either by policyholders, intermediaries, or by employees within an insurer in order to get a cheaper premium or improve results.
- There should be a way of verifying the exposure provided by the policyholder. This measure will form the basis of many calculations so it is vital that it is not wrong!
- It needs to be acceptable to policyholders, regulators, and the market more generally.

Example 4.2 A motor insurer creates their own exposure measure, called 'risk points'. This is calculated by the insurer by giving the make and model of vehicle a number of points on an internally-created scale from 1 to 100 (100 being the riskiest vehicles), and then multiplying this by the time in years on cover.

This is measurable, verifiable, and widely acceptable, but this exposure measure is not ideal.

After an influx in new car models, the account managers decide to re-allocate all vehicles on the scale. This means that historic exposure is now changed, so historic data, such as the price tariff, can no longer be used without a complicated conversion.

The measure also causes a perverse incentive for **underwriters**, who have targets to write a certain amount of exposure per year, to write risky business. Because risky business is worth more risk points, they can write 1 risky policy or 5 low-risk policies for the same amount of risk points. But the former is quicker, despite not aligning with the business' strategy. Equally, if the business decides to de-risk their portfolio (i.e., it wants to write less risky business), it will appear that their volumes have decreased, even if number of policies stays the same, so chief executives may not want to take that prudent action.

The table below outlines common exposure measures for popular lines of business.

Exercise 4.1 For each exposure measure, decide how many of the ideal qualities it possesses.

Product	Exposure measure	Comments
Motor	Vehicle years	1 vehicle, insured for 1 year = 1 vehicle year
Property/home insurance	Sum insured	This is usually an amount specified by the policyholder, and is usually the maximum amount that the insurer would pay out in the case of a total loss
Pet/health insurance	Number of lives	Some pet insurance companies use number of tails instead, which avoids any debate about how many lives cats really have!

(continued)

(continued)

Product	Exposure measure	Comments
Marine	Sum insured, or dead weight (or gross registered) tonnage	For total loss coverage, sum insured is usually the preferred metric. However, for non-total loss, the size of the vessel is more informative. To measure this, the weight of the ship is used—for cargo ships, one metric is Dead Weight Tonnage (i.e., the maximum weight of cargo that can be carried); the passenger ship equivalent is Gross Registered Tonnage
Liability lines	Wage roll/number of employees	The more employees you have, the more chances there are for one to have an accident. It is usually the average wage roll or the number of employees over the previous year, to allow for natural fluctuation. This can underestimate exposure if a business grows rapidly in a year
Cyber insurance	Annual revenue/number of employees	As well as reputational penalties, companies can be fined for data breaches. The fine is often a percentage of annual revenue, hence the use of annual revenue as a metric. Number of employees or number of unique customers could also be used, to assess how vulnerable or attractive a company is as a target

4.4 The Importance of Being Earnest (and Honest): GDPR and Ethics in Actuarial Work

Organizations can extract a lot of information—and therefore a lot of profit—from individuals' data. As a customer, there often isn't much choice about this. You can't even buy a pencil online without giving your name, address, email, and credit card details, let alone buy insurance. For a customer, this leads to two questions: firstly, what is the organization doing with my data, especially if they are sharing it with other parties, and why? Secondly: what happens if the data leaks from the organization into the public sphere, either through cyber-crime or by accident?

As more and more data on us is collected by organizations—what we watch and buy, what we eat, and where we go-, customers are increasingly nervous about the data being misused, or passed into the wrong hands. Whether we binge watch reality TV or had to Google how to boil an egg, there is data that we wouldn't want being linked to us (at least publicly!).

To reassure customers, a raft of data protection legislation has sprung up around the world, with the majority of countries having some form of legislation. In May 2018, the European Union brought a particularly comprehensive set of rules and consumer rights into force, called the European General Data Protection Regulation (GDPR). To give all the detail would be a book in itself—the regulation is 261 pages![3]—but there are some important features which impact actuarial work which we will highlight here.

4.4.1 Who Cares About GDPR?

Unlike previous privacy regulation, GDPR applies to all businesses who hold the data of EU residents, no matter where the company is based. This means that almost every big multinational must protect their EU customers in the way set out in the regulation, so it is globally important. The consequences of not abiding by GDPR are potentially very expensive: the fines for breaching rules can be up to $20 million USD, or 4% of annual global turnover, whichever is higher. For a multinational such as Amazon, based on their 2020 turnover the fine could be up to $12 billion! Unsurprisingly, companies have been quick to comply.

[3] Regulation (EU) 2016/ 679 of the European Parliament—on the protection of natural persons with regard to the processing of personal data and on the free movement of such data, and repealing Directive 95/46/EC (General Data Protection Regulation).

In the US, states are starting to make their own legislation based on the principles of GDPR, in particular the California Privacy and Protection law which was effected in January 2020, and the Washington Privacy Act in July 2021. Other countries such as Iceland, China, Brazil, Argentina, Switzerland, and Malaysia also have GDPR-style regulation, so it is globally important as a benchmark.

Did you know?
In 2019, Google was fined $50 million for not being transparent about how customer data was being collected and used for targeted advertising.

4.4.2 What Are the Broad Aims of GDPR?

GDPR defines some data as 'personal', meaning it could be used to identify an individual. Some of this is obvious, such as full name, a unique user ID or your IP address, but it could also be data that if used in conjunction, could be identifying, such as demographic information or social media posts. One team of researchers found that 99.98% of Americans would be correctly re-identified in any dataset using just 15 demographic attributes, among them age, gender, and marital status.[4]

Example 4.3 There is only one ginger person living in a certain street. By combining the street name and hair color, this person could be uniquely identified. Therefore, hair color and street name are personal attributes.

After a string of high-profile data leaks, consumers were becoming very concerned about their data privacy. One goal of GDPR was to protect customers from these leaks, while also increasing transparent about how and why data is collected and used.

4.4.3 What Are the New Consumer Rights?

Under GDPR, consumers have increased rights over their data, including:

[4] Estimating the success of re-identifications in incomplete datasets using generative models.

- The right to informed consent

Previously, terms and conditions about the collection and use of customer data were often long, full of technical jargon and legal terms. They were often vague about what data analysis would be used for. Under GDPR, consent must be sought for each specific purpose. EU citizens might have noticed that there are more Terms and Conditions pop ups when they are online: as companies change the data they wish to collect and the analyses they wish to conduct, new consent must be sought for that specific purposes.

Example 4.4 "The company reserves the right to use data for advertising" should be swapped for "Data will be passed to third-party advertisers".

- The right to access data

Under GDPR, customers have the right to ask a company if their individual data is being used for a specific process, and what data the company holds about them as an individual. So, a customer can receive a copy of whatever data a company holds about them, for free. For a social media platform such as Facebook, consumers can download archives of every post, photo, and place they've ever posted or checked in at, as well as every friend request they have accepted, declined, or even everyone who has ever even sent a friend request. It also tells you which (and how many) adverts you have clicked on, and which advertisers have your data. A fascinating read!

- The right to be forgotten

If a customer needs to give consent for a company to have their data, it makes sense that they should also be able to revoke that consent. This extends to data the company already holds as well: a customer can request that a company 'forgets' their data, as long as it is not against the public interest to do so.

Example 4.5 After requesting a copy of their location data from their smartwatch provider, a customer requests that historic location data is deleted. This data is not important to anyone other than the customer, and after confirming this with relevant authorities, the company deletes it.

Conversely, a politician is accused of holding illicit meetings at a certain hotel with a friend who owns a business. This business was subsequently awarded a large government contract, paid for by public money. The work was not completed despite full payment, and it made the business a large profit. The politician wishes that their smartwatch company delete the location data that would confirm or deny their location during these meetings. After considering the details, the company decides that deleting this data would not be in the public interest, and they retain the information in case of future investigation.

4.4.4 What Does GDPR Mean for Actuaries?

Actuaries are unlikely to be customer-facing and dealing with data access requests, but GDPR affects how actuaries should handle data. For example, they should de-personalize any personalized data they receive (e.g., by removing names and phone numbers of policyholders). They also need to advocate for de-personalization, or pseudonyminization, at source: at the earliest point that datasets don't require personal information, it should be de-personalized to reduce the number of colleagues with access to that information. Office workers may have noticed an increased business focus on data security: not emailing datasets to your personal email address for working at home; locking hard copies of data away; and password-protecting datasets.

GDPR (and responsible data use more generally) has additional importance for actuaries, as unlike data scientists, data engineers, and data governance officers, actuaries are part of a profession. Being part of a profession requires not only strong technical and communication skills, but also a commitment to uphold certain ethical and professional standards. Actuaries are therefore people who have made a commitment to these standards and as such, have a duty to their clients and the public to act responsibly. As GDPR and data privacy regulation are deemed to be in the public's interest, it is an actuary's duty to not only comply, but also follow best practice and encourage others to do so too.

Actuaries tend to have a certain mystique about them. Even before the creation of professional codes of conduct, actuaries were trusted because they were actuaries, and that was enough. However, a few high-profile cases of misconduct[5] led long-established institutes to require their members to act

[5] The most notable of these was the Equity Funding scandal in the 1973, in which the Equity Funding Corporation of America created numerous fictious insurance policies and engaged in accounting fraud. Over 100 employees were involved in the deception, including the company's actuaries.

ethically and professionally and follow a code of conduct. It is important for actuaries to maintain public trust, for several reasons.

Firstly, the public generally lack knowledge about what actuaries do, which means that actuarial work receives less scrutiny than other professions.

Exercise 4.2 Compare how often doctors, teachers, and actuaries are mentioned in the news. Is it positive or negative? Is the article about the quality of their work, the ease of entering the profession, or something else?

Secondly, traditional actuarial stomping grounds of pensions and insurance work on the principle of uberrima fides (utmost good faith). When the industry trades on promises, it is important that customers perceive it to be trustworthy.

Thirdly, actuaries inform and make very important decisions, both in terms of financial value and in terms of impact. Actuaries are frequently involved in mergers and acquisitions, valuing large assets or liabilities, advising governments, and financial products that have a material impact on an individual, so it is critical that they are not only competent but also trustworthy.

Institutes have slightly different codes of conduct, but the principles are largely the same globally. The Institute and Faculty of Actuaries bases their code around six points. Actuaries must:

- Act professionally, with integrity and honesty
- Be impartial, and avoid and disclose conflicts of interest
- Ensure competence at any work undertaken, and take care in their work
- Ensure compliance with all relevant local guidance and regulation
- Communicate in way that is clear and appropriate to the audience
- Disclose any information that is relevant, and any action that is unlawful[6]

On top of a commitment to adhere to the code, actuaries have to undertake continuous professional development and professionalism training every year. This helps members ensure they are competent and up to date with any changes to guidance.

Actuarial integrity extends much further than responsible data use. Honest valuation of assets and liabilities is often used as an example where actuarial integrity is tested, e.g., if board members put pressure on actuaries to reduce

[6] IFoA's Actuaries' Code.

their estimate of ultimate claims in order to make the company finances seem better than they truly are. It is of course very important that actuaries do show integrity in these circumstances, but as actuaries move into a wider variety of roles, areas, and industries, applications of the code of conduct increase as well. Here are just three examples of tricky ethical dilemmas for actuaries.

4.4.5 Price-Walking

Pricing is a traditionally actuarial pursuit, but with great power comes great responsibility—and between the advances in data and statistical techniques, there is certainly a lot more analytical power these days! But it's not just the use of data in pricing that can be deemed unfair. In 2018, a 'supercomplaint' was brought to the UK's Financial Conduct Authority (FCA) regarding how insurers offer discounts to new customers, but increase the price for customers who renew every year. This practice is known as price-walking, as the price starts low for new customers but 'walks' upward every year. This was deemed unfair and unethical of insurers, as people who don't switch are more likely to be vulnerable in some way (note that lack of financial literacy or ability is a vulnerability in itself). Insurers were also deemed to be engaging in practices that make it more difficult for customers to switch, e.g., by auto-renewing policies without timely reminders to the customer that the policy is renewing, or by not providing a fair comparison of the previous year's price and the new price. The prescription from the FCA is for improved communications with customers to increase transparency about their premium, and to put limits on new customer discounts.

4.4.6 Discriminatory Data

The insurance industry naturally discriminates. And that's not a problem, as long as it is only discrimination based on risk. With data, we can create increasingly sophisticated and accurate statistical models to determine that risk, so the judgment of **underwriters** or pricing actuaries is less important. This is a good thing, as an individual's judgment is subjective and sometimes biased. Nonetheless, it is simultaneously reassuring and disheartening that the majority of states and countries have laws to stop the explicit use of protected characteristics in insurance pricing exist. For example, in North Carolina:

> No insurer shall refuse to insure or refuse to continue to insure an individual; limit the amount, extent, or kind of coverage available to an individual; or

charge an individual a different rate for the same coverage, because of the race, color, or national or ethnic origin of that individual.

However, census data often tells you the racial makeup of a neighborhood or postcode district. It also might tell you the outcome of a vote in a certain district. This may be allowable as a **rating factor**, since the feature relates not to the individual themselves but the wider geographic area. These proxy factors might be statistically significant and even explainable. For example, there may be a correlation between people who live in an area represented by a conservative politician and a higher risk of heart attacks. This could be explained because people who live in an area represented by a conservative politician might be more likely to be conservative themselves. Voting conservatively is correlated being older, and over 50s are more likely to suffer a heart attack. But here, the risk comes from their age, not voting for any specific party! The policyholder might vote differently to their neighbors, or be a young conservative voter. So although the voting outcome in the district may be correlated with risk, it would be more transparent and ethical to use the direct risk factor here: the age of the policyholder. More generally, it should be remembered that data can reflect and perpetuate biases of the population and the environment from which it was captured.

A question for pricing actuaries is what to do with indirectly discriminatory factors, such as home address. Clearly, home address for motor or home insurance will be related to the risk: some areas have higher rates of carjacking or theft, or are more likely to flood. But some neighborhoods are largely monocultural, and so including home address may indirectly (and inadvertently) capture the effect of race on risk.

This is an area where pricing actuaries face real dilemmas currently: some say that fitting the factor of race of policyholder in the initial model helps to avoid it being captured by other factors (and then leaving all other factors unchanged, race can be removed from the model entirely), but this requires policyholders to give data about their protected characteristics. Some practitioners seek to ignore it entirely, so there is no data with which the link between risk and race could be ever investigated thoroughly (although as mentioned, census data that captures '% of Residents from X Ethnicity' can be found free online). To what extent are we responsible for discrimination if the data justifies it? How do we remove historic bias against certain groups from our data? And how do we deal with non-risk-based discrimination when we find it?

4.4.7 Environmental, Social, and Governance (ESG) and Sustainable Investment

Some actuaries work in investment, primarily focused on matching liabilities with assets so that risk is neutralized. According to the textbooks, the considerations for which assets are best suited to certain liabilities are based around their financial characteristics: their liquidity, value, duration, currency, etc. In practice, the industry is acknowledging that some investments that would traditionally be considered 'good', are not so good for the environment, or for people or animals. For example, oil or tobacco companies were traditionally thought to be good investments if a company wanted to invest in shares, but there is a growing acknowledgment that this may not be what companies or their customers wish to support any longer. The role for investment actuaries now is to consider another dimension to asset matching: whether the ethics of the asset match the buyer's, and whether value can be defined in a way that takes people, planet, and profit into account. There is also the risk of stranded assets, which is the risk that an asset which has been invested in (such as a company, or a physical asset such as an oil rig) will become worthless due to public sentiment about that industry or item, before the investor has a chance to sell it—or that they are forced to sell at a loss. The risk of stranded assets, as well as the reputational risk of investing in oil rigs or tobacco companies, is just one consideration for investment actuaries. While the industry and regulation catches up with public sentiment on this issue, it is up to actuaries in this field to be forward-thinking, act with integrity, and set their own ethical standards in this area.

4.5 The Future of Data: Insurtech and AI

The extra security, administration, and responsibility of data protection and data ethics are well worth the costs, as more data can lead to better understanding and quantifying risk. This section discusses how data could revolutionize actuarial work, focusing on three technologies that are already among us: telematics, smart appliances, and machine learning. These, along with technologies that automate aspects of insurance business, are sometimes referred to as **Insurtech**.

4.5.1 Telematics

Telematics has been the hot topic of car insurance since the early noughties, frequently discussed as the future of car insurance. And yet it hasn't particularly taken off for the majority. What's the problem?

Telematics is sometimes known as black box technology: it's a GPS system in your car that monitors and transmits your location to your insurance provider. By knowing exactly where you are—and these systems usually work to within 2 metres—not only can your insurer know your location, but also your speed and your acceleration and braking habits. It's clear why this is useful for insurers: how and where you drive are the main risk factors for causing an accident. If your insurer knows you are driving at 55 mph in a 30 mph zone, or that you frequently drive incredibly quickly around town centers late at night, then it is likely that they will deem you a higher risk.

Telematics devices have been adopted by younger drivers in search of lower premiums. If they can prove that they only drive in daylight hours, at reasonable speeds, and with gentle acceleration then it is likely that they will be offered a lower premium. However, uptake by the wider population hasn't happened due to a combination of privacy concerns and a lack of incentive: for experienced drivers, the risk of being deemed a bad driver by your own car is more costly (and embarassing!) than the possibility for a small discount at renewal. But telematics are inherently opaque: they are called black boxes for a reason! Without transparency about how they affect premium, customers are unlikely to want additional monitoring.

That's not to say that the past two decades have all been hot air from insurers. A small group of telematics users has allowed the industry to work through some tricky situations while limiting exposure to risk, so in some ways it has been fortunate that it hasn't been more popular thus far. The main problem has been that it is very difficult to harness the unstructured data into a useful rating structure. Looking at each set of data indvidually, it is easy for a person to decide if someone is a particularly good or bad driver by considering what time of day they drive, whether they speed, how aggressively they change speed, and in what areas they drive.

Initially, this information was reduced to structured data in a simple way: have they ever driven past 11 pm? How often do they travel faster than 60 mph? Do they drive more in cities or countryside? However, that meant that unusual behavior would sometimes get categorized as usual. For example, a parent who usually drives very cautiously in suburbia might be considered a good risk. One night, their child is ill and they take her to the city hospital at midnight. If this data was captured using a binary data capture, they might be wrongly captured as a poor risk. It would be better to consider driving late night as either a frequency or as a percentage of total miles (i.e., 1 in 240 trips started after 11 pm, or 0.2% of miles driven in the past year were driven after 11 pm). However, this might unfairly penalize shift workers, so perhaps the relevant behavior is trips that start and end at the same location, that begin after 11 pm. As every few seconds a new data point will be captured, identifying these trips is much more complex than a boolean operator applied to the journey start time! This type of manipulation is called feature engineering: making a useful variable from other variables that, in their current state, are not easy to work with. Over the past twenty years, part of the learning process of telematics has been learning how to deal with unstructured data: how much we can simplify, and how best to do that.

Did you know…?
Apollo 11 (the spacecraft that landed on the moon in 1969) had two computers on board. Each weighed 32 kg and had 4 kB of RAM. Today, all of that memory could fit on a small USB stick.

The role of telematics systems in driverless cars may boost its adoption in insurance. Of course, other ethical dilemmas will then be debated (such as whether a driver should be ultimately responsible for the vehicle, even if it is on autopilot mode), but the possibility and potential keeps telematics on the emerging tech list.

4.5.2 Smart Appliances and Wearable Tech

Smart appliances are those that connect to the internet (and often a smartphone or tablet) to give the customer control, convenience, or more information. For example, a regular central heating system may be set on a timer to control when the heating comes on or off, via buttons in the house. But a smart central heating system would connect to the internet and could be set using an app on your (smart)phone. This gives more control, because a customer can set the heating wherever they are; more convenience, because it can be set from the comfort of the sofa; and more information, because it is easier to display energy usage information and other statistics on a phone than on the small display of a boiler. Of course, this also has benefits for your electric company as they can read your meter virtually, and by looking at general trends in heating usage, better plan and predict future demand.

But what relevance might smart appliances have for insurers and pension providers?

On the face of it, these appliances are a bit of a nuisance for insurers, as they are more expensive than 'dumb' equivalents and therefore more expensive to repair or replace in the case of a claim. However, the additional information they generate could also be shared, and the additional data and control may mean that claims can be averted or minimized.

Example 4.6 A family goes away for a two-week winter holiday. On the second day they are away, their pipes burst (this often happens if the pipes freeze then thaw quickly, so this type of claim is more prevalent in cold weather). This starts to slowly flood their home. A burst pipe can release as much water as 50 bathtubs a day, so after 12 days, they arrive home, totally unaware of the damage that awaits them, to realize their whole ground floor is ruined. Their insurer pays tens of thousands of dollars to pump out the water, fix the house, and replace their furniture.

Contrast that with a family with a smart water system. On their second day, their pipes burst. The smart water meter registers the abnormally low water pressure in their home, and notifies the homeowner immediately. They call a neighbor to turn the stopcock off immediately, and a plumber to fix the pipes in their absence. They return home nervously, but to little damage other than the wooden floor in the area of the leak. The insurer pays the cost of the emergency plumber and $5,000 for new floorboards.

The additional information also has operational benefits for insurers. Insurers who inspect and survey equipment as part of their service (sometimes offered as part of engineering cover to ensure that equipment is safe to use) would naturally want to prioritize checking the equipment that is most likely to be faulty. Smart electrical systems may flag that a certain area of wiring is getting too hot. While this isn't an immediate issue, a surveyor would know to check that area as a priority to reduce the risk of future fire claims.

Another operational benefit is that the additional information collected could be easily passed to insurers for pricing purposes. Perhaps an actuary thinks that the amount of electrical wire in a home would be a good risk factor for domestic fire insurance, but no homeowners know that information off the top of their head. Their smart electric system might have that information, and it could either be input by the individual or perhaps the customer could permit the insurer to access the information from the app as a one-off for pricing purposes.

Similarly, smart watches and smart fitness trackers are technologies that health and life insurers are very interested in. These devices can track step count, body temperature, heart rate, and much more, so could be useful to assess how much exercise someone does and if they have a high resting heart rate, which can be a sign of—often undiagnosed—tachycardia or other cardiovascular issues. Fitness apps where the user tracks their diet or inputs other health-related data could also be useful for pricing purposes. As with other technologies, the main blockers to widespread adoption of these as rating factors are that these technologies are still expensive and not used by all age groups; the data is difficult to wrangle as it can be unstructured; and the cost involved in cleaning and analyzing these large datasets is currently not worth the additional benefit the data gives to insurers. However, they can give positive opportunities for customer engagement, and prompt positive behaviors and proactive risk mitigation.

4.5.3 Machine Learning

Actuaries have traditionally had mathematics and statistics backgrounds in addition to the technical actuarial training they undertake. Machine learning, being from the domain of computer science, has therefore been often missed in actuarial science as a probabilistic approach is usually the more comfortable for the analyst! Machine learning and statistical techniques often aim to solve the same problems of prediction, classification, and regression, so in one sense there hasn't been much loss due to the lack of diversity in approach. However, with the introduction and adoption of unstructured data sources and Big Data,[7] statistical techniques—which often boil down to parameter estimation for a specified model structure—can be too time-intensive and reliant on actuaries' data exploration. In contrast, machine learning suggests model structures, including interactions, so is all but necessary for deriving value from unstructured and vast datasets.

However, statistical modeling has the advantage in situations where there is some understanding about the factors underlying the modeled variable, which is the case in most insurance products at the moment. Further, statistical models are much easier to explain and understand than models created by machine learning, which is very important when these models are implemented and have the potential to affect millions of customers. This is not just an issue of fair pricing but also the calibration and implementation of the models, and how they deal with extreme and unlikely events.

---

[7] Big data is usually defined as data that has sufficient volume, velocity, and variety. These attributes are known as the 3V's of Big Data. Some people prefer to state 5V's, with the inclusion of veracity and value but this seems superfluous: all data should always be of good quality and of some value in order to warrant collection and storage.

Machine learning is already being embedded within large insurance companies, typically alongside traditional statistical modeling. For example, the use of unsupervised machine learning to classify customers into driving categories, and the use of supervised learning in determining whether a theft was due to keyless car technology or not. Some companies are using machine learning to do the data preparation, create their pricing models, and deploy the tariff from start to finish. Actuaries with a good knowledge and capability in these approaches are already in high demand: within a decade this skillset might even be essential. Deep learning, which is a subset of machine learning that uses artificial neural networks to improve the model, and Artificial Intelligence, where the computer effectively mimics the human brain, are the next level of sophistication. Applications in insurance or pensions include catastrophe modeling and **reserving**, as well as credit scoring or spotting fraudulent claims. As more actuaries develop in-depth knowledge of these techniques and with the right data and regulatory approval, they could soon be considered powerful and essential tools for actuarial work, in the same way that **Generalized Linear Models** are today.[8]

One significant downside of Machine Learning is that it depends solely on the data. Therefore, if the data is biased in any way, the computer will replicate that bias in the model it creates. This is called systematic bias. For example, a local council creates a smartphone app to monitor if a driver drives over a pothole, with the aim of discovering where the most potholes are then fixing them in a fair way. They apply machine learning to this, and fix the potholes. However, the majority of the funding goes toward more affluent areas. What's the problem? Well, the data was collected on a smartphone app for drivers. Data on the experiences of people without smartphones, or those who don't drive, weren't collected so the results were biased toward the neighborhoods where lots of people do have smartphones and cars. Another example might be sitting in your home: does voice recognition software (including Siri, Alexa, and Google Home) often fail to recognize your voice? If so, it might be because of the systematic bias in the learning algorithm. These devices are trained to understand speech based on a database of clips of thousands of people speaking. What's the problem? The database of clips is 69% male voices, so these devices are trained to predominantly understand men's speech. These systematic biases mean that machine learning still requires a lot of guiding, validation, and review from data scientists and actuaries alike.

---

[8] In fact, an IFoA Working Party specifically evaluated neural networks against GLMs for pricing insurance in 1996. The report is very readable and provides an excellent introduction to neural networks. https://www.actuaries.org.uk/system/files/documents/pdf/0417-0438.pdf

As seen with telematics, the move to using unstructured data sources to give worthwhile insight to insurers is likely to be slow. However, cloud data storage, which can reduce the upfront infrastructure costs of data storage might make it more appealing, as will an increased awareness of and comfort with machine learning techniques. Actuaries are not immune from being replaced by Artificial Intelligence, but it seems more likely that actuaries will become shepherds and interpreters of AI insight, rather than being replaced entirely. This is because of the interdisciplinary nature of actuarial work—it is not only mathematical calculation (the area in which we have already been replaced with computers!), but considering if the data is appropriate, what modeling is required, eliciting expert opinion, synthesizing relevant market and environmental trends, using actuarial judgment, and the application of ethics and integrity. Combined with new forms of insurance, new modeling techniques, new claim types, and new risks, it seems likely that actuaries will have an even more important role in harnessing the power of data in years to come.

5

Annuities

Chapter Objectives

- Create a cashflow model
- Find the present value of a certain future cashflow
- Find the accumulated value of a deposit
- Use effective and nominal interest rates

The father of this pleasant grandfather, of the neighbourhood of Mount Pleasant, was a horny-skinned, two-legged, money-getting species of spider who spun webs to catch unwary flies and retired into holes until they were entrapped. The name of this old pagan's god was Compound Interest.

—Charles Dickens, Bleak House

5.1 Interested in Interest?

If you've ever played 'Heads I Win, Tails You Lose', you will already know the difference between certain and uncertain events.

Actuarial science is primarily concerned with the quantification of uncertain events. But in any introduction to actuarial science, it would be remiss not to include some detail on the quantification of certain events too; after

© The Author(s), under exclusive license to Springer Nature Switzerland AG 2022
C. Townsend, *A Risky Business*,
https://doi.org/10.1007/978-3-031-11673-5_5

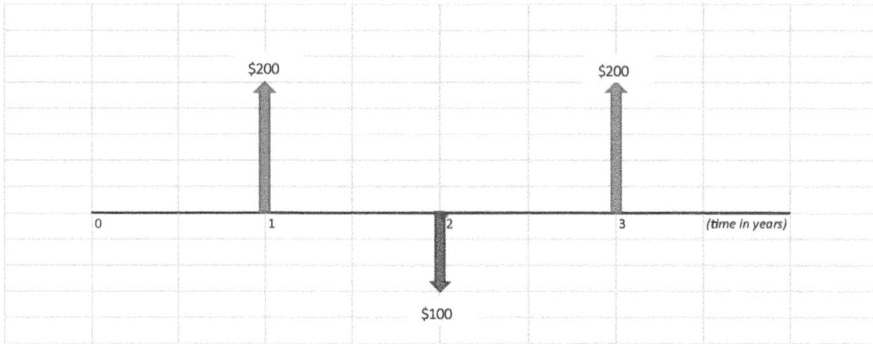

Fig. 5.1 An example of a three-payment cashflow model. The first and third payments are income, the second payment is outgoing

all, if we can't quantify certain events, what hope have we for the complexities of maybe?

In this chapter, we will consider monetary transactions that we are certain will happen at a given time in the future. Actuaries often use the term **cashflows** to refer to monetary transactions. When money is received, that is called a positive cashflow. A negative cashflow is when money is outgoing. We always want to know the value and the timing of every cashflow—that is, how much has happened and when it happened. While we could show that information in a table or list, an annotated timeline can help clarify the chronology. Actuaries call this a cashflow model.

Figure 5.1 represents a positive cashflow (we receive money in) of $200 after 1 year, a negative cashflow (we pay money out) of $100 after a further year, and a positive cashflow of $200 after one more year. Cashflow models such as this are useful for lots of reasons: to assess the benefits of a project, to model viability (e.g., to identify if a business will become insolvent), or to decide how much to sell a business for. **Cashflow** models are also used for **reserving** large claims or for modeling **reinsurance** recoveries.[1]

At any point on the timeline, we can easily see what money we have already received or paid out, and also what cashflows are expected to come. By considering what is yet to come, we can work out how much the future cashflow stream is worth. Note that the start of the timeline usually represents now—also known as time 0: analyzing historic cashflows that have no bearing on future cashflows is more the realm of accountants, not actuaries. But valuing a cashflow model is not as straightforward as just totting up the numbers on the timeline!

[1] See Chapter 10 for more on reserving, and Chapter 13 for reinsurance.

There is a 'time value' to money. Would you rather have a dollar in your pocket now, or a dollar in a year's time? Most people would think a dollar today would be better.

But how much better?

The answer depends on the interest you will be able to earn on your initial sum, and how far in the future you will receive the money. If the dollar was given to you tomorrow, you might feel indifferent. To calculate the value today of receiving \$1 in 1 year, we divide the dollar by $(1 + i)$, where i is the expected interest rate.

If the future dollar was in n years' time, where n is sufficiently large number of years, you would likely want to grab the cash now! In mathematical terms,

$$\text{Net Present Value} \times (1 + i)^n = \text{Value in } n \text{ years}$$

The **net present value (NPV)** is the actuarial term for the value today of a future cashflow. A rational person should be equally happy to receive either the NPV today or the future cashflow—they are equivalent. In the equation above, the future cashflow is received in n years, and the annual interest during this period will be i%. As you might expect, the NPV is very sensitive to the $(1 + i)^n$ term, so it is common for it to be called the accumulation factor. Similarly, the value in n years is often called the **accumulated value** at n. Our equation above therefore simplifies to:

$$\text{Present Value} \times \text{Accumulation Factor} = \text{Accumlated Value at time } n$$

This conversion of the invested sum into a larger sum is usually based on some important assumptions:

1. First, note that this assumes that we are using compound interest. Compound interest is when both the initial sum and the interest itself earns interest. For the rest of this chapter, we'll always assume we are using compound interest (unless explicitly stated otherwise).

   The alternative to compound interest is simple interest, where interest is only earned on the initial sum. This is uncommon because it involves remembering how much the initial deposit was, as well as keeping track of how much is in the account in total. Of course, it is also disadvantageous to the customer, who could earn more interest on the same money by withdrawing the earned interest and then immediately depositing it with the same bank.[2]

[2] Customers would then be incentivized to withdraw all their previously earned interest, and then re-deposit it so that the bank wouldn't know it was previously earned interest. This increase in

2. Secondly, the interest is assumed to be paid at the end of defined time periods, such as every quarter year or every year. It is commonplace for interest to be paid at the end of the year, for *n* years. The alternative is that interest is paid in advance. If the interest were to be paid upfront, a customer could then withdraw the money and then immediately deposit it again, at which point the bank would presumably pay the interest upfront and then the customer withdraws it and so on. If circumstances dictate, the bank may then require the customer to repay the advanced interest, but that is a lot more administratively difficult. Because of this, in practice, interest payments are never made in advance!
3. Finally, note that the time and the interest rate must 'match' in terms of time. In the above equivalence, we have assumed interest is only paid once a year, at the annual interest rate. This is called the effective rate: interest is paid once per time period.

Further, the above assumes fixed or stable interest rates however in reality, shocks and variability can and do occur. Shocks or events can even change the outlook overnight. After 9/11, interest rates dropped from 6.5% to 2% in less than a year. Anyone calculating the **accumulated value** of a lump sum based on 6.5% in early September would have overestimated the future value of their money. Even in these simple cashflow models, we can see that there is still some uncertainty: even if we are certain of the amounts and the timings, there is still uncertainty related to the interest rate.

So far, we have considered situations where interest is paid once per period, specifically, once a year. However, high street banks usually pay interest every month. But does it matter how frequently interest is paid?

Let's assume an account has a 12% annual effective rate. Is that equivalent to paying 1% every month?

The graph below compares 12% interest paid annually to 1% interest every month. If we start by investing $1000, in an annual payment, there is one payment of $120. In the other account there are 12 payments of interest, but the amount varies: initially it is smaller—the first payment is only $10—, but as the interest compounds, in later months the payment gets bigger: the final payment is $11.16. At the end of the year, the account that pays 1% every month has earned $6.83 more than the account paying 12% yearly. The space between the two lines shows the accumulated value of the interest (Fig. 5.2).

transactions would be operationally difficult for the bank. This practical reason is why simple interest is uncommon, and would only ever be feasible for accounts with limited withdrawals.

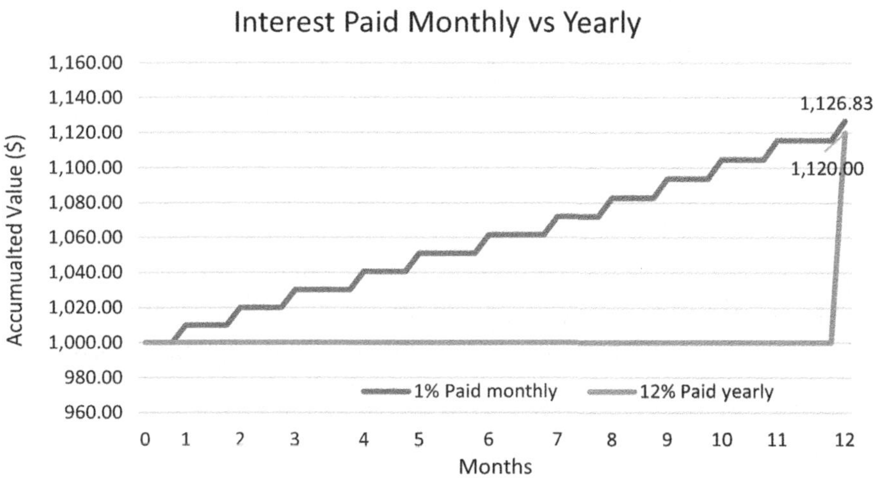

Fig. 5.2 1% monthly vs 12% annual interest

However, there must be an interest rate, paid every month, such that the two accounts will be equal at the end of the year.

> **Exercise 5.1** Using trial and error, calculate to 4 decimal places what that might be. [The answer will be calculated in Example 5.1 using another method.]

To ensure the two methods end up at the same value, the monthly interest rate must be, in percentage terms, smaller than 1%. To find this effective monthly rate, which we will denote as *m*, we use the following calculation, where *i* is the effective annual rate.

$$(1+m)^{12} = 1+i$$

> **Example 5.1** What is the effective monthly rate for an annual interest rate of 12%?
>
> $$\sqrt[12]{1+0.12} - 1 = m = 0.9489\%$$

So, the effective monthly rate in this example is less than 1%, but the amount of interest paid increases every month. In the first month, you only receive $9.49 in interest (or 0.9489% of your initial deposit), but by the final month you receive $10.528 (0.1053% of your initial deposit) in interest.

Did you know…?
To give a rough guide as to how long a lump sum doubles in value under compound interest, divide 72 by the annual interest rate. For example, with an interest rate of 8% a year, a lump sum will take 72/8 = 9 years to double. Strange but true!

It can be easier to pretend that the customer will be paid equal installments (in this case, equal installments of £10.57 per month, or 1.057% a month). This is called the nominal interest rate. The nominal monthly interest rate is denoted $i^{(12)}$. The following conversion can be used:

$$\frac{\text{Nominal Monthly Rate}}{12} = \text{Effective Annual Rate}$$

The benefit of nominal rates is that it is much easier to calculate the total interest (simply multiply one interest payment by the number of times it is paid). The downside is that it is hard to compare monthly nominal rates with quarterly nominal rates. For this reason, banks must clearly show the AER—annual effective rate—when advertising products. Because nominal rates are only really a simplification, they can't be used in accumulation calculations: convert to the effective rate as soon as possible.

Converting from nominal to effective rates allows us to calculate an accumulation factor. Using accumulation factors instead of individual interest rates saves us calculation time. It may avoid a calculation altogether, as they are listed in actuarial tables.

AER or APR?
In the UK, banks are required to give a customer an easy way to compare interest rates.

For savings products, they are required to give the *AER*, the annual effective rate: this is the same as the annual effective rate to be used in present value calculations.

However, for loan products, they are required to give the *APR*, the annual percentage rate: this is the annual effective rate of interest rounded to the nearest 0.01%.

Ultimately, the benefit of using a present value to evaluate the worth of cashflows (rather than their value at the time of the payment) is that they are directly comparable. For example, would you rather have £50 in a year's time, or £55 in 2 years' time? Even when told the effective annual interest rate is 5%, most people wouldn't be certain which is worth more. Contrast this with being offered £47.62 today or £49.89 today: the choice is obvious.[3]

5.2 Introduction to Annuities

An annuity is a very serious business.

—*Jane Austen, Sense and Sensibility*

In the last section we calculated the present value of a certain future cashflow. Now, we will extend this methodology in order to evaluate the present value of a series of regular cashflows.

A further benefit of calculating the present value of a cashflow, rather than the value at the time of payment, is that present values can be added together without further calculation. Therefore, even if the model has multiple cashflows (as shown in Fig. 5.1) , the **net present value** of the whole series can be calculated. A negative net present value means that the value of outgoing future payments is more than the value of future income and is often used as a quick way of assessing whether a business or project is viable.

Net present values also act as a 'fair price' (excluding transaction costs, profit or expenses, etc.) for a series of payments. This allows cashflow streams to be exchanged; so, a lump sum could be traded for a series of regular payments. Such a series is called an **annuity**.

Did you know ...?
In Ancient Rome, it was against the law for moneylenders to calculate interest using compound interest.

Who would buy an annuity? Why not just put the lump sum in a bank and withdraw the money like an income in future? There are lots of reasons why a customer may have this preference: perhaps a retiree doesn't want to burn through their lump sum too quickly and be left with no money in later life; perhaps someone doesn't trust themselves with that much money in one go, or perhaps (if annuity payments are guaranteed) they don't want to risk

[3] Assuming a rational person who subscribes to the economic assumption of monotonicity, that is, *more is more*. Remember that a good actuary always states their assumptions!

that interest rate changes will mean investing their lump sum won't grow as much as they expect. Perhaps the annuity rates (i.e., the price of an annuity) is just too good to miss. For whatever reason, the customer would prefer to trade in their lump sum for a regular series of payments in the future.

There are three key features of an annuity:

1. The payments are made at regular intervals.
2. The payments are of an agreed amount.
3. The payments continue for an agreed length of time (denoted by n).

In this chapter, we will only consider annuities where the payments are certain; the next chapter considers the case where the agreed length of time is contingent on an event. This chapter will also assume that effective interest rates are constant throughout the whole length of the annuity and denoted by i%.

A cashflow model of an annuity might look something like Fig. 5.3. This annuity will pay $100 at the end of every year for 3 years. The present is denoted by time 0, so the first payment will be made in one year's time.

The **net present value** (**NPV**) of an annuity is simply the sum of the present values of the payments.

Example 5.2 The net present value of the annuity in Fig. 5.3, assuming a 3% annual effective interest rate, is

$$\text{NPV} = \frac{\$100}{1.03} + \frac{\$100}{1.03^2} + \frac{\$100}{1.03^3} = 97.087 + 94.260 + 91.514 = \$282.86$$

While summing the NPV for individual cashflows is fine for short annuities, it is not uncommon for an annuity to last 25 years and so these

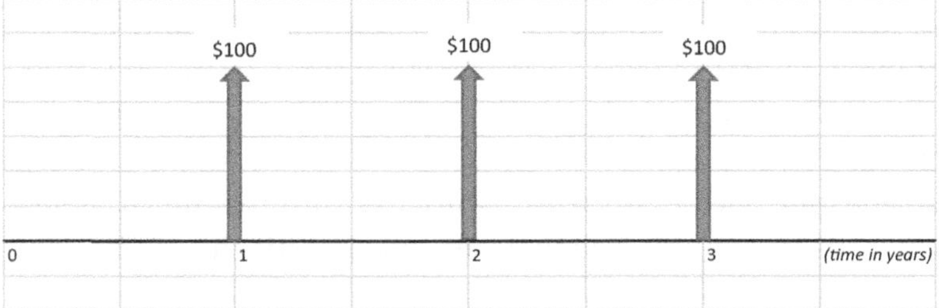

Fig. 5.3 A 3-year annuity of $100

calculations will quickly become tedious. Fortunately, using a bit of algebra, a simplification can be found:

$$\begin{aligned}\text{Net Present Value} &= \text{NPV of 1st payment} + \text{NPV of 2nd payment} + \cdots \\ &\quad + \text{NPV of } n\text{th payment}\end{aligned}$$

$$\text{Net Present Value} = \text{Payment Value} \times \left[\frac{1}{1+i} + \frac{1}{(1+i)^2} + \cdots + \frac{1}{(1+i)^n}\right]$$

You might recognize this as a finite geometric series. Using the result that

$$a + ar + ar^2 + \cdots + ar^{n-1} = a\left(\frac{1-r^n}{1-r}\right)$$

this simplifies to:

$$\text{Net Present Value of an Annuity} = \text{Payment value} \times \frac{1 - \left(\frac{1}{(1+i)^n}\right)}{i}$$

This final term is called the **annuity factor**, denoted a_n for a given interest rate.

Exercise 5.2 Use this formula to confirm the NPV in Example 5.2.

The formula for accumulation factors for annuities is so frequently used by actuaries that even inputting it into a calculator gets tiresome! For commonly used interest rates and values of n, the **annuity factor** is calculated in the *Formulae and Tables for Examinations of The Institute and Faculty of Actuaries* book, or your local equivalent. For interest rates of 3% and 5%, an accumulation and annuity factor table has been created in this book's appendix.

The problem with a present value—and the present more generally—that it doesn't last long enough! In 3 months, what we are today calling the present value will be outdated. Will we have to redo all our calculations? Fortunately, we can apply an accumulation factor to our present value to move it forward or backward in time with only one quick calculation.

Top tip
For the exercises in this book, you don't need a copy of the Actuarial Tables as the appendix contains a table of accumulation and annuity factors, and a small mortality table.

However, if you plan to sit actuarial exams it is well worth borrowing or buying a copy of the full version. At first, they seem difficult and unnecessary, but spend time upfront getting comfortable with using them. Get to know what is tabulated (and what is not), and practice, practice, practice!

This is because present values can be found and summed at any chosen point in time, as long as they are all referring to the same point in time. For example, two values at time 4 can be summed, but not a value at time 1 and a value at time 4. Using a timeline can help clarify which payments can be most easily grouped, reducing the amount of calculation.

In Fig. 5.4, instead of calculating the NPV for each payment in turn, it is equivalent to calculate the value at time 1 for all the payments (which is only four calculations not five, because the payment at time 1 is already at time 1), and then restate the NPV at time 1 as an NPV at time 0 using just one calculation. This means that once the NPV has been calculated, it can quickly be moved to any other point in time.

A similar logic can be used to find the present value at any time period, including at the time of the final payment. This gives us the following

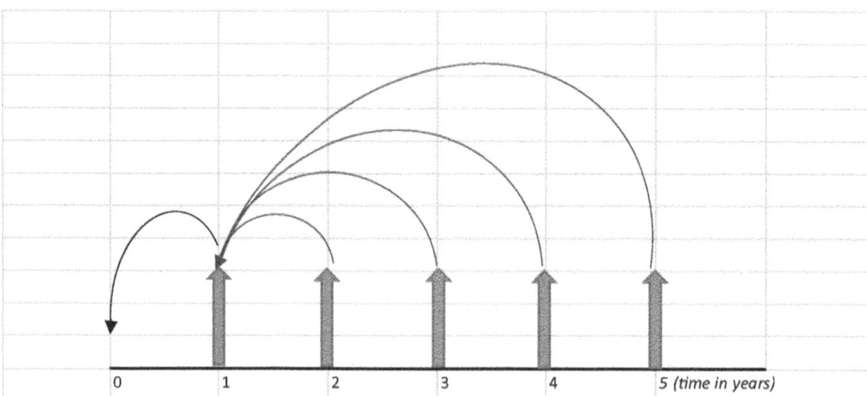

Fig. 5.4 Once the NPV of a series of payments is found, it can be quickly restated as an NPV at any point in time

equivalence:

$$\text{NPV at time } n = \text{NPV at time } 0 \times (1+i)^n$$

Example 5.3 The net present value at time 0 of the annuity in Fig. 5.3: A 3-year annuity of \$100, assuming a 3% annual effective interest rate, is calculated to be \$282.86. Using this information, calculate the net present value at time 0 of an annuity that pays \$100 at the end of every year for six years, at a 3% annual effective interest rate.

The given result is for a 3-year annuity of \$100. We can split the annuity in the question into two consecutive three-year annuities: one that is exactly like the one in Fig. 5.3 (that has a first payment at time 1 and an NPV at time 0 of \$282.86), and one that is a 3-year annuity of \$100 but where the first payment is at time 4, not time 1. We know that the NPV at time 0 for the first three payments is \$282.86. But this result also tells us that the NPV at time 3 for the final three payments is \$282.86.

So we have one NPV at time 0, and one at time 3. We need to roll the NPV at time 3 back to time 0:

$$\$282.86 = \text{NPV at time } 0 \times 1.03^3$$

Then, once they are both at time 0, we can add them together to give the NPV at time 0 for the whole six-year annuity:

$$\text{NPV at time } 0 = \$282.86 + \frac{\$282.86}{1.03^3} = \$541.72$$

How to Use Actuarial Tables

Actuarial tables are a regular companion for all actuarial students. Actuarial tables have several sections which each relate to a different type of calculation that you might be doing. Get to know your copy of the tables by having a good read through, because after some practice you will find that looking up values from the tables is quicker than typing it into your calculator. Try to spot the following sections:

- Formulae: This section contains information on statistical distributions (such as their mean, variance, moment generating function, probability density function and parameters). These formulae will be most useful to you in the statistics-based technical exams. Also in this section are mathematical formulae for the Newton–Raphson method, Integration by Parts, Taylor Series, and Bayes' Theorem. Day to day you won't be using these 'first principle' type formulae, but they are nice to have as a back-up in an exam situation.

- Compound Interest: This section of the tables will be the most relevant to this chapter. Each page is a set effective interest rate, and down the side is the length of the annuity in years. The table is full of the calculated values at the set interest rate: the accumulation factor $(1 + i)^n$, the accumulated values, the net present value of annuities, and the present value of an increasing annuity. Also calculated are the 'paid monthly' nominal rates.
- Population mortality tables (for standard lives, assured lives, and pensioners): The values in these tables will be explained further in Chapter 6, but the gist is that the ages of people are down the side, and the table contains the probability of them dying. It's a bit morbid, so after reading the pensions chapter, try looking up your own age and gender to discover the likelihood you'll die in the next year!
- Pensions (relating to present values of pensions, including rates of sickness, etc.): These sections relate to various regular payments that are conditional on the health or age of the recipient, for example, the present value of an annuity that starts at age 65, or a sickness benefit that pays weekly for the duration of a sickness. This section contains relevant assumptions and the corresponding calculations.
- Time Series: This basically contains some examples of useful time series: Retail Prices Index, the FTSE-100, bank base rates and even (in the UK), the National Lottery bonus ball number, among others. In an exam on time series, the question might point you toward one of these examples as some of the long-winded summary statistics are already calculated for you (you would usually have to show your working for the first part of the calculation to prove you know what you are doing, and then be allowed to just look up values from the tables).
- Statistical tables: This section contains the probabilities for commonly-used distributions, such as Gaussian, Chi-squared, Poisson, and the F distribution, and some pseudo-random values. You may already be familiar with such tabulations if you have sat a statistics exam before.

If everything is calculated in the book, what's the point of knowing the formulae? Well, in an exam they are unlikely to ask a straightforward question that can simply be looked up! You may be given an interest rate that isn't in the tables, forcing you to do it from scratch, or something like Example 5.3 where the annuity needs to be manipulated into something that can be looked up in the tables.

The other useful thing to know about actuarial tables is that they aren't cheap, but if your employer is sponsoring your actuarial exams, they may also be willing to pay for the tables!

5.3 Variations on Annuities

Because the accumulation factors and annuity factors are already tabulated in the Actuarial Tables, the skill is in understanding the situation and choosing the correct parameters for i and n. This becomes more challenging when the **annuity** is not straightforward.

The previous section considered a very specific type of annuity, in which:

- Payment is made once per year.
- The first payment is made at the end of the first year.
- Each payment is the same amount.
- The **cashflows** continue for a finite term.

If you're happy to simplify in this way, feel free to skip ahead to Sect. 5.4, as in this section we will relax each of these assumptions in turn. For annuity calculations that are less standard, drawing a timeline is recommended. Drawing a timeline will help in spotting exactly how long to accumulate for—draw the 'jumps' in yourself—and will help to ensure that no payments are missed. When doing present value calculations for annuities, remember that there is usually more than one way to get to the answer. For example, as we saw previously, the present value of each cashflow can be found and summed or the value of the cashflows at time 1 can be found. Values can be found and summed at any chosen point in time. As the annuities become more complex, there will be many ways to split them up. As with many actuarial problems, there is not a single correct method, but some smart thinking and splitting of the problem can be much more efficient.

5.3.1 Annuities Paid Monthly

If interest can be paid monthly, so can annuity payments. Specifically, if an annuity pays \$12 over the year, then if paid monthly, a payment of \$1 is paid every month. As we've seen previously with interest paid monthly rather than yearly, the annuity paid monthly will be worth slightly more than its annually paid equivalent, as the earlier payments will attract interest.

To quantify the value of the monthly payments compared to the yearly equivalent, we can convert our monthly payments into an annual payment

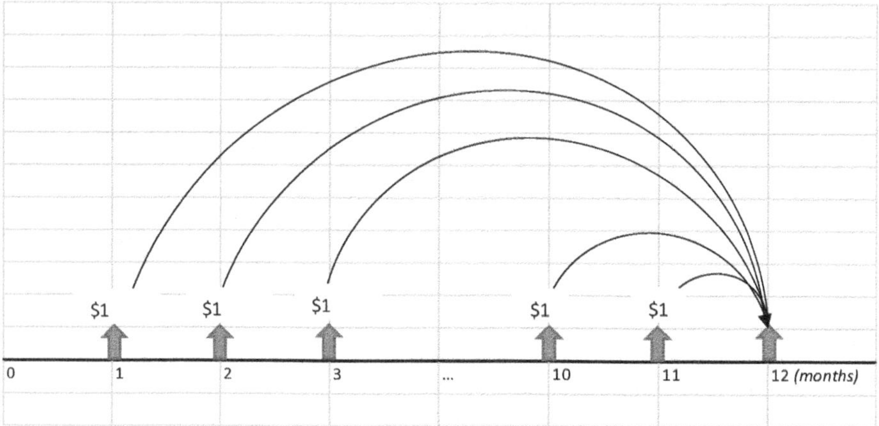

Fig. 5.5 Jumping between monthly payments and annual payments

by 'jumping' them forward to the end of the year, as in Fig. 5.5. What would the value of that annual payment be?

If the monthly payments are accumulated to the end of the year they relate to, we can see that this is equivalent to one annually paid annuity that pays $(1+i)^{11/12} + (1+i)^{10/12} + (1+i)^{9/12} + \cdots + (1+i)^{1/12} + 1$ at time 12.

The present value of a monthly paid annuity can be calculated either with reference to an annually paid annuity (and therefore can be found in the Actuarial Tables), or can be calculated from scratch. Using the logic above, we can simplify our calculation to:

$$\text{Monthly Paid Annuity Factor}_{i,n} = \frac{1 - \left(\frac{1}{(1+i)}\right)^n}{i^{(12)}} = \frac{i}{i^{(12)}} \times a_{i,n}$$

where $i^{(12)}$ is the nominal interest rate paid monthly (i.e., the interest rate pretending that every interest payment is equal) and $a_{i,n}$ is the annually paid annuity factor for an annuity of n years at interest rate i.

5.3.2 Deferred Annuities

A **deferred annuity** is one where the first payment is delayed until a specified future time. These would commonly be bought by people who would like to purchase an annuity now—perhaps because the prices of annuities are particularly cheap—but don't want or need the income stream for a while. The length of delay is denoted m, and as before, n is used to denote the duration of the payments. Therefore, the final payment of a deferred annuity is at $m + n$ years in the future.

Exercise 5.3 Draw a timeline of a 3-year annuity of $100, deferred for 5 years.

At time m, the deferred annuity is no different to a standard annuity! Therefore, we can calculate the Net Present Value of a cashflow as at time m, using the methods we are already familiar with, then jump that value back m years to time 0, as we did in Example 5.3.

$$\text{Deferred Annuity Factor}_{i,m,n} = \text{Annuity Factor}_{i,n} \times \left(\frac{1}{1+i}\right)^m$$

Exercise 5.4 Can you think of another way to calculate the present value of a deferred annuity? [Hint: instead of being a deferred annuity, think of it as a regular annuity that is missing some payments.]

5.3.3 Increasing Annuities

An increasing annuity is simply an annuity where the payments increase over time. Specifically, the annuity pays 1 at time 1, 2 at time 2, 3 at time 3, etc. One reason why a customer might want to buy a product like this is if they suspect their expenses will increase over time; for example, they anticipate that they will need an increasing amount of care or money for medical expenses as they age (Fig. 5.6).

Summing the present value of each payment gives us the present value of the increasing annuity.

$$\begin{aligned}\text{Increasing Annuity Factor} &= \frac{1}{(1+i)} + \frac{2}{(1+i)^2} + \frac{3}{(1+i)^3} + \cdots + \frac{n}{(1+i)^n} \\ &= \sum_{t=1}^{n} t \frac{1}{1+i}^{t}\end{aligned}$$

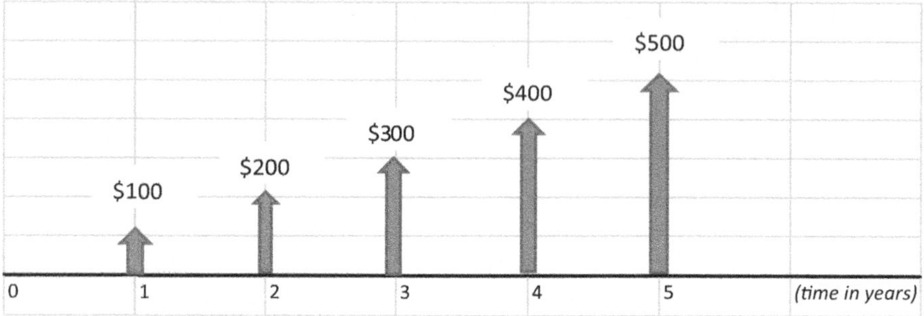

Fig. 5.6 An increasing annuity

Again, using our knowledge of geometric sequences, this can be written as:

$$\text{Increasing Annuity Factor}_{i,n} = \frac{\text{Annuity Factor}_{i,n-1} + 1 - n\left(\left(\frac{1}{1+i}\right)^{n}\right)}{i}$$

Increasing Annuity Factors are already calculated in the actuarial tables, so fortunately it is not necessary to recall this formula very often!

Of course, an annuity could be decreasing instead of increasing. These are less common than increasing annuities in real life, but prevalent in exam questions! One way of calculating the present value of a decreasing annuity is to break it up into a level annuity (where all payments are equal and of the value of the first payment) plus an outgoing increasing annuity, as in Fig. 5.7.

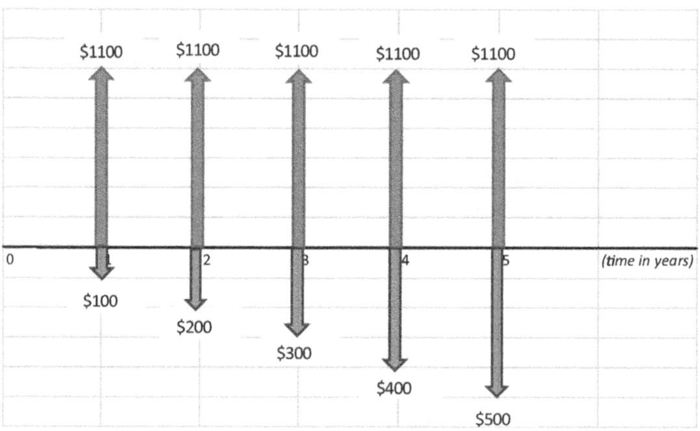

Fig. 5.7 A 5-year decreasing annuity that starts with a first payment of $1000 and decreases by $100 a year

Combining or splitting up complex annuities into more standard parts is usually quicker than calculating the present value of each payment individually, particularly as you get more practice.

5.3.4 Perpetuities

It is ironic that the final type of annuity we will consider in this chapter is one that never ends: it is payable forever. With an infinite number of payments, the present value of a **perpetuity** may seem like a difficult thing to evaluate, but in fact, perpetuities are very straightforward for two reasons:

1. Firstly, perpetuity factors can be found by letting n increase until it tends to infinity, which simplifies neatly.

$$\text{Perpetuity Factor} = \lim_{n\to\infty} \frac{1 - \frac{1}{1+i}^{n}}{i} = \frac{1}{i}$$

2. They aren't very common. There aren't many reasons to buy a perpetuity from a customer's point of view: most customers would prefer to have a larger payment for a shorter time (even shortening the term to 100 years is likely to cover the customer's lifetime!). But companies don't want to be continually paying out forever either, so perpetuities have historically been issued by governments, mostly in desperate attempts to raise money for war.

Did you know…?
In 2015, the British Government bought back £218 million of consols: perpetuities issued in order to raise money for war. The majority of those bought back were issued in 1927 by Winston Churchill to refinance debt from the First World War, but the oldest was from 1853.

Example 5.4 What might the annual payment have been for the consols that were bought back by the UK Government, if their net present value in 2015 was £218m, assuming an interest rate of 3%.

£218 million = Annual Payment × $\frac{1}{0.03}$, so we could estimate the annual payment as £6.54 million.

5.4 IOU: Calculating Loan Repayments

Although the terminology and notation of the previous chapter may seem niche, in reality most people are involved with an annuity. Unfortunately, it's not usually in the way they'd like: if you have a mortgage or loan with regular payments, you are paying an annuity to your lender.

In order to repay the loan by the end of the term, each payment must first cover the interest due and then some of the outstanding capital. Some loans don't require the loan to be paid until the end of the term, so the regular payments only cover the interest. Unsurprisingly, these are called interest-only loans. Interest-only mortgages were very common before 2008, but after the financial crash it became apparent that a lot of customers wouldn't be able to ever pay off the loan and would be forced to sell their home to pay the loan at the end of the term. Because of this, nowadays interest-only loans are not very common: lenders must check that the customer has a means of repaying at the end of the term and often require large deposits. In this section we will therefore assume that all loans are repayment loans.

It is unnecessary to calculate the present value at time 0 of repayment loans, as that is the amount initially borrowed. However, we may want to calculate:

- How much of the loan is outstanding (owed) at any given time, or
- How much of the regular payments go toward paying off the capital?

Given that the present value of all future payments at the start of the loan is equal to the amount initially borrowed, it is logical that the present value of all future payments at a given time t is equal to the amount borrowed—that is, still owed—at time t. So, we can use our knowledge of how to calculate present values of annuities to calculate how much of the loan is outstanding.

Example 5.5 A customer has five years until their mortgage is fully paid off. Their monthly repayment is £1,000, payable at the end of the month. The annual effective interest rate is 3%. What is their balance outstanding?

This is an annuity paid monthly, with parameters $n = 5$ and $i = 0.03$. Recall that the relevant annuity factor is therefore

$$\text{Monthly Paid Annuity Factor}_{i,n} = \frac{1-\left(\frac{1}{(1+i)}\right)^n}{i^{(12)}} = \frac{i}{i^{(12)}} \times \text{Annually Paid Annuity Factor}_{i,n}$$

where $i^{(12)}$ is the nominal interest rate paid monthly.

We can look up the nominal monthly interest rate and the annual annuity factor for $n = 5$, $i = 0.03$ in the tables, so without further ado (and remembering that the annual payment is £12,000, not £1,000), we can use this formula to give us an NPV of:

$$£12,000 \times \frac{i}{i^{(12)}} \times \text{Annually Paid Annuity Factor}_{3\%,5} = £12,000 \times 1.013677 \times 4.45797$$
$$= £54,227.30$$

So, net present value calculations can be used to find the balance outstanding. Until the loan is paid off, the monthly repayments can feel interminable and the homeowner may want to find out how much of each payment goes toward paying off the capital. Before we attempt this calculation, let's first consider how a mortgage balance changes over time.

At the end of the month, interest is applied to the remaining balance. For example, suppose someone has a \$100k mortgage initially. Then, at the end of the month, interest of \$100 is added to the balance. The homeowner now owes \$100,100. They then make a payment of \$500, bringing the balance to \$99,600. Here, the homeowner paid \$500 but the balance only reduced by \$400, as the repayment had to also cover the \$100 of interest accrued over the month. Notice that if the regular payment is more than the interest accrued, then eventually the mortgage will be paid off.

If the amount of interest accrued per month is known, it's easy to work out how much of the repayment goes toward paying off the capital owed—it's simply the mortgage payment less the interest accrued. However, assuming compound interest, the absolute amount of interest paid each month will change: as the outstanding balance decreases over time, the amount of interest added every month will also decrease over time.

As long as the interest rate stays the same, as the balance outstanding decreases, the amount of capital paid off each month increases. It's much quicker to pay off the last \$10,000 than it is to pay the first \$10,000, as you can see in Fig. 5.8. The higher the interest rate (and in Fig. 5.8, a very high interest rate of 3% per month has been selected), the more concave the graph looks. Specifically in this example, it takes 51 months to pay the first \$10,000, but only 4 months to pay the final \$10,000. It is also worth noting the proportion of the payment that is interest repayment. In this example, 97% of the first payment is interest repayment, but only 46% of the 100th payment is interest.

So, to calculate how much of a monthly repayment goes toward paying off the capital, we must do the following:

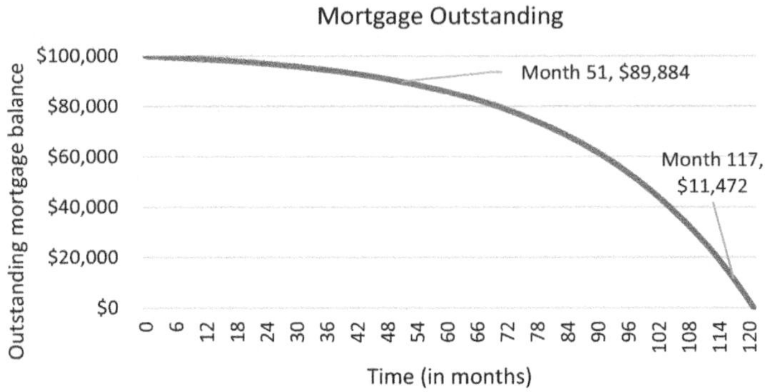

Fig. 5.8 Mortgage outstanding for a $100,000 mortgage, 3% interest per month, and a 10-year term. The two data points marked are the 51st month, when the balance first drops below $90,000, and Month 117, the last month before the balance drops below $10,000

1. Calculate the loan outstanding immediately after the previous payment.
2. To the loan outstanding, add the interest that would be applied to that amount over the month.
3. Reduce the balance by the repayment, then calculate how much the capital has reduced by.

This calculation is nothing new: it is simply a combination of calculations already covered in this chapter. Once you can find the accumulation factor and net present value of an annuity, there are lots of other loan-based calculations that can be performed, such as:

- How much will I owe on my student loans when I am 40?
- If the government makes a one-off payment of $10,000 immediately, when will I finish paying the loan off?
- If the interest rate increases to 3% but my payments remain the same, how will that affect the payment term?
- How much will I need to pay a month in order to be debt-free by 2030?

The possibilities are endless and slightly addictive, and can be helpful in turning a large and intangible loan into a series of smaller milestones.

This chapter will finish with a slightly longer mortgage calculation example. The next chapter considers annuities where the payments are somewhat uncertain—or does it?!

5.5 Worked Example

A homeowner takes out a 20-year mortgage for $250,000, at an annual interest rate of 5%, paid monthly. Mortgage payments are made monthly.

a. *How much is their monthly repayment?*
b. *If the homeowner pays an additional $5,000, how much sooner will they be mortgage free?*
c. *What proportion of their first payment (if they didn't pay the $5,000 lump sum) is interest?*

a. To calculate their monthly payment, we can use that the NPV is the amount owed. Therefore, we have that:

$$250,000 = \frac{0.05}{i^{(12)}} \times a_{20} \times \text{Annual payment}$$

By looking up $i^{(12)}$ and a_{20} in the actuarial tables at 5%, or otherwise,[4] we can substitute the terms in our equation to be:
$250,000 = \frac{0.05}{0.048889} \times 12.4622 \times \text{Annual payment}$, so by rearranging we find the annual payment is $19,615.106. This is paid evenly throughout the year, so dividing by 12 we can find the monthly payment of **$1,634.59**.

b. If the homeowner pays $5,000 immediately, this reduces the net present value of the annuity to $245,000, but leaves the monthly payment the same. Then, we can solve the following for n.

$$245,000 = \frac{0.05}{i^{(12)}} \times a_{20} \times \$19615.10631.$$

Recalling that $a_n = \frac{1}{i}\left[1 - \frac{1}{1+i}^n\right]$, this simplifies to:

$$0.6106418599 = \left[1 - \frac{1}{1.05}^n\right],$$

[4] When math textbooks say "or otherwise", it usually means that there is no other way to do the calculation. This is not a math textbook, and so the alternative method of calculating using the formula on page 103 is also welcome, although very slightly longer. Or, you could have looked up $\frac{i}{i^{(12)}}$ in the actuarial tables, as it is also typically tabulated.

$$n \log\left(\frac{1}{1.05}\right) = \ln(0.3893581)$$

$n = 19.3329$ years, which is 8.005 months sooner. So, the homeowner would be mortgage free **8 months sooner** than they would otherwise, just by paying an initial sum of $5,000.

c. To calculate the proportion of the first payment that is interest, note that their first payment interest payment is due at the end of the first month and calculated on the full $250,000. We use the interest rate given the question and, noting that this is an annual effective interest rate, we can apply it for 1/12th of the period. The outstanding balance at the end of the first month is therefore:

$$250,000 \times 1.05^{\frac{1}{12}} = \$251,018.50$$

So, the interest applied in the month is $1,018.50. This means that out of our first payment of $1,634.59, $1,018.50 is interest; that's **62.3%** of our first payment.

Top tip
Don't round your calculations midway through a question – make use of all the decimal places and only round at the very end.

6

Life-Dependent Products: Your Money or Your Life

Chapter Objectives

- Become familiar with life assurance terminology, notation, and common products
- Be able to create appropriate cashflows for uncommon circumstances
- Find the life expectancy of single and joint lives
- Find the present value of life-based annuities products

We all die son, it's just a question of when.

— Frank, Blue Bloods S1 E22

The previous chapter focused on the value of a series of **cashflows** that are given at certain times—that is, we knew exactly when the money would be paid. It is clearly going to be more difficult to evaluate the value of a series of cashflows that are paid at an uncertain time in the future, for example those that are paid if (and only if) something happens.

As we saw previously, annuities are very popular for those who have retired as the **annuity** income effectively replaces a salary. It is easy to see why a pensioner may want to buy an annuity, but it is less obvious what the term of such an annuity should be. If a longer term is chosen, the regular payments are smaller; if a shorter term is chosen then the recipient may be without any income at all after the term has ended. Ideally, someone in this circumstance

© The Author(s), under exclusive license to Springer Nature Switzerland AG 2022
C. Townsend, *A Risky Business*,
https://doi.org/10.1007/978-3-031-11673-5_6

would buy an annuity that lasts exactly the length of their remaining life: this gives them the maximum regular payment with minimum worry about the future. Therefore, if we knew when a person was going to die, we could use the techniques we learned in the previous chapter. All that remains is to determine when, exactly, an individual is going to die.

6.1 When Will They Die? (And Other Morbid Questions)

It is important (and surprisingly necessary!) to clarify that an actuary can't actually tell you the exact day or year you will die. However, they can make estimates about the typical life expectancy for people of different demographics: age and sex being the most commonly used, and the ones included in the Actuarial Tables.

Life expectancy is a familiar concept. It is widely used by governments and non-governmental organizations to compare countries, for example. Even without analyzing the statistics, unfortunately most people know someone who died 'too young'. However, life expectancy as used in these metrics or in common parlance usually is calculated as the expected lifetime for someone born today. Therefore, life expectancy in the traditional sense is not that useful when performing calculations, as the person will have already lived a bit of their life. An actuary is more interested in the future expected lifetime. For a person aged x, we denote this T_x.

But annuities aren't the only products that are concerned with the policyholder's death: another type of product is a whole life assurance. Unlike an annuity, a whole life assurance pays nothing while the policyholder is alive. However, on their death, it pays out an agreed sum. Also, unlike an annuity that has a specified term of n years, a whole life assurance will pay out on death whenever that occurs; it is more akin to a perpetuity in that no time period has to be specified upfront.

The cashflow model for a whole life assurance is a payment of the specified amount when the policyholder dies.

Using the techniques of the previous chapter, the value of the whole life assurance is the net present value of the payment, that is:

$$\text{NPV of a Whole Life Assurance} = \left(\frac{1}{1+i}\right)^{T_x}$$

Here, the benefit is paid exactly at the time of death. However, that's not very realistic for three reasons:

- There is almost always a notification delay—a delay between the death occurring and the assurance provider being informed of the death as the policyholder's family manage urgent practicalities and come to terms with their loss.
- Assurance companies require proof of death in the form of a death certificate. A doctor will provide this for the family, but there may be a delay of a week or so while the cause of death is established.
- For administrative reasons, the assurance provider often takes a few days to release the funds. For example, in order to check that the money would be sent to the correct account.

Insurance or Assurance?

Assurance and insurance, from a customer's point of view, are largely similar: they are products that pay out when a specified event happens. The distinction comes in the characteristics of the risk of the event occurring:

- Insurance products consider events that are unlikely to happen. For example, your house may flood, but it is more likely that it won't. Assurance products pay out for events that definitely will happen or are very likely to happen.
- For this reason, assurances are usually limited to life products (as death is certain, but there is no moral hazard to insuring it).
- Insurance aims to remove the financial risk of the event: if your house floods, your insurance will reinstate you to your previous financial position. This is called the principle of **indemnity**. The policyholder is not financially any better or worse off if the house floods, than if it does not flood.
- Assurance pays out a specified sum, though this value is not meant to represent the value of the life lost. The policyholder's estate (or beneficiaries of the assurance policy) is financially better off after the event than they were before it.

Payment immediately on death is also problematic for providers as they need to keep a larger proportion of funds liquid. This is a problem because in exchange for the liquidity (the ease of access to the money), they may receive a lower interest rate on the money, reducing their returns. For this reason, whole life assurances are often payable at the end of the year of death. We denote the value of a whole life assurance payable at the end of the year of death as

$$A_x = \left(\frac{1}{1+i}\right)^{(K_x+1)}$$

Effectively, we want to round the value T_x up to the nearest integer. We can write that as $K_x + 1$, where K_x is the curtate future lifetime of a life aged x—that is, K_x is the integer part of T_x. A_x, the value of a whole life assurance payable at the end of the year of death for a life aged x, is tabulated for some interest rates.

Whole life assurances were very popular post-war, until the 1980s when interest rate variability made providers jittery about taking on risks for such a long time. The present value of whole life assurances is highly sensitive to the interest rate, so estimating this accurately (which is incredibly difficult in periods of high volatility, and even more so over long periods of time) is vital to avoiding large losses. For this reason, after the 1980s, another variant emerged: Term Assurance.

A term assurance is a life product that pays out if the policyholder dies within a specified term. For example, a 25-year term assurance might pay £100,000 if a policyholder dies *within* those 25 years. If the policyholder dies *after* the term ends, they receive nothing.

Exercise 6.1 For a customer seeking to purchase an assurance product, which would you expect to be cheaper: a whole life assurance or a 20-year term assurance? Would your answer depend on the age of the policyholder?

For actuaries, the difference in value between a term assurance and a whole life assurance depends on the likelihood of a policyholder dying during the term, and the value of a term assurance is very sensitive to this assumption. Specifically, in a whole life assurance, a policyholder dying sooner than expected only has the effect of bringing forward payments, whereas in a term assurance, this has the effect of giving rise to a payment that otherwise would not have occurred at all. In term assurances, dying one year sooner can change the profitability of the product for the provider significantly, if that year is the final one in the term. For term assurances, actuaries need accurate and precise estimates of not only future life expectancies, but also the probability of death in each year.

6.2 What Life Tables Tell Us About Life

So what is the probability of death in any given year of age? National statistics bureaus can give us this information, as they count how many deaths occur per year for each year of age. By dividing this by the number of people who reached that age in the first place, they can give an estimate for the probability of death in the next year. For example, if 1000 people reached their 50th birthday in a country, but only 980 reached their 51st birthday, the probability of survival between the ages of 50 and 51 would be 0.980, and therefore the probability of death between the ages of 50 and 51 would be 0.02. This probability is denoted q_{50}, and more generally, the probability of death between the ages of x and $x + 1$ is denoted q_x (Fig. 6.1).

Unsurprisingly, there is a much higher likelihood of dying in the next year if you are 90 than if you are 49: older people are most likely to die of heart disease, cancer, flu, Alzheimer's, and almost every other disease, not to mention frailty leading to falls and other accidents. The trend of increasing mortality with increasing age for adults is consistent across most developed nations (this dataset is from Australia) and probably not new information for you. However, the relatively high probability of death at older ages obscures the trend for younger ages (Fig. 6.2).

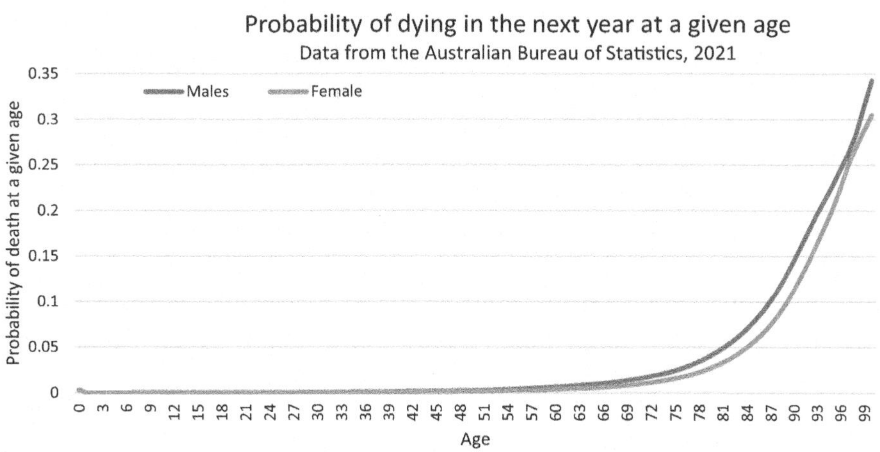

Fig. 6.1 Probability of dying in the next year at a given age

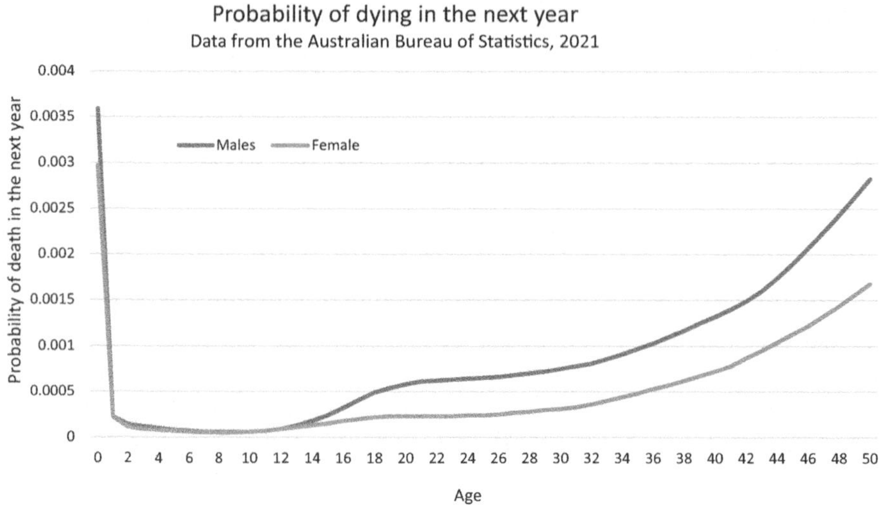

Fig. 6.2 Probability of dying at a given age, zoomed into mortality for under 50s

By zooming in to the younger ages, a much more interesting (albeit no less morbid) graph emerges: mortality does generally increase with age, with the exception of babies between the age of 0 and 1.

As a cohort, babies (under 1) have the highest future life expectancy of any age group. However, the risk of death for this group is unfortunately relatively high; genetic illness or complications during birth may mean that a newborn won't survive. Post-birth, new parents will be worried about sudden infant death syndrome (previously called cot death), and with their immature immune systems, infants are highly susceptible to infectious diseases.

Mortality then drops again, only to increase in the late teenage years. This increase in mortality is often attributed to risk-taking behavior such as reckless driving, drink-related deaths, recreational drugs, and other preventable causes, and is known as the 'accident hump'. Although unintentional accidents are the leading cause of death for teenagers in all developed countries, tragically, suicide is the second highest cause of death for this age group.

It is for this age group that the largest difference is seen between male and female lives. Traditionally, male and female lives have been enumerated separately due to the differences in causes of deaths (manual working accidents versus maternal mortality) and more recently, due to the difference in prevalence of diseases such as heart disease, cancers, or other late-life illnesses. However, not much has been made of the difference in mortality between young men and women. The graph below is data from the US Centers for Disease Control and Prevention, although there are similar trends seen in

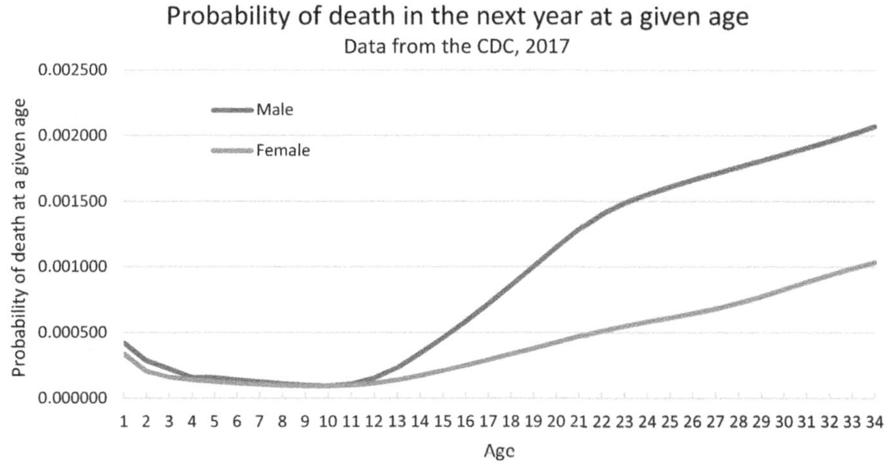

Fig. 6.3 Probability of dying at a given age. Notice how male mortality is much higher than that of females due to risk-taking behavior

other developed countries; male teenagers have a much steeper accident hump than females of the same age (Fig. 6.3).

Interestingly, though most developed countries exhibit the same trends and causes of death for infants and adults; the US is a bit of an anomaly for preschooler (aged 1–4) and child (4–15 years) mortality. Here, among the lowest mortalities of the life table, it is possible to see some of the most tragic deaths of all—the preventable deaths (Fig. 6.4).

In the UK, the leading cause of death for 1–15-year olds are childhood cancers [ONS,[1] 2016], whereas in the US, the leading cause of death for this age group is drowning [CDC, 2017], predominantly in swimming pools. Death by shooting (homicide) is the 4th highest cause of death for 1–9-year olds in the US: in the UK there have been no school shootings since handgun control was introduced in 1996. The absolute level of mortality is low for under-21s and they are unlikely to buy life insurance products themselves. However, analysis of their mortality is still important, albeit from a public health perspective.

Life tables are free and publicly available for most countries and are full of interesting statistics. Life insurers do not solely use national life tables in their pricing as they are not very granular, although they are normally provided for male/female and any provinces/states/territories. However, they can give a good indication of general trends in the causes and numbers of deaths.

[1] The UK's Office of National Statistics. All data in this section is publicly available from the relevant government source for each country.

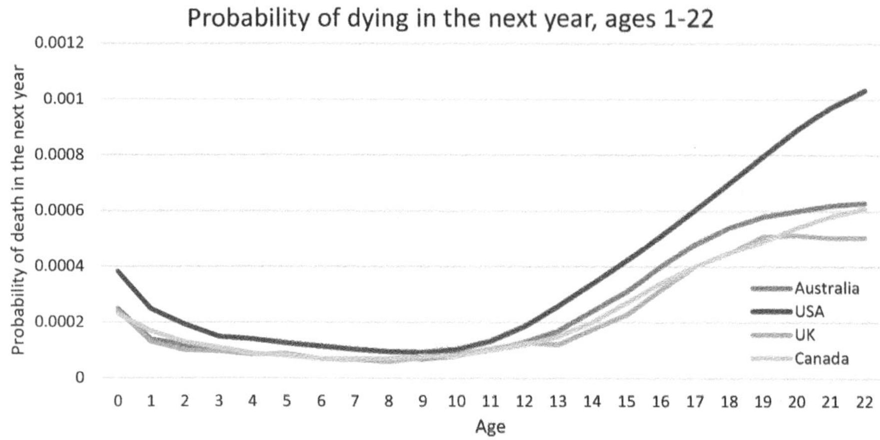

Fig. 6.4 Mortality for a subset of developed countries (Data from the Australian Bureau of National Statistics, CDC, the Office of National Statistics, and Statistics Canada[2])

They are usually compiled by actuarial teams, so their production and analysis are one example of a public sector actuarial role. Public sector actuaries also advise policymakers on demographic changes, the state pension, and much more. Such analysis is both morbid and influential as by comparing life tables with other countries, other provinces, or other genders we can start to infer how societal trends and government policy affects how all of us live—and die.

6.3 Calculations with Life Tables

With a general feel for using life tables, we can now start some technical calculations.

We have already introduced q_x, the probability of death between the ages of x and $x + 1$. The other useful (and slightly more optimistic) probability is p_x, the probability of survival between the ages of x and $x + 1$. This is calculated as $1 - q_x$, and is also tabulated in actuarial tables.

Let's return to the expected present value of a term assurance. Remember, term assurance is a product that pays a lump sum at the end of the year of death if the policyholder dies within the term (of n years), otherwise it

[2] Here I've chosen the four developed nations that produce the most actuaries. For a comparison of all countries, the World Health Organization has collated the relevant data here: Life tables by country (GHE: Life tables) (who.int).

pays nothing. The expected present value can be viewed as the probability-weighted combination of $n + 1$ outcomes:

- Outcome 1: the policyholder dies in the first year. NPV of this outcome = Lump sum$/(1 + i)$
- Outcome 2: the policyholder dies in the second year. NPV of this outcome = Lump sum$/(1 + i)^2$
- ...
- Outcome n: the policyholder dies in the nth year. NPV of this outcome = Lump sum$/(1 + i)^n$
- Outcome $n + 1$: the policyholder dies after the end of the term. NPV of this outcome = 0

Using the life tables, we now are able to accurately assign probabilities to each outcome. If the policyholder starts the term assurance on their xth birthday, the probability of outcome 1 is simply q_x. However, assigning outcome 2 the probability q_{x+1} overlooks the fact that to die between $x + 1$ and $x + 2$, our policyholder must survive until they are $x + 1$ in the first place. Consequently, the probability of outcome 2 is $p_x q_{x+1}$.

Then, to achieve outcome 3, our policyholder must survive the first two years and then die so this outcome has probability $p_x p_{x+1} q_{x+2}$.

For outcome $n + 1$ there is a shortcut: the probability of a policyholder dying at any point after n years is simply the probability of them surviving at least n years: $p_x p_{x+1} p_{x+2} \dots p_{x+n}$. However, first recall that the net present value of this outcome is 0, as no lump sum will be paid, so the probability of this occurrence is unimportant.

Example 6.1 What is the expected present value of a five-year term assurance which pays out \$100,000 at the end of the year of death, which begins on the policyholder's 60th birthday, assuming an interest rate of 3%?

Outcome	Probability of occurrence	1/accumulation factor	Probability /accumulation factor
Dies in 1st year, aged 60	$q_{60} = 0.005774$	$\frac{1}{1.03} = 0.970874$	0.005606
Dies in 2nd year, aged 61	$p_{60} \times q_{61} = 0.006396$	$\frac{1}{1.03^2} = 0.942596$	0.006029
Dies in 3rd year, aged 62	$p_{60} \times p_{61} \times q_{62} = 0.007077$	$\frac{1}{1.03^3} = 0.915142$	0.006476
Dies in 4th year, aged 63	$p_{60} \times p_{61} \times p_{62} \times q_{63} = 0.007821$	$\frac{1}{1.03^4} = 0.888487$	0.006948
Dies in 5th year, aged 64	$p_{60} \times p_{61} \times p_{62} \times p_{63} \times q_{64} = 0.008631$	$\frac{1}{1.03^5} = 0.862609$	0.007445

The total expected present value is the sum of the right-hand column, multiplied by the lump sum; in this case, 0.032504 × \$100,000 = **\$3250.43.**

These calculations are repetitive, and even in the short example above some shortcuts can be made. For example, perhaps one might evaluate $(p_{60} \times p_{61}) \div 1.03^3$, save that answer on your calculator and use the saved value in your calculations of outcomes 3, 4, and 5, instead of rekeying. A spreadsheet is the obvious choice, although not usually available in exam conditions. As with annuity calculations, there are multiple ways to get the correct answer.

Term assurances are popular for a few reasons. The first is that the nearest alternative (whole life assurance) is not commonly sold. Secondly, a term assurance is much cheaper than a whole life assurance for a policyholder the same age, the difference in value for a lump sum of 1 being A_{x+n}. Thirdly, life assurances are usually bought in order to service debt that the policyholder has, or costs that would arise in the case of their death. This is most commonly a mortgage with a fixed term. Life assurances are also often bought by parents of young children in order to provide for their children in the case of their death. If a parent of young children died, not only would the remaining spouse be the only earner, but also the sole caregiver and solely in charge of the many domestic duties that arise with small children. A life assurance could enable the remaining parent to get a nanny or cleaner, for example. Single parents are particularly keen on having life assurance, fearing their children being left with nothing and no one. These motivations also explain why term assurance is more common: hopefully, both mortgages and children eventually cease to be financial liabilities! Term assurances are often purchased with the term either equal to the remaining mortgage term, or until the policyholder's children are expected to be financially independent.

These motivations also explain the popularity of another product: decreasing term assurance. Like a term assurance, it pays out a lump sum at the end of the year of a policyholder's death. However, the lump sum decreases in absolute terms over time. For example, in the case of death in that year, \$100,000 might be paid at the end of the first year, but only \$95,000 in the second year, \$90,000 in the third, etc., until the term ends. The price of such a product can be determined using a similar logic to decreasing annuities. Here, the value of a decreasing term assurance can be written in terms of a level term assurance less an increasing term assurance.

6.4 Life-Dependent Annuities

The previous section covered whole life and term assurances: products that pay the benefit after the death of the policyholder. This section will cover the opposite; products that pay the benefits until the death of the policyholder. These products are popular with retirees as they create a regular income stream, but they eliminate the difficult decision about how long the annuity term should be. Recall that given a specific lump sum, a longer term means a lower regular payment, but a shorter term could leave the policyholder with no income at all. By buying an annuity until death, the risk of living longer than expected and outliving a fixed term (called **longevity risk**) is passed to the annuity provider. Note that longevity risk is the opposite of the mortality risk that was held by a term assurance provider; the mortality risk is the risk that a policyholder will die sooner than expected. On a practical level, buying a life-dependent annuity is also good for the customer, as most people don't have the knowledge nor the data with which to estimate their own future expected lifetime—not everyone is an actuary after all! The wider benefits of large-scale pension schemes and the relative merits of passing or retaining longevity or mortality risks are discussed in the next chapter. The rest of this chapter will focus on calculations pertaining to life-dependent annuities.

From first principles, the expected present value of an annuity that pays \$1 at the end of every year providing that the policyholder is still alive can be expressed as the sum of the expected value of each payment. The graph below shows the present value of every potential payment (assuming a 1% interest rate) and the probability of 1-year survival for each age, given that the policyholder is 65 today (Fig. 6.5).

There are, of course, many potential payments, although there are none past the age of 100 in the graph above. Many life tables stop at age 100: there simply isn't enough data to make any reliable conclusions past that age. In effect, these life tables assume that a person will die once they reach 100.

Evaluating the expected value of a life-dependent annuity in this way would be incredibly time-consuming for medium or large values of n. To alleviate some of the pain of these calculations, the expected present value of a life-dependent level annuity paying $1 a year is calculated in the actuarial tables using the notation a_x.

A life-dependent annuity may be deferred; that is, the payments do not start being paid until some point in the future. This might be used by a person who receives a lump sum while still of working age, but wishes to use the money to increase their income in retirement. A cashflow timeline can help identify the differences compared to an immediate annuity; at time m, it is simply an immediate life-dependent level annuity that we already know how to evaluate. However, there is 'extra' timeline before that point. To roll back to time 0, the immediate life-dependent level annuity value must be converted to a time 0 present value by dividing by $(1+i)^m$. The additional consideration for life-dependent deferred annuities is that the policyholder

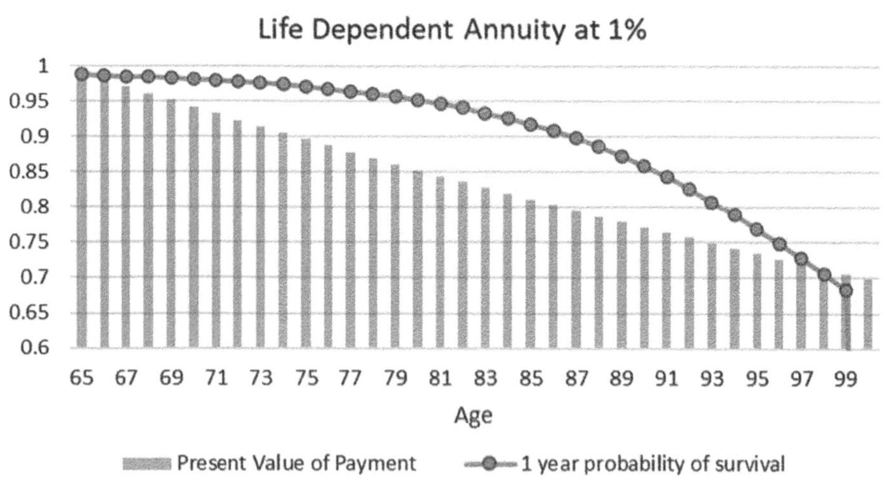

Fig. 6.5 Comparing the present value of each yearly annuity payment with the probability that the payment will be made, given that the person survived the previous year

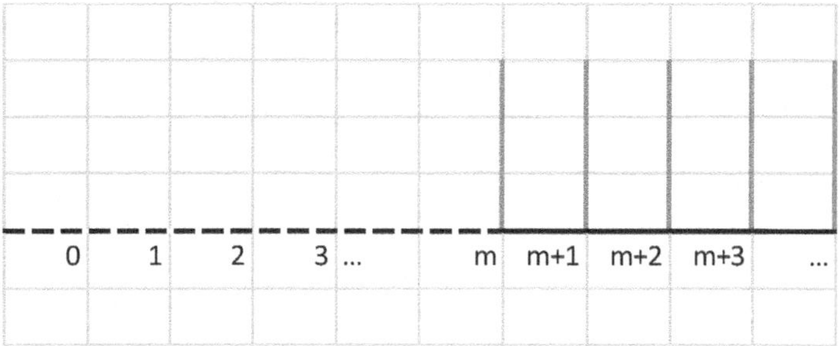

Fig. 6.6 A cashflow model of a deferred life-dependent annuity. Here, it is deferred for *m* years, and ceases upon death

must be alive at time m, so we must also multiply by the relevant p_x probabilities. A deferred annuity is denoted $a_{m|n}$, i.e., the net present value of an annuity for a life aged n, deferred by m years, paying \$1 per year from age $n+m$ (Fig. 6.6).

Example 6.2 A life-dependent annuity for a 60-year-old has an NPV of \$250,000 from the point of the first payment. A customer wishes to buy this annuity, but have it start in 3 years' time, instead of immediately, as he is currently 57. What is the NPV of the annuity he wishes to buy, assuming a 3% interest rate?

As in the previous chapter, we want to roll back the NPV 3 years. But we also need the customer to survive for three years! So, the NPV at time 0 (that is, three years before the annuity payments begin) is:

$$\frac{\$250,000}{1.03^3} \times p_{57} \times p_{58} \times p_{59} = 228785.4 \times 0.995829 \times 0.995351 \times 0.994818$$
$$= \$225,596.80$$

6.5 Variation in Mortality Rates

So far, this chapter has assumed that people of a certain age and gender in a country all have the same life expectancy, or the same probability of surviving a set time. However, there are many other factors that influence mortality. There are two motives for examining these in detail. Firstly, to accurately price life-dependent assurance or annuity products; and secondly, to reduce their effect on mortality. These are two very different goals: the former is to appreciate the impact of each factor but ultimately to change the behavior of the life assurance company, whereas the latter is to learn about the **risk**

factors in order to change the behavior of the population. This might be the role of a public health department, an actuarial advisor to the government, or (on a smaller scale) individuals concerned with their own health.

Exercise 6.2 Other than age, gender, and country, can you think of any risk factors contributing to differences in mortality?

There are many risk factors, and some could be difficult to distinguish from each other. Some include:

- Occupation
- Housing
- Access to healthcare
- Level of poverty
- Nutrition
- Smoking status
- Hobbies (especially high-risk activities such as sky-diving)
- Genetics

We'll discuss the effects of some of these risk factors below.

6.5.1 Occupation

The Bureau of Labor Statistics has found that the average American spends 44 hours a week at work, so it's no surprise that your job has a large impact on your overall life and health. Occupation can directly or indirectly impact mortality. A direct effect on mortality might be a job such as active-duty military, where death is an ever-present threat. However, most people in developed countries don't have such dangerous roles, and yet their job still affects their mortality indirectly. For example, as an actuary, it is unlikely that a spreadsheet will kill you (although there may be occasions where it feels like it!), but doing such a sedentary job may lead you to gain weight and increase your risk of cardiovascular disease. For indirect effects on mortality, particularly sedentary roles, it could be argued that occupation is not important if a person is active in their leisure time. Is there a discernible difference between a Zumba instructor who spends all day being active and then watches seven hours of television every evening, and an office worker, who spends seven hours working in front of a screen, but is particularly active in their leisure time? Here, it is really screen time and lack of activity that is in fact the cause of the impact on mortality, and occupation is being used as a quick way of assessing that. As the popularity and prevalence of wearable fitness trackers continue to rise, perhaps occupation will become less useful and assurance companies will instead ask for a monthly average of your daily step count. They could also quantify the level of occupational stress that an individual is under by tracking heart rate or blood pressure.

There are practical issues with assessing the risk of a profession. Firstly, people may over- or understate their job, making it difficult to group accurately. For example, a company CEO seems like a very low-risk office role, but the title may be used by a sole trader in the scaffolding or roofing industry. Another difficulty is obtaining sufficient data for some professions, particularly those such as actuarial which have high entry requirements and therefore low numbers of practitioners.

Jobs that have frequent medical checks, such as pilots, might attract a healthier type of person. Crucially, these checks eliminate individuals who do not meet the required health standard, leaving an individual in poorer health to move to another profession. These individuals (who likely have similar skillsets) may enter a new profession in sufficient quantities to mean that the new profession experiences higher mortality. A similar effect was particularly acute in the UK when coalmines were shut down, leaving many thousands of men out of work and possibly with damaged respiratory systems. In these cases, recording a full employment history would be the ideal solution, but may lead to issues in choosing which profession a person should be categorized as—does a three-month stint waitressing in college count? What about a three-month stint working with asbestos? Also, collecting all this information would be logistically difficult for both life assurance companies and public health officials, especially for those individuals who change jobs frequently.

6.5.2 Smoking

Most people are aware that smoking cigarettes is bad for your health, and certainly for life assurance products smoking dramatically alters the price of the product you buy.

Exercise 6.3 How would smoking status affect the price of (a) a whole life assurance, (b) a 20-year term life assurance, (c) 20-year annuity certain (i.e., paid to the estate regardless of survival), and (d) a life-dependent annuity?

However, since their recent introduction, e-cigarettes have increased in popularity, particularly with younger demographics. Insurance companies needed to decide how much to charge e-cigarette users relative to non-smokers and traditional tobacco users. In 2018 the Association of British Insurers surveyed several companies, who all said that the price for e-cigarette users would be the same as traditional tobacco users. This is the prudent choice, as medical evidence suggests that e-cigarettes may be better, but they are certainly worse than no tobacco at all.[3]

6.5.3 Income-Related Risk Factors

Poverty, nutrition, housing, and access to healthcare are often correlated and can be grouped together as effects of poverty; those who can least afford safe and suitable housing are also those who will be least able to afford a balanced and nutritious diet and good quality, comprehensive private healthcare.

The effects of obesity and malnutrition on mortality are well-documented. Heart disease is the biggest killer in the US, with diabetes being listed seventh. Chronic liver disease (usually caused by excess alcohol consumption) is the tenth biggest killer in America today. Assurance providers will usually capture weight or Body Mass Index (BMI) data about a policyholder as a proxy for diet, and it is easy to see the relevance. Of course, not everyone who is overweight is in poverty and not everyone who experiences poverty is overweight, but poor quality, highly processed food lacking in nutritive value can be cheaper than individual fresh ingredients, especially when the cost of travel to the store, any food waste, and cooking fuel or utensils are factored in.

Poor quality housing often conjures images of overcrowded, unsanitary Victorian slums and there is no doubt of the effect of such living conditions on the mortality of the residents.[4] Unfortunately, poor quality housing still exists but in different and sometimes invisible ways. Damp and spore-based fungi can cause and aggravate asthma and respiratory health, and overcrowding can spread bouts of coughs, colds, and other infectious diseases such as typhoid. While typhoid outbreaks seem a world away, in 2020 the most overcrowded London boroughs such as Newham (where 25% of people live

[3] Open Research: Electronic cigarettes and health outcomes: systematic review of global evidence (anu.edu.au).

[4] In the late 1890s, Bethnal Green in East London was an infamous slum, where 25% of babies died in their first year and the average (total) life expectancy for males was 19 years. The charity Action For Children has found that in 2020, 59.6% of children in Bethnal Green are living below the poverty line, the highest concentration of child poverty in London.

in conditions that the council deem overcrowded[5]) were hit hardest by coronavirus. Covid-19 modeling will be covered in more detail in Chapter 17, and it is difficult to precisely link the spread in these areas to housing explicitly, but lack of space in which to isolate would have contributed.

6.5.4 Genetics

Historically genetics has not been considered a risk factor for life assurance (beyond genetic illnesses that are often spotted at birth), as genetic testing was prohibitively expensive. Even now, family medical information is usually collected as a proxy. Some issues (both practical and ethical) with collecting at-home genetic testing information was discussed in Chapter 4, but this is an area where future actuaries may have to wrangle both with large, complex datasets, and their own ethical code.

6.5.5 Selection Bias

Selection bias occurs throughout science, normally used to mean that the sample was not randomly chosen and therefore extrapolating results from this group to the wider population is invalid. In actuarial science, this term means that the customers an insurer or pensions provider attracts are not randomly selected from the population.

It has already been noted that customers may struggle to accurately estimate their future life expectancy, whereas a life assurance provider will have plenty of survival data for people with the same risk factors. However, when it comes to life and health, a customer arguably has insider knowledge. Assurance providers will of course try to quantify the key factors such as age, sex, occupation, weight, medical history, and smoker status, but they will not be able to capture everything. For example, a person might have multiple aunts, uncles, and grandparents that survived to grand old ages and this might give them the impression that they too will live well into their nineties. An assurance provider may ask about medical history or diagnosed medical conditions, but it won't be able to collect data on general and mysterious aches and pains; a knee that sometimes hurts a bit, the regular headaches treated with over-the-counter paracetamol, or the chest pain that you get occasionally but which goes away untreated.

---

[5] Overcrowded Households by Borough—ONS. https://data.london.gov.uk/dataset/overcrowded-households-borough.

Someone who perceives they are in worse health than the average for their age may make different choices, bringing forward big trips and eschewing saving. This difference in behavior includes buying different life assurance products: a person who perceives that their health is worse than average may opt for a term assurance, whereas someone with expectations for a long life might buy a life-dependent annuity on the basis it will give them 'better value for money'. Although not known for her actuarial prowess, Jane Austen is right to note that "*People always live forever when there is an annuity to be paid them.*"[6] This is not merely a determination to wring the last penny from assurance providers, but due to this effect of selection bias.

6.6 Joint Life Annuities

In this chapter we have considered life assurance products that are dependent on one life. However, spouses often organize their financial arrangements together, so products dependent on two lives are also popular. A joint life product is—as the name implies—dependent on the status of two lives. There are four possible states here: both alive, both deceased, and the two combinations of one person being alive and the other deceased.

One key assumption in these calculations is that the two lives are independent. This is purely for mathematical convenience. In fact, it is highly unlikely that two married individuals' lives will be independent. They will probably have similar diets, breathe in the same cigarette smoke, and they are more likely to be in an accident together. As alluded to previously, they are also likely to contract the same illnesses, which can have lethal consequences. However, for expediency, let's assume their lives are independent.

There are two statistics of interest here. Firstly, the joint life status, denoted xy, which means that both people must be alive. So, an annuity a_{xy} means an annuity that is paid annually while both people are alive. It stops when the first death occurs.

The second statistic of interest is the last survivor status, denoted $\overline{xy}$. This refers to the time until both people die. In this way, an annuity $a_{\overline{xy}}$ is paid annually until the second death occurs and both people have died.

Exercise 6.4 Which is bigger, A_{xy} or $A_{\overline{xy}}$, that is, the Net Present Value of a lump sum paid on the joint life status or the NPV of a lump sum paid on the last survivor status?

[6] From Sense and Sensibility.

Both statistics can be evaluated using single-life functions. The probability that the joint life status survives the year is equal to

$$p_{xy} = p_x p_y$$

But care needs to be taken with the inverse; the probability of joint life status failing in a year is NOT equal to $q_x q_y$. This is because joint life status can fail in three ways: person x dies, person y dies, or they both die. It is easier to write

$$q_{xy} = 1 - p_{xy}$$

For last survivor probabilities, we are interested in the event that both people have died. The probability the last survivor status fails is therefore:

$$q_{\overline{xy}} = q_x q_y$$

Exercise 6.5 Write an expression for $p_{\overline{xy}}$.

For calculations involving joint life status, it is usually easier to work with the probabilities of survival, i.e., in terms of $p \ldots$, whereas for calculations involving last survivor status, it is often easier to use the probabilities of death, i.e., $q \ldots$ functions. Calculations involving two lives are very similar to their single-life equivalents, although more care needs to be taken in assigning the probabilities to each outcome.

Example 6.3 Calculate the expected present value of a joint life five-year term assurance with a lump sum of \$250,000. The policyholders are a 30-year-old male and a 28-year-old female, and the effective annual interest rate is 3%.

The key terms in this question are 'term assurance' and 'joint life'. This product will pay a lump sum when the first person dies. Next evaluate each possible outcome:

Outcome 1: Lump sum is paid at the end of the first year—at least one person died that year.
Outcome 2: Lump sum is paid at the end of the second year—both survived the first year but one or both died in the second year.

Outcome 3: Paid at the end of the third year—both survived two years, then one or both died.
Outcome 4: Paid at the end of the fourth year—both survived three years, then at least one death occurred.
Outcome 5: Paid at the end of the fifth year—both survived four years, then at least one death occurred.
Outcome 6: No payment is made—they both survive five years.

Remember that in every outcome, there are three events that could trigger the payment in that year: he dies, she dies, or they both die. In the table below, the probability of occurrence will list the three events in that order. Here, order matters as the lives are different sexes and therefore to enumerate the probabilities you will need to use both a male and a female life table.

Top tip: in an examination, avoid making mistakes in enumerating by using two different color pens. The pens that have four different colored inks are really useful: assign one color to the male life and one to the female life probabilities. In the table below, red is used to denote the female life probabilities (Color figure online)

Outcome	Probability of occurrence	1/Accumulation factor	Probability / Accumulation Factor
1	$q_{30}p_{28} + p_{30}q_{28} + q_{30}q_{28}$	$\frac{1}{1.03} = 0.970874$	0.000658 x 0.970874 = 0.000639
2	$p_{30}\,p_{28}\,[q_{31}p_{29} + p_{31}q_{29} + q_{31}q_{29}]$	$\frac{1}{1.03^2} = 0.942596$	0.0006433
3	$p_{31}\,p_{30}\,p_{29}\,p_{28}\,[q_{32}p_{30} + p_{32}q_{30} + q_{32}q_{30}]$	$\frac{1}{1.03^3} = 0.915142$	0.0006506
4	$p_{32}\,p_{31}\,p_{30}\,p_{30}\,p_{29}\,p_{28}\,[q_{33}p_{31} + p_{33}q_{31} + q_{33}q_{31}]$	$\frac{1}{1.03^4} = 0.888487$	0.0006605
5	$p_{33}\,p_{32}\,p_{31}\,p_{30}\,p_{31}\,p_{30}\,p_{29}\,p_{28}\,[q_{34}p_{32} + p_{34}q_{32} + q_{34}q_{32}]$	$\frac{1}{1.03^5} = 0.862609$	0.0006734

We sum the final column and multiply by the lump sum: 0.0032665 × \$250,000 = **\$816.63**.

6.7 Reversionary Annuities

Historically, **reversionary annuities** were also popular. Reversionary annuities are annuities that are payable to a person from the point that another person dies. As with most products involving two lives, these are most commonly taken out by spouses, and one version of this is a widow's pension. When fewer women had their own careers or pensions, they were dependent on their husbands'. However, if their husband died, some women would have no income. For this reason, many pension providers started to include a provision for the widow of the policyholder. These widow's pensions would usually be a smaller amount than the husband's full pension, but enough to keep the wolves from the door. As more women entered the workplace and started their own pensions, and due to the decline of **Defined Benefit** pensions (which we will discuss in the next chapter), widow's pensions are a lot less common nowadays. However, a large number are still in operation due to the long run-off time between starting and winding up a pension scheme.

In the previous section, the notation q_{xy} was introduced as meaning that the joint life status of xy fails in the next year, i.e., either x or y or both die. With a bit more notation, we can extend this to cover the necessary probabilities for calculations involving reversionary annuities:

q^{1}_{xy} is the probability that in the next year, x dies, and does so before y.
q^{2}_{xy} is the probability that in the next year, x dies, and does so after y.

Note that it is not important to the calculation when y dies, only that they do so in the correct order. Specifically, in q^{1}_{xy}, y can die any time after x, even outside of the one-year period under consideration. The important part is that person x dies within the year period, and person y waits their turn.

If we knew which person was going to die first, the expected present value of a reversionary annuity payable to y on the death of person x would be a simple calculation: it would be of value 0 if person y dies first, and if person x dies first then it would have the value of a regular annuity lasting $K_y - K_x$ years.[7]

To save listing all the (many) years for which a widow could survive, it is easiest to calculate this in terms of the expected present value of annuities from the annuity tables. Specifically, the expected value of a reversionary annuity, paid to person y on the death of person x, can be evaluated as $a_y - a_{xy}$.

Reversionary annuities are most commonly taken out by spouses, but that doesn't have to be the case. As with other life-dependent products, a policyholder could buy a policy that is dependent on the life of a parent, a sibling, or a friend. However, for non-spouses, the policyholder must prove that they have an **insurable interest** in the death of another, that is, that they lose out financially in the case of that person's death. Spouses do not have to prove this as in a marriage, assets will be shared, but for all other relationships it

[7] This notation was defined in Sect. 6.1.

must be proven to the product provider. There are many valid reasons, for example, suppose someone owes you a substantial sum of money. If they died, you wouldn't get it back. In this scenario, life assurance bought by the lender on the life of the debtor would be preferable to life assurance bought by the debtor themselves, as the life assurance would pay directly to the person owed the money. This avoids any delay in payment due to the will being contested, etc. The proof required here would be the debt agreement. Having to prove insurable interest also stops the lender from receiving a vastly inflated lump sum in the case of death of the debtor: the sum assured would be the amount owed.

Dying of a Broken Heart?
In this chapter we have assumed the lives of spouses are independent. However, Broken Heart syndrome (also called Takotsubo cardiomyopathy) is a real medical condition that affects mortality. Specifically, if a spouse dies, the remaining spouse has a much higher rate of mortality than would be otherwise expected, particularly for the first year. It is fortunately quite rare, although it is likely that some Takotsubo patients are misdiagnosed as heart attack sufferers.

A reversionary annuity might be bought on behalf of a child on the life of their parent. The person acting in the interest of the child would have to prove that the child would lose out on child maintenance payments or similar income if their parent died. In this scenario, the reversionary annuity might only last until the child is an adult (perhaps 18 or 21 years). It would pay out yearly if the parent died within the 18 years and if the child was still alive. As with other fixed-term annuities, it would stop paying out after the 18-year term ended. A similar situation would be the death of a business partner. If it could be proven that their contribution to the business was a regular sum of money and on their death that would cease, then there may be an insurable interest there and a reversionary annuity may be appropriate.

As with other products, insurable interest is designed to stop a person profiting from another person's misfortune. It is even more important for life products, as the event in question is a death. Without insurable interest, the policyholder would be able to benefit from the death of someone that they don't even have to know! And that would make the policyholder both very rich, and potentially the prime suspect in a homicide.

This chapter has covered the calculations based on life-dependent annuity and assurance products, as well as the general trends and causes of variation in mortality. The next chapter will consider how aggregating life and health dependent products affects society as a whole, the financing of such arrangements, and the advantages and disadvantages of transferring mortality and longevity risk.

7

Pensions: Who Pays?

Chapter Objectives:

- Understand the difference between Defined Benefit (DB) and Defined Contribution (DC) schemes, and the pros and cons of each
- Understand the main funding methods for DB schemes and their advantages.
- Consider the challenges and possible solutions to the problem of social care

pension (noun): Late Middle English (in the sense 'payment, tax, regular sum paid to retain allegiance'): from Old French, from Latin pensio(n-) 'payment', from pendere 'to pay'.

The previous chapter covered life-dependent assurance and **annuity** products. One commonly held annuity is a pension; a regular income paid to someone because they no longer work due to age or poor health. However, whereas in the last chapter we considered individual pensions or individual life assurance policies, in this chapter we will see the benefits and challenges of scaling this up to a system that covers thousands or even millions of people. But before we look at the challenges of the present, let's start at the beginning.

© The Author(s), under exclusive license to Springer Nature Switzerland AG 2022
C. Townsend, *A Risky Business*,
https://doi.org/10.1007/978-3-031-11673-5_7

7.1 A Short History of Pensions

One of the first examples of a pension arose in 13 BC, when Augustus Caesar created a pension for veterans of the Roman Army. After 25 years of service in a legion, soldiers would be given a lump sum of 13 years' full pay. However, this was not designed with geriatric care in mind—it was no more than a state-wide incentive, to protect Caesar from being overthrown by disgruntled and impoverished ex-soldiers (and it worked). Annuities did exist, as evidenced by a third-century Roman annuity pricing table, but they were most commonly given as part of people's wills, rather than bought as a commodity. So, while the Romans might have done many things for their descendants, giving us aqueducts, sanitation, and roads, it is hard to argue that they created a life insurance market.[1]

After the fall of the Roman empire, no further large pension schemes were set up. There wasn't much demand for life-dependent annuities as those who could afford the upfront cost could usually live off their wealth in old age, and the poor couldn't afford them. There were some attempts at selling such annuities in Tudor England, but the lack of demand and a poor annuity rate meant that sales were low.

Meanwhile in the Dutch Republic of the 1500s, the government had been selling insurance products in order to raise funds. These policies were not pensions, but **perpetuities** and life-dependent annuities sold at the rate of 14 times the annual income of the annuity. It may seem an anathema to twenty-first-century actuaries, but no data were collected by the Dutch government to quantify the death rate and ascertain if these annuities were reasonably priced.

Across the English Channel, plague had ravaged the land for several years. The British government wanted to reassure the public that the plague was not as fatal as it was generally believed, and so in 1592 the London Bills of Mortality were first collected. These were weekly records of the number of deaths and christenings in the parishes of London, and were published sporadically until 1603. Unfortunately, the bills were initially only collected in plague years, so with no baseline comparison available, they provided little comfort to Londoners.

In 1653, there are records of the first employer-sponsored pension scheme, administered by the British Royal Navy.[2] They were the first large company to provide pensions for those in their employ, albeit in very small numbers until

[1] Although they did give us 'actuaries' in a linguistic sense—the word 'actuary' comes from the Latin *actuarius*, who acted as a mix between a secretary and an accountant in the Roman Senate.

[2] Royal Navy ratings' pensions seventeenth-twentieth centuries - The National Archives.

the 1670s. However, the scheme was generally limited to officers, although pensions could also be awarded to those who were injured or died on duty in particularly heroic circumstances, and who had over 20 years of service. In the seventeenth century, there were two main forms of Royal Naval pensions; a pension for a wounded or injured serviceman, received as a monthly income, and a 'Royal Bounty', which was a lump sum of one year's wage paid to the widow or dependent children of someone killed in service. However, these pensions were not particularly generous and many ex-servicemen had to find alternative employment to supplement their income. The life expectancy at this time for a seaman was only 48 years, so unlike today's pensioners who can expect to draw a pension for a few decades, payments were usually short-lived. These schemes were known as pensions, but a person often could not maintain a reasonable standard of living on them alone. The intention was not to provide financially but to boost employee engagement, as it was reasoned that a well-remunerated serviceman was more likely to show more courage.

Commercial life insurance remained largely unavailable in England until King William III—formerly known as Prince William of Orange—needed to raise £1 million (about 25% of the national budget at the time) after the 'Glorious Revolution' of 1688. Stealing the fundraising idea from his homeland, his government sold life annuities for £7 per £1 of annuity income. His idea was incredibly unpopular, with only £110,000 being raised and only 1,002 lives insured, but King William was at least bright enough to collect data on his customers. Interestingly, half of the lives insured were under the age of 11, demonstrating that the British public understood the obvious flaw in the government's flat pricing, i.e., that if all annuities are priced the same, you get a lot more value for money if the holder is likely to live a long time.

Several mathematicians had created mortality rate estimates based on the London Bills of Mortality data, but the government seemed unbothered by their flawed annuity pricing structure. This may have been less problematic than it would appear to us today. Firstly, the number of lives was so small that it may not have been worth the effort of applying new-fangled probability theory, particularly if the true objective was to raise funds initially, not make a long-term profit. Secondly, although age had a bearing on mortality, factors such as poverty and gender were much more important in determining your fate. Thirdly, an outbreak of plague or any other contagious disease would vary the mortality rates so much that a differential due to age would seem trivial. Regardless of what was implemented, mathematicians across Europe were pushing the boundaries of what was theoretically possible.

At this point, a set of mathematical tools previously used in pure mathematics and astronomy proved potent in calculating probabilities and accumulation factors. This was calculus, invented by Newton and Leibnitz. With the creation of annuity tables and calculus, the time seemed ripe for a private company to launch a life insurance product. In 1698, the London-based Mercer's Company launched two products: a widow's pension (also called a **reversionary annuity**) which gave a widow an annual income after the death of her husband; and a one-year life assurance policy that paid a lump sum on death. The former was horrendously loss-making; in 1738 Mercer had to reduce the amount of its annuity payments and in 1747 it required a partial government bailout. The Mercer life assurance product was not loss-making, but only because it didn't offer a fixed lump sum on death. Instead, Mercer was acting as a manager of a pool of policyholders, where the total premiums (less expenses) for the year were divided equally among those who had died. Therefore, Mercer was not actually taking on any mortality risk at all on this product—which, given the losses for their other product, was probably a very good thing.

The first successful examples of risk transfer in a life insurance product were products designed by a mathematician called James Dodson. Unfortunately, Dodson never lived to see his success and died (uninsured) at age 52. In 1762, a company called Equitable Life used Dodson's work to do two rather revolutionary things. Firstly, it offered whole life assurance, rather than single year cover. This increased demand substantially, as single year cover was only usually taken out by wealthy individuals who were usually about to embark upon something risky. A corollary of this multi-year cover was that the premium was paid annually, rather than a single premium upfront. Now the tables had turned, and the policyholder was effectively providing an annuity to the company.

Secondly, the premium depended on the age of the policyholder; the older you were, the higher the premium. Further, James Dodson had used data to determine the premium. He had carefully studied the London Bills of Mortality from 1728 to 1750 (in 1728, the Bills started including the decade age of the deceased, which is one reason why earlier years were unsuitable for this analysis). However, he didn't use the average mortality for each age group, as he very reasonably concluded that by using the average, half of the years would experience worse mortality than assumed in the pricing. Instead, he used the death rates from 1741, during which England experienced unexpected frost. The resulting famine caused many extra deaths in England (as well as in Ireland, where this event is known as the Bliain an Áir—the year of slaughter). As a result, Equitable Life not only made a profit, but had managed to create an appetite for life assurance products. Not only that, but it had subtly shifted the center of life insurance innovation from Antwerp to London.

Equitable's innovation continued under the leadership of Edward Rowe Mores. In 1773, it spent four months conducting the first valuation of the assets and liabilities of the company. It also analyzed the mix of business, to ensure that an influx of new policyholders did not distort the overall age profile of the policyholders, and it applied rigorous underwriting to ensure that policyholders were not "*of feeble constitutions*". Mores also invented the position of 'actuary', declaring that the chief officer of Equitable should be known by the title. Despite the marine insurers of Lloyd's jumping on this new life insurance bandwagon, Equitable continued to be a leading UK life insurer until the 1990s, managing £26 billion of funds at its peak. Commercial life insurance markets were not just created, but profitable.

Once it had been shown that life insurance was not always loss-making, more private companies and institutions started offering pensions to their workforce, including the East India Company and the Bank of England. While these schemes helped to widen access to pensions, access was not universal and the poorest in society most often were left without. The idea crossed the Atlantic and in 1875, American Express was the first to offer a pension scheme in the US. In 1889, this was Europe's hot topic, and while Britain was busy setting up a committee (called the National Pensions Committee), in the German empire the Old Age and Disability Bill was introduced by Otto von Bismarck. This legislation provided a pension in the form of an annuity for all workers from their 70th birthday onward, paid for entirely by the state, funded by a tax on workers. Although it is often thought of as the first universal state pension, its universality was restricted, rather predictably, to men. While this was undoubtably a leap forward for the welfare state, Bismarck's motivation appears to have been political; to create legislation to appease the public and therefore neutralize the perceived threat of Marxism. The scheme was a success, both for Bismarck and the German population. And, by 1908, the British National Pensions Committee had finally come up with legislation for a British pension, which was remarkable similar to the German offering. Having created universal state pensions, governments of the next 120 years would grapple with the problem of paying for them. And the rest, as they say, is history!

London Bills of Mortality

The London Bills of Mortality were ground-breaking data collection exercises due to the sheer amount of data needing to be captured in a world without computers or even wide-scale literacy. However, these tables were nowhere near ideal for their purpose. The first limitation of the Bills was that they were only performed in plague years initially, which gave public officials no clear idea of how fatal the plague actually was! From 1603 they were performed every year, but still had significant limitations including:

- The Bills only actually captured the number of burials in Church of England churchyards, and not deaths themselves, so systematically underestimated fatalities (particularly of Jewish, immigrant, and Catholic families).
- They didn't record the number of people in the parish, so death rates couldn't be calculated.
- They didn't record the reason for death—a particular limitation given the reason they were created! Cause of death started being captured in 1629. In 1632, cause of death was also used to create another piece of data: whether the cause of death was from a disease spread by human-to-human transmission. In particular, clerks were told to specifically identify the plague, smallpox, measles, tuberculosis, and syphilis (listed as French Pox in the Bills).
- They didn't record the age of the deceased, so the data couldn't be used to estimate different death rates for different ages. This started to be captured in 1728.
- They didn't allow for the substantial amount of immigration into London, and the ages of immigrants.

The Bills of Mortality showed the potential power of large datasets in the public sector, and paved the way for the first UK-wide census in 1841.

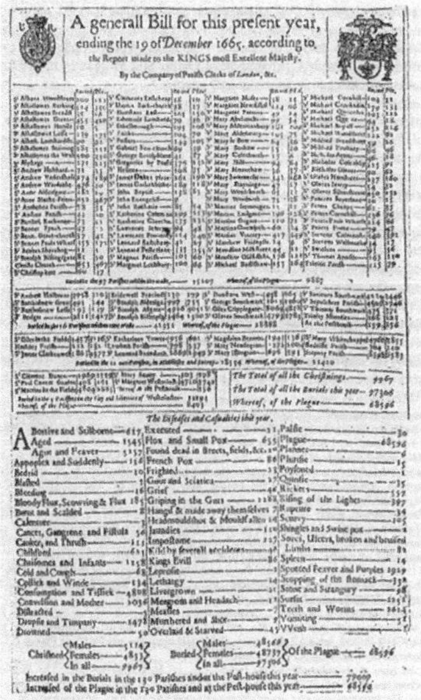

A generall Bill for this present year, ending the 19 of December 1665. according to the Report made to the KINGS most Excellent Majesty.

By the Company of Parish Clerks of London, &c.

Buried in the 97 Parishes within the walls — 15207 Whereof, of the Plague — 9887

Buried in the 16 Parishes without the walls — 41351 Whereof, of the Plague — 28888

Buried in the 12 out-Parishes, in Middlesex and Surrey — 18554 Whereof, of the Plague — 21420

Buried in the 5 Parishes in the City and Liberties of Westminster — 12194 Whereof, of the Plague — 8403

The Total of all the Christnings — 9967

The Total of all the Burials this year — 97306

Whereof, of the Plague — 68596

The Diseases and Casualties this year.

Abortive and Stilborne	617	Executed	21	Palsie	30
Aged	1545	Flox and Small Pox	655	Plague	68596
Ague and Feaver	5257	Found dead in streets, fields, &c.	20	Plannet	6
Appoplex and Suddenly	116	French Pox	86	Plurisie	15
Bedrid	10	Frighted	23	Poysoned	1
Blasted	5	Gout and Sciatica	27	Quinsie	35
Bleeding	16	Grief	46	Rickets	557
Bloody Flux, Scowring & Flux	185	Griping in the Guts	1288	Rising of the Lights	397
Burnt and Scalded	8	Hanged & made away themselves	7	Rupture	34
Calenture	3	Headmouldshot & Mouldfallen	14	Scurvy	105
Cancer, Gangrene and Fistula	56	Jaundies	110	Shingles and Swine pox	2
Canker, and Thrush	111	Impostume	227	Sores, Ulcers, broken and bruised Limbes	82
Childbed	625	Kild by severall accidents	46	Spleen	14
Chrisomes and Infants	1258	Kings Evill	86	Spotted Feaver and Purples	1929
Cold and Cough	68	Leprosie	2	Stopping of the stomach	332
Collick and Winde	134	Lethargy	14	Stone and Stranguary	98
Consumption and Tissick	4808	Livergrown	20	Surfet	1251
Convulsion and Mother	2036	Meagrom and Headach	12	Teeth and Worms	2614
Distracted	5	Measles	7	Vomiting	51
Dropsie and Timpany	1478	Murthered and Shot	9	Wenn	1
Drowned	50	Overlaid & Starved	45		

Christned: Males — 5114, Females — 4853, In all — 9967

Buried: Males — 48569, Females — 48737, In all — 97306. Of the Plague — 68596

Increased in the Burials in the 130 Parishes and at the Pest-house this year — 79009

Increased of the Plague in the 130 Parishes and at the Pest-house this year — 68590

7.2 Defining Defined Benefit Schemes

Some pensions, including the original Royal Navy and American Express pensions, are described in terms of the amount that will be paid on retirement; the pensioner might receive a percentage of their salary, or a set monetary amount. This is called a **Defined Benefit** (DB) scheme, as the benefit (the amount paid out) is defined in advance and is usually guaranteed for the rest of the pensioner's life.

But what is the amount that they will receive?

This comes down to the terms and conditions of the specific scheme, but it will usually be an amount linked to the individual's salary. However, an individual will usually have many salaries over their lifetime. Specifically, one hopes that their salary will increase over time. There are two main reasons for this increase. Firstly, it is usual for experience or promotion to be recognized with a salary increase: an intern is not paid the same as the CEO. But secondly, salaries increase over time to balance inflation and reflect changes to the cost of living. A worker who is many years from retirement might want to understand the value of their defined benefit pension, and to do so needs to estimate their salary that their pension will be based upon. This is called their pensionable salary.

There are lots of potential options for the salary used in defined benefit schemes. Some are more commonly used: final salary, 'average of final three years'' salary, 'average of final five years'' salary, and 'career-averaged' salary, which is the average salary of your tenure at the company. As ever, there are advantages and disadvantages to each choice.

Exercise 7.1: Which of these options would you expect to give the highest amount?

Final salary is simplest to use administratively; there is no need to keep records of previous salaries other than the current one for individuals. As the final salary is expected to be higher than the average of the final three years' salary (due to inflation if nothing else), it is more generous to the pensioner. For businesses, a final salary scheme creates the risk that older employees will achieve a promotion and then retire immediately, costing the business more in pension costs but receiving little to no benefit from their promotion. For this reason, a final salary scheme can affect an individual's chance of being promoted at older ages, despite their intentions. They may also penalize career changes or part-time working in the late stages of a career. For example, someone turning 65 may not wish to retire but instead work part-time, but a final salary scheme would mean that this would dramatically decrease their pensionable salary and therefore they are more likely to retire altogether at 65.

An average of the final three years' salary protects the business against this risk, as it is usually calculated based on the amount of time at each salary.

Example 7.1: Matilda receives a salary of $50,000 and normally receives a 1% inflationary increase at the start of each year. She gets promoted six months before she retires. Her new role receives a salary of $60,000.

The pensionable salary for her average of final three years' DB pension scheme is:

$$\left(\$50k + \$50k \times 1.01 + \$50k \times 1.01^2 \times 0.5 + \$60k \times 0.5\right) \div 3 = \$52,000.83$$

If Matilda was in a final salary scheme, the relevant salary would be the full $60,000.

However, if Matilda defers her retirement in order to stay for three years in her new role, her pensionable salary would be $60,602, the average of final three years. This is slightly higher than $60 k because of the yearly inflation applied.

In a final salary scheme, after staying for three years', her pensionable salary would be $61,206 (that is, the $60,000 inflated for three years at 1%).

Schemes usually pay a percentage of a person's pensionable salary, and this is usually based on length of service at the company. In the UK, a common structure for a defined benefit pension is to receive $\frac{n}{60}$ × Pensionable Salary, where n is the number of years' service at the company. This is usually just the integer part of the number of years' service, to avoid fractions upon fractions. This is usually the amount for a level annuity, that is, one that does not increase over time. Another option is for the pensioner to receive

$\frac{n}{80} \times$ Pensionable Salary initially but for this amount to increase in line with a standard measure of inflation, such as the national retail price index. The important thing to note here is that the option is for the employer, not the employee. The only choice for an employee at a company with a defined benefit scheme is whether to opt in or opt out.

Example 7.2: Matilda's pensionable salary is \$60,602. She has been with her company for 32 years and 6 months. Her company gives a pension of 1/60th her pensionable salary per full year of service. Her annual pension is therefore

$$\frac{32}{60} \times 60,602 = \$32,321.07$$

The benefits are clear, but what is the cost to an individual of joining a defined benefit scheme?

Usually, the contribution is fixed: an individual either contributes a certain percentage of their salary to be part of the scheme, or they opt out and pay nothing. The important thing to note is that they are putting money into the whole scheme's pot, rather than their own. In a defined benefit scheme, an individual does not have their own pension pot, but trusts that the scheme will have enough to pay out what is promised at the right time.

One advantage of a defined benefit scheme for the recipient is that they will find it easier to budget as they know what income they will receive, and it is guaranteed for life. The retiree won't have to worry about living longer than expected, as the pension will pay an annual amount. This is because in a defined benefit scheme, the scheme owner holds the longevity risk—the risk that participants will live longer than expected. The longevity risk originally sat with the individual, and was a financial risk in that if they lived longer than expected, they would run out of money. But in a DB scheme, the longevity risk is passed to the scheme owner.

In theory, the scheme owner should be much better equipped to deal with longevity risk than an individual. The company provides pensions for a pool of people, some of whom will die early than expected and some of whom will die later than expected. There are efficiencies of scale here, and the scheme can employ or consult a pensions actuary to calculate an expectation for future lifetimes of groups of participants. They can calculate the expected **net present value** of the annuities for each group, then aggregate to find the expected present value of the whole scheme's liabilities. The scheme should hold assets greater or equal to the value of their liabilities to be considered solvent. If not, the scheme owner may need to put some more assets (i.e., money) into the scheme 'pot'. This exercise would be repeated every year at least, and this is usually the reason that laypeople hear anything about actuaries at all. A written statement is usually sent to participants of the scheme to let them know the actuarial valuation of the pension scheme liabilities, with the intention of reassuring them that there is sufficient money for it to continue to run. It's not just participants that would be interested in this report though, as shareholders of the company would also be interested if the value of the liabilities are greater than the value of the assets, in which case the company may need to put money into the scheme, potentially reducing dividends or the share price.

Companies contributing to pension funds is not in itself a bad thing. Pension schemes were initially set up with the intention of attracting and retaining employees, with the pension scheme acting as an additional form of remuneration. The issue for companies is only if it needs to make additional payments to fund shortfalls in the scheme. There are some other options for the company:

- Increasing employee contributions. This option wouldn't be particularly popular with current employees, who would effectively be paying more for the same pension benefits. However, it might be combined with additional benefits of lesser value than the increase (such as death in service benefits or increased ill-health pensions) to make it more palatable. This is effectively a pay cut for employees, but might be perceived as being worth it for the continuation of the defined benefit scheme.
- Reducing the benefits in the scheme—although this would be unpopular with employees and ex-employees, potentially negating the morale-boosting effects of the scheme in the first place. This action might also be contrary to pension legislation and fairness, as participants may have contributed but may not receive the promised benefits. This would be pretty much the worst-case scenario for a pension scheme: the company would take a hit to their reputation, but employees (and especially ex-employees who are already in receipt of their pension) might have no choice but to accept a reduction in their pension.

Here, actuaries may make themselves unpopular with the scheme owners. If the scheme is significantly underfunded, that is, the expected value of its liabilities is much greater than its assets, then an actuary may be the first person to discover this unfortunate fact, and therefore often the first person to communicate this to all the stakeholders. Therefore, writing pension scheme valuation reports involves not only actuarial understanding, but also a large dose of integrity and the ability to manage stakeholder expectations. The report must also be very clear and free of technical jargon, as company executives, scheme participants, and company shareholders are unlikely to be experienced pension actuaries themselves.

In a defined benefit scheme, the scheme owner holds the longevity risk. However, if lots of participants live longer than expected, then the scheme owner may find the scheme unexpectedly expensive. This is exactly what happened in the twentieth century. Due to improved scientific and medical understanding, life expectancy increased in the US from 48 to 79 years, with similar increases in almost all developed countries. Defined benefit schemes, invented in the mid 1800s, were not expecting these improvements and so quickly became underfunded as the assumptions used to calculate the contributions required from the employees became increasingly out-of-date. Within 50 years, the life expectancy of Americans had increased by 20 years (although there was a sudden dip in life expectancy in 1916, this quickly recovered post-war). Instead of supporting pensioners on average for a few years, now the same pension pot would have to stretch to fund them for approximately 15 years. At the same time, the pensions provided grew more generous, particularly in the boom years post World War II (Fig. 7.1).

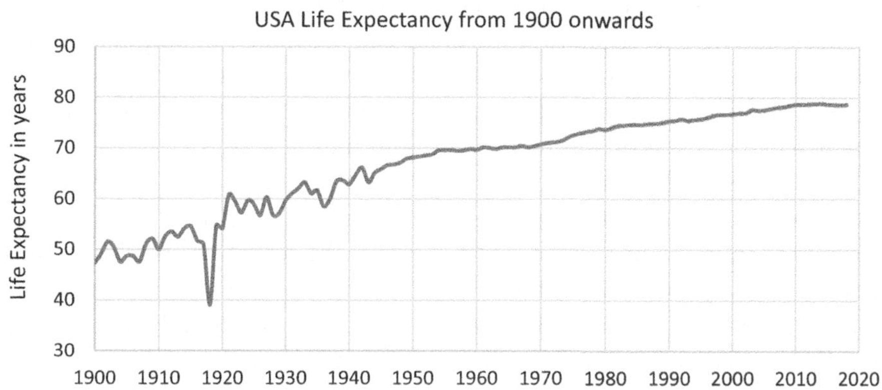

Fig. 7.1 Life expectancy at birth over the past 120 years. Notice the dip in 1918 due to Spanish Flu. Data from the CDC, 2020. An interactive version of this chart can be found at NCHS Data Visualization Gallery - Mortality Trends in the United States (cdc.gov)

7.3 The DB Disappearance

America introduced a new Act of Congress called the Revenue Act of 1978. The main aim of this act was to simplify tax law, reducing corporation and individual taxation, but it also made deferred remuneration tax-free. Put simply, employees could receive part of their wage tax-free, if they receive it sufficiently far into the future. The idea was that by taking money out of the system for now, the 'Great Inflation' of the American 1970s would be eased. But it wasn't long before companies identified that by using this Act, they could remunerate their employees without the burden of running defined benefit schemes: employees could defer their income themselves for retirement (tax-free), and employers could add to this contribution as they wished. By 1983 these '401k' schemes[3]—more generally known as **Defined Contribution** (DC) schemes—were popular with American corporations. They soon migrated across the Atlantic, where the political shift toward individual rather than collective responsibility meant that DC schemes were seen as fashionable (if ever a pension scheme could be called that).

The main reason these schemes became popular with corporations is because they don't involve any transfer of risk from individual to scheme owner. The individual holds their own longevity risk. In the face of a trend of increasing life expectancy, a defined contribution scheme means an employer can provide remuneration to employees but without ongoing exposure to longevity risk. The main reason that many employees joined the schemes was that they didn't really have a choice; if your employer does not offer a defined benefit scheme then you are not able to join it!

There are other benefits of DC rather than DB schemes to companies too (which also generally shift the risk to the individual), such as the lack of exposure to investment risk. In a DB pension, the scheme puts all the contributions into a fund. This fund is invested, apart from the amount that is required to pay any benefits due immediately, in order to enhance its value. Like life expectancy, actuaries will make assumptions about the investment return on the fund. If it is lower than expected, then the value of the fund will be lower than expected and, again, the scheme owner may have to inject some additional money into the scheme. In a DC pension, the individual is not promised any specific benefit on retirement, other than the value at the time of their deferred income. So, if the market gives a lower return than expected, the individual has a smaller fund than expected, and has to

[3] Nowadays there are lots of American variants of DC schemes. Although 401(k) is the most well-known, there is also the Roth 401(k), (which is a post-tax scheme), 403(b) schemes for education and charitable sector workers, 457 schemes for government workers, and many more!

cut their cloth accordingly. Most employers work with a pension provider for their employees, who will look after all the individual funds, invest at the employee's request, and provide information and yearly valuations. This means the individual doesn't have to do the leg-work of investing themselves, but they still bear the risk—that is, the financial consequence—if investment return is lower than expected.

Exercise 7.2: Suppose that a new combined drug mixture massively and rapidly reduced the incidence of cancer and vascular disease. What would be the implications for the pensions industry? [Hint: You might want to consider DB and DC pensions separately, and think about the different stakeholders here. For example: the scheme owners, the customers, the fund investors, and the government.]

That is not to say there are no advantages at all to the individual. One is that DC pensions don't have the same level of risk relating to pension provider failure or bankruptcy, because a DC pension isn't guaranteed. But the main advantage for individuals of DC schemes is that the individual has more choice: over the amount contributed, the amount and timing of money withdrawn, and the types of investments invested in. This is an advantage, but only really for those who are particularly financial literate or who have extreme or unusual circumstances. For example, someone in particularly poor health may choose to withdraw the majority of their pension soon after retirement, rather than buying an annuity. They might benefit more from a DC rather than DB pension, as they are unlikely to live long and may prefer to receive a higher amount of money for a shorter time. Recall that in a DB pension, the employer sets the amount of the benefit, so the individual would not have a choice in this. For those would are extremely financially literate, a DC pension offers a wider choice: they could choose to receive a higher amount and invest it in something, or do something even more exotic that they feel will be profitable.

The real issue here is that some individuals will make bad choices, primarily due to financial illiteracy or because they are vulnerable in some other way, including being the recipient of bad advice. They might receive too high an income initially, leaving them with too little to finance the rest of their days. As they have control of a very large amount of money, they might find themselves a target for scammers, wishing them to part with their pension—and they may find themselves an easy target. They might invest a large amount with the hope of growth but be bitterly disappointed. Then, society has a choice; does it pick up the tab for these poor or unlucky decisions, or does it allow people to descend into poverty?

Although most companies no longer offer defined benefit schemes, a huge number of schemes are still paying out. The problem with defined contribution schemes is that most people don't put in enough, if any contribution is made at all. Part of the issue may be a communication issue, and the higher level of financial literacy required for a DC pension is also a barrier. Part of the issue of low contribution levels may be psychological; the concept of tax-efficient deferred remuneration is framed in a way that means that employees are deferring income, which leads to feelings of loss aversion and encourages lower contributions. More practically, when the cost of living is high, people might need to take all their remuneration immediately to pay their current bills.

As DB schemes come to an end and pensioners start retiring with inadequate funds in their DC pot, care for the elderly will become a critical issue. Many countries have a state pension, which is a back-up to prevent terrible hardship but is not always enough to live well on.[4] Instead, some governments have put in place rules around defined contribution pensions. Examples include:

- Limits on the total amount of draw-down (lump sums taken early in retirement), e.g., a limit of 20% of the individual's total pension pot can be taken out in the first five years.
- Rules around what can be bought. For example, in the UK there used to be a law that pensioners under the age of 75 had to buy an annuity (although this is no longer the case).
- Taxing pensions in the same way as income. As income tax is usually a percentage of income, and the percentage often increases at certain income levels, this will deter individuals from taking a high income.
- Requiring a minimum employer contribution, and/or a minimum individual contribution. This was introduced in the UK as part of the UK's Pensions Act 2008, which not only sets out a minimum percentage of salary that must be put into a pension, but also states that employees must be automatically enrolled, so it is opt-out rather than opt-in.

Let's summarize this section with a direct comparison of defined benefit and defined contribution schemes.

---

[4] The OECD's Pension Report 2017 ranks state pensions according to their generosity versus the country's average working-age salary. The Netherlands, Turkey, and Croatia were the most generous, with pensioners receiving more than the average working salary. In contrast, South Africa provides around 18% of working salary, and the UK has the lowest rate of all developed countries, providing only 29%. In these countries, it is even more critical that young people contribute to their pension, as the alternative may be a very poor retirement indeed.

Defined Benefit Scheme	Defined Contribution Scheme
Participant pays a % of their salary, chosen by the company	Participant pays any amount they choose
Pension received is defined by the number of years' service and some function of participant's salary	Pension received is dependent on the amount of contributions made, investment returns, and the amount withdrawn
Individual has little control of benefit received	Individual has total control of the benefits they receive
Employer may be required to top up fund assets	Individual is responsible for their own pension pot
Employer/scheme owner holds the longevity and investment risk	Individual holds their own longevity and investment risk
Scheme must run until all members are deceased	Scheme offers no long-term commitment for employers

7.4 The Fundamentals of Funding

The previous sections covered the difference between defined benefit and defined contribution pension schemes. We saw that in a defined contribution scheme, the final benefit amount is not bound by any agreement so can flex to fit the value of the fund at the point of retirement. In DC schemes the fund is assigned to a specific individual, so is akin to setting up a bank account. However, in a defined benefit scheme, there is an agreement in advance between the scheme owner and the recipient about the level of pension that will be received. The scheme owner needs to meet this agreed cost, so setting the right amount of contribution is important. Recall that a DB pension fund is for the use of all scheme members and is not assigned to any one person, so the scheme also needs to decide how much each person should pay.

This section considers some different systems for funding for DB schemes, and their advantages and disadvantages. The three types we will mainly consider are regular contribution, terminal funding, and pay-as-you-go.

7.4.1 Regular Contribution

As the name suggests, contributions are taken regularly and gradually build up to the expected amount. Payments are made by individuals in the time between joining and the first benefit being paid. If the contribution was set at a level amount (in monetary terms), we could use the formulae from Chapter 5 to calculate the size of the payments. Although it can be difficult to calculate the exact level of contribution accurately for benefits that will be paid decades in the future, this method is simple to understand. The other benefit of this method is that the contributions paid early on will attract significant interest.

7.4.2 Terminal Funding

The alternative to regular contributions is a lump sum. This is usually done at the start of the pension scheme (unimaginatively referred to as 'lump sum in advance'), or more commonly, at the point that the benefits start being payable, called terminal funding. The lump sum should be big enough to cover the expected cost of all future benefits. This funding method is a bit risky as there is no fund until benefits for the cohort become payable, and if someone retires earlier than expected, or the company goes bust, there would be no money set aside (although a government appointed pensions regulator would likely be proactive in that situation—see the chapter on capital to understand how). The benefit of terminal funding is that if the scheme is solely funded by an employer, they may find it simpler to pay a lump sum than regular amounts. A company may also want to hold capital in the company (rather than the pension fund) for as long as possible, perhaps to achieve growth ambitions.

7.4.3 Pay As You Go

In this method, today's workers pay for today's pensioners. This is often called unfunded as the idea is that no fund is ever really created but rather that the annual contributions of the scheme are equal to the annual outgo of the scheme; or as my grandma would say, 'robbing Peter to pay Paul'.

Example 7.3: A company has 1000 current employees and 50 retired employees who receive a DB pension from the company's scheme. Last year, the company paid out $1,000,000 in benefits to the pensioners.

If the expected benefit cost this year is $1 million, the company needs to raise $1m in contributions this year. The company's total wage roll is $40 million, so it sets the level of contribution at $1m/$40m = **2.5%** of each employee's salary (on average, $1,000 per current employee this year).

In a hypothetical example, this works rather neatly as the contributions can be set to exactly equal the required benefits. It saves the company the administration and related costs of running a fund, and so the company is not holding any investment risk either. However, in reality, a small fund does need to be maintained to allow for slight variations in the amount of benefits paid or wage roll. The reasons for this are purely practical; employees will find it harder to budget if their pension contribution keeps changing their take-home pay. From an employer's point of view, increasing the contribution will likely negatively affect morale and may incur employee or union opposition. It is usually more practical to set a contribution level for the next few years to avoid unnecessary fluctuations, but this does require a bit of a 'float' fund. A small fund also means that if the benefit bill is slightly lower than expected, there is a pot in which to keep the small surplus. Equally, if the benefit bill is slightly higher than expected, the float fund may mean that there is no urgent scramble for cash mid-year.

Example 7.3 continued: This year, due to an internal reshuffle, 10 experienced employees retired and they weren't replaced. As they were experienced employees who had worked at the company for many years and received high salaries, their pension benefits total $400,000 per year. Now, the company has 60 pensioners with a total benefit bill of $1.4 million and now has 990 current employees. The average salary has decreased to $38,000 due to the higher earners leaving, so each employee now must make a contribution of $\frac{1,400,000}{38,000 \times 990} = \mathbf{3.72\%}$ of their salary.

The following year, a particularly fatal strain of flu circulates through the area and kills eight of the company's pensioners. The total benefits bill is now back to $1 million which, split between the 990 employees, equates to a contribution of **2.65%** of their salary. This fluctuation in contribution is inconvenient for the employees, and they feel that talking about pensions is very tiresome, so they agree to pay **3.25%** of their salary for the next five years.

A pay-as-you-go system works well when the ratio of pensioners to current workers is both stable and low; that is, there are plenty of current workers compared to pensioners. For example, DB pension schemes worked well for American Rail companies and the British Navy as there were plenty of current employees but relatively few pensioners due to the low life expectancy of ex-naval officers or ex-railroad laborers. However, there is a risk in companies using a pay-as-you-go funding method. If the company goes bankrupt, today's workers will have been paying into a scheme to fund others' pensions, but no one will be able to fund their pension. More immediately, there will be no workers to fund the current pensioners' benefits, so those pensioners would be left with no income at all.

The graph below compares the value of a pension fund under the methods discussed above. Note that the value of the fund under lump sum methods

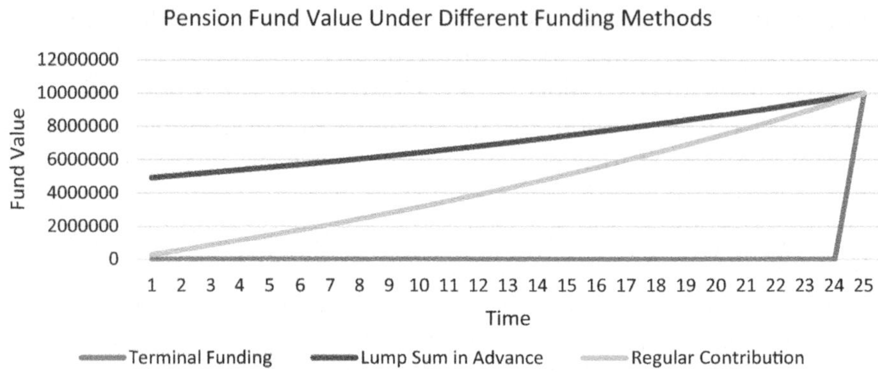

Fig. 7.2 Fund Value for different funding methods. The benefit with NPV of $10 million is due to be paid at time $t = 25$

and the regular contribution method are the same at the point of first benefit payments (Fig. 7.2).

There are lots of examples of a pay-as-you-go schemes, as universal state pensions can be viewed as a pay-as-you-go DB scheme. Funding state pensions in this way does not pose the same issues around default risk as it does in the private sector, as a state government (in some form or another) is certain to exist; however, provision could be subject to change.

Since Bismarck's universal pension act, state pensions have been created all over the globe. As state pension effectively tends to be a defined benefit scheme, where the government is the scheme owner and the contributions are a percentage of income applied as a tax. Like other scheme owners, the government can vary the amount of benefit received, the age of eligibility to receive a pension, and the percentage contribution required. Unlike employer schemes, state pensions do not usually relate directly to an individual's final salary. State pension schemes are invariably funded using the pay-as-you-go method for political reasons; election cycles are short, so using an upfront funding method would deny a government immediate public spending, but a just-in-time method may draw criticism for being financially irresponsible.[5]

State pensions are usually the largest part of the state's welfare packages. For example, in the UK two-thirds of welfare claimants—just under 20% of the entire population—are claiming a state pension, totaling almost £100 billion a year. It is a similar story in most developed countries, and opponents of a welfare state may object to the size of the bill. This is set to increase as more

[5] Besides, Bismarck's initial scheme was funded using a pay-as-you-go method, and was the original system that other governments copied.

of the population ages, and is leaving countries with a bill that is something of a ticking time bomb.

Consider how the demographics of the US are predicted to change by 2050. Almost a quarter of the population is expected to be over 65 and claiming a state pension (currently it is around 16%). Children are expected to comprise approximately 20% of the population. As such, that leaves approximately half of the population (as some adults do not work/pay income tax) to support the other half. Remember, the state pension is currently a pay-as-you-go system, so today's workers fund today's pensioners. A higher proportion of pensioners makes it very expensive for today's workers. Very simplistically, that means two working adults have to fund one state pension.

This is going to mean some tough decisions for governments. Should the state continue as is, continuing to fund state pensions at the same level, even though it would mean much higher taxation for the working population? Or should it reduce the pension eligibility or amount, and risk many older people (who might have planned their retirement around the state pension) struggling? There are a few obvious solutions, although few are politically easy.

- Increase taxation, which will predominantly fall on those of working age, to fund the current level of pension.

  Although no politician likes increasing taxes, this is actually probably the easiest option in a practical sense and requires no pension reform. The risk is that younger people may see this as unfair, particularly if the state pension is higher than a minimum wage or increases faster than wages and could cause intergenerational division. Increasing taxes could have other knock-on effects, such as more people opting out of work or working part-time, or may even mean that young people struggle financially and decide to have fewer children, or none at all. This would then exacerbate the problem long term, as then fewer children would grow into fewer working-age adults.

- Decrease the amount of the state pension.

  The main risk with this is that it leaves many pensioners struggling financially, particularly if the state pension is their sole income. Equivalently though, it is currently common for the state pension to increase with inflation,[6] so if this wasn't done for a period it would decrease the real

[6] Increases such as these are not guaranteed however. Despite making it a key election pledge in 2019, the UK government suspended the 'triple lock' on pensions in 2021. The triple lock guaranteed that

value of the pension without decreasing the nominal value. This is particularly effective during periods of high inflation, although it can increase pensioner poverty if the pension was not very generous to begin with.

- Increase the age of eligibility.

  To reduce the number of claimants, a government could increase the pension age. This is likely to be unpopular with the cohort affected, but has some precedence (The UK increased the state pension age to 66 in 2020, and Australia is increasing theirs to 67 in 2023). The downside with this is that it cannot be implemented immediately—it needs a few years' notice for the age group affected to be able to either change their retirement plans or to put more into their private pension. The other issue this option raises is that life expectancy is not homogeneous throughout the population: those on lower incomes usually have a lower life expectancy than their wealthier counterparts and the difference can be stark. An MIT study[7] showed that rich American men live 14 years 7 months longer than the poorest. Increasing the pension age may mean that the poorest in society are only allowed a very short retirement—or none at all.

- Increase the taxation base.

  Some taxes apply only to younger workers. This could be changed to increase the amount of tax raised, which could be used to pay for the state pension. For example, in the UK, so-called national insurance (in reality, a tax of 13.25% of income) is not paid by working or retired individuals once they are old enough to become eligible for the state pension—although like all tax laws, there are lots of exceptions, such as some exemptions for the self-employed. This age limit could be scrapped, increasing the tax base.

- Increase the number of working-age adults.

  If the state pension bill must increase, then at least it can be made more palatable by spreading the cost across more workers. One way to increase this demographic is to encourage people to have more children. There are

pensions would rise by the highest of inflation, wage increases or 2.5%. Without this promise, pensioners are exposed to inflation risk.

[7] New study shows rich, poor have huge mortality gap in U.S. | MIT News | Massachusetts Institute of Technology. https://news.mit.edu/2016/study-rich-poor-huge-mortality-gap-us-0411

some obvious disadvantages with doing this. Firstly, it is a rather slow way to increase the number of working-age adults, with a lag time of more than 18 years until the children of this policy enter the workforce. In the short term, it may even exacerbate the issue by taking some people out of the workforce for a few years and demanding increased funding elsewhere. Secondly, it may be difficult to dismantle the systems that discourage families from having more children. For example, an expensive, over-heated housing market, a lack of paid parental leave, or prohibitively expensive childcare. Encouraging families to have more children might be practically difficult, let alone the political or ethical questions it might raise.

This option has been chosen in South Korea, which offers a payment to the parents of children born from 2022. A payment of 2 million won ($1850 USD) is received at birth, then 300,000 won ($270) a month until the child turns 1. This is not South Korea's first foray into payments per child. It has spent the equivalent of $1 billion USD since 2007 but is still experiencing decreasing fertility rates.

Germany is also trying to increase the number of working adults in their population, but by encouraging immigration. This has had the desired effect, as 17% of people living in Germany are first generation immigrants. While this solution is practically easier and seemingly more effective, it can come with political considerations for governments.

Collective Defined Contribution Schemes (CDC schemes)

Contrary to popular belief, there are lots of innovations in the pensions industry. One to watch is the beginnings of Collective Defined Contribution schemes. Noting that in individual DC schemes, individuals hold all the longevity and investment risk, CDCs look to share those risks in a pool.

A member joins, and their contribution is set. The contribution can be set in ways similar to those of DC schemes, i.e., to target a set benefit, or as a percentage or a dollar amount of their salary. An employer might add to this contribution. Then the expected investment return will be calculated, to give an estimated benefit. So far, so like DC schemes.

Now, lots of members join the same fund. The individual no longer has their own pot of money, but is in a pooled scheme. However, unlike a DB scheme where the benefit is guaranteed, in a CDC the benefit is not guaranteed. This is good for employers as they won't be liable for having to top up the fund when it gets low. Instead, in a CDC, the benefit is variable, dependent on investments and longevity. However, like a DB scheme, this is usually an annual pension, rather (or as well as) a lump sum on retirement.

This is more efficient as instead of individual members buying an annuity at the prevailing rate, incurring transaction costs and advice costs, the scheme provides it automatically.

But what happens if investments are so poor that the guaranteed minimum is not reached? In these cases, the contributions of working members would have to be increased (or conversely, their benefits would be decreased). So the subsidy comes from within the scheme, instead of from employers. At the moment CDCs are fairly new and research is still ongoing to quantify the downside risk from this sharing: what is the worst likely scenario for a member? From this, the quantification of intergenerational fairness with CDCs can be evaluated. It is already clear that the more certainty you give over benefits, the more risk sharing, and therefore the more potential for unfairness. However, there could be upsides: Willis Towers Watson estimates that CDC pensions could yield a pension that is 70% higher than the average DC scheme, and 40% higher than DB schemes.

Some companies are already creating CDCs. The UK's Royal Mail, a privatized company that is unionized for historic reasons, decided that CDCs might be answer to for them, as it balances the needs of the unionized members (who want an income) with the needs of the company (who wants to be absolved of pension guarantees). As of November 2021, the contribution rates were set at 6% of employee's salaries, plus a contribution of 13.6% from the employer. It guarantees a minimum of 3/80 of pensionable salary per year of service. For an optional extra 1% contribution, employees can also access a lump sum on retirement.

7.5 Social Care

As if funding state pensions is not challenging enough, governments are increasingly having to deal with the question of funding social care as well.

Social care is an umbrella term for non-medical care required for a basic quality of life. It includes assistance getting out of bed, going to the shops, dressing, getting meals, and going to the bathroom, but encompasses a wide range of activities. Social care is most often required by those with disabilities and the elderly. Providing social care was previously not a big issue for two reasons. Firstly, as medical care was more limited, the frail and disabled didn't often survive long. Secondly, where it was required, it was expected that women (daughters or daughters-in-law) provide the care, unpaid. As more women have careers out of the home and young people are geographically more mobile, they often can't drop in to provide daily care. Successive rounds of changing government guidelines and structures, plus the challenge of reduced budgets, have taken a toll on social care. As such, social care is now often outsourced to paid carers and therefore needs funding.

Funding social care is much more challenging than funding a universal state pension or a personal pension as not everyone will require social care, whereas everyone who retires at 65 will require an income. It is also difficult to assess who is at greater risk of needing it, as scientists still aren't sure of the **risk factors** for developing illnesses such as Alzheimer's and dementia. Like a pension, requiring social care in old age seems a very long way off for today's workers. Numerous studies have shown that willingness to save for something declines the further in the future the event is. Social care is also subject to the uncertainty of needing it. People's optimism about their own future reduces their willingness to save for it—and they mistake hope for certainty.

Willingness to save for social care is often also clouded by confusion over who pays for it. In the UK, the NHS will pay for treatment and care if you have a physical disability such as cancer. Consequently, people often think that the NHS will pay for care for those with a medical diagnosis of dementia (but it does not). Local councils often provide social care for adults with disabilities, and some people mistakenly believe that they will pay for social care for older people. Unfortunately, the eligibility criteria are so high that very few people qualify, or they qualify for a level of care below what they require. The costs of social care are huge. At-home care—where a carer visits the person's home—costs on average $4,300 a month in the US.

Every country has a slightly different approach to funding social care, as every country has a different healthcare system. There doesn't seem to be a 'right' way—no country seems to have a financially sustainable social care system that is robust enough to cope with the aging population but is sufficiently affordable for the majority to have access to it. Actuaries are well equipped to wrangle with this problem, as it combines the traditional actuarial areas of pensions and population modeling. For some countries, setting up a new system to fund social care would potentially create a sizeable new industry which would require actuarial expertise.

Let's consider some approaches to funding social care. There are many variations on these approaches, so consider for yourself how each could be flexed.

Exercise 7.3: For each approach, consider whether it could be pay-as-you-go, terminally funded, or funded by regular contribution, and how robust it is to an aging population.

1. The Ostrich Model

   Individuals fund social care (or not) as they see fit—no accusations of a nanny state here! This method creates no state burden of funding. Proponents may argue that it encourages personal responsibility and more freedom of choice, although low take-up may mean that private products are inefficient and costly. Disadvantages are that it may leave a lot of people without any care, particularly those who are already retired. Young people in low-income families may be required to give up work to be a carer full-time for their relatives, perpetuating inequality. This approach is common in developing countries.
2. The Fully State-Funded Model

   This is the most comprehensive approach: the state pays for care based on an assessment, including any accommodation costs -often called 'hotel fees'. This is the cost of living in a nursing home rather than your own home. The main advantage of this system is the universal coverage. The government bears all the risk under this model.

   Case study—Japan: This was the approach taken by Japan in 2000, where the government started a relatively generous and universal social care system. To fund this, they increased general taxation but also levied an additional tax on those aged 40 and above. Pensioners also contribute and receive a card entitling them to social care after the state-defined threshold of need is met. This created an influx of participants in the social care market, so pensioners had more choice in their care providers. However, Japan had underestimated the needs of its population; 10% of pensioners met the threshold in 2000, but 16% met it in 2005. Because of this, Japan started to means-test the payment of accommodation fees in 2005. Demand continues to rise, so Japan is considering doubling the percentage contribution from pensioners who are currently receiving social care. Japan recognized that due to cultural expectation, women were more likely to give up work to fulfill caring responsibilities and specifically wanted to protect their female workforce, which is why they are keen that benefits adequately cover the cost of care.
3. The Mandatory Private Insurance Model

   Under this approach, individuals would have to buy insurance from private providers but would have choice over the provider and level of cover, subject to a minimum level. This is similar to car insurance in most countries; it is mandatory to have third-party liability cover, but many people often buy accidental damage cover too. The main advantage is universal coverage, but the downside is that private providers are often for-profit, leading to higher costs for the population.

Case study—The Netherlands: The Netherlands was the first country to introduce universal state care in 1968, using Bismarck's pay-as-you-go funding approach. However, due to rising costs, in 2006 the system was overhauled to one where citizens must purchase long-term social care insurance from a private provider. To ensure this was affordable, the market was heavily regulated. The premium must be a percentage of the person's income, and employers must make a contribution as well. There are different levels of coverage and different benefit options—people can choose from direct provision of care, or cash which with to buy care. The problem with this system is that many people, particularly those in the poorest sector of society, opt to receive cash and then purchase a lower level of care than they actually require, leaving them vulnerable. For poorer families, working-age women often give up their own careers and provide informal care for their relatives, so the relative can then keep all of the cash benefit. The poorest pensioners are therefore at risk of receiving inadequate care. The other issue with this scheme is that, like car insurance, providers have been locked in a price war, rather than focusing on the level of benefits or quality of care. This in turn means that consumers are less likely to receive anything more than the most basic care.

4. The Capped Cost Model

   Under this system, the government limits its own exposure to risk by only providing a certain level of benefits to an individual, with the acknowledgment that this may not be sufficient to cover their needs. This could either be done as percentage of care costs, or as an absolute amount. The advantage is that this provides a minimum level of care but at a much lower cost to the state.

Case study—France: In France, the government pays a percentage of care costs, but the percentage is means-tested and varies between 0 and 90%. To pay their share, individuals are encouraged—but not required—to have private long-term care insurance. The most popular insurance provides a monthly payment once the individual reaches the threshold for care and the premiums for these are not taxed to increase affordability. The French government has made employing home care tax-deductible to encourage people to use their payments for carers.

The Institute and Faculty of Actuaries (based in the UK) made the opposite recommendation for social care in the UK and suggested that instead of the government paying up to a cap, the government should pay the amount over the cap. Specifically, they suggested that individuals should pay for their own care up to a cumulative amount of £100,000, but after that the government should pay the rest. The French model limits risk for the government but means it pays a small amount for a lot of people. The IFoA's suggestion would limit the risk for individuals but keep costs manageable for the government as they would only be paying any costs for less than 10% of the population. It is important to note that in the UK there is a particular political aversion to the idea of having to sell one's home to pay for care in later life, so under this arrangement those with £100,000 saved would not have to endure that outcome. In both variations, the government and the individual (perhaps with the help of insurers) share the risk.

The issue of funding social care will become more urgent in the next decade as countries struggle with an aging global population. Which method is chosen depends on the healthcare system and demographics, but also government objectives, risk appetite, and willingness to be open about the risk individuals are bearing. Actuaries are sounding the alarm about the risk and making relevant recommendations. Will they be heard before it is too late?

8

Pricing: The Science of Estimating the Risk Cost

Chapter Objectives:

- Find the risk cost using first principles
- Understand the problem of anti-selection to insurers, how to spot it, and how to guard against it
- Understand the basics of how to use Generalized Linear Models to estimate risk cost
- Find the risk cost of a large risk by using Increased Limit Factors.

Estimating is what you do when you don't know.

—Sherman Kent

Have you ever been in a store and found an item which is missing its tag? You take it to the register, and wait with trepidation (and a little bit of dread) while they ring it up: it could bust your budget, or it could be a bargain.

It is a similar circumstance in actuarial work: at the point of selling the policy, the insurer doesn't know if it will be a bargain or regrettably expensive for them, as the exposure to risk hasn't even started. And so—as seen in Chapter 2—it's not clear if the price charged will be enough to cover the cost of the policy or not, and so the insurer is not sure whether the policy will lead to a profit or not.

© The Author(s), under exclusive license to Springer Nature Switzerland AG 2022
C. Townsend, *A Risky Business*,
https://doi.org/10.1007/978-3-031-11673-5_8

At first glance it may seem that a profit is made if the insured event doesn't occur. That is true, at least for a while. If you insure for a year against an event that doesn't occur in that year, then the **premium** (the price paid by the customer) is more than the cost of the policy, and so a profit is made. Surely then, if an event never occurs, an insurance company will make a big profit?

Unfortunately for insurance companies, that is unlikely. After a while, customers will re-evaluate the risk of the event happening and may no longer deem it necessary to buy insurance. The demand for that particular insurance product will reduce to zero and therefore no longer generates a profit.[1]

And what about life assurance companies? They are very certain that their event in question will occur, as "*nothing can be said to be certain, except death and taxes*".[2] So, avoiding the insured event cannot be the main path to profitability, or whole life assurance would cease to exist.

Profit occurs when the premium charged is more than the cost of the policy.[3] Therefore, no matter what the cost is, if the premium charged is higher, an **underwriting profit** can be generated. Bear in mind that, at an individual policy level, most policies will be profitable, as they will earn premium but not incur any claims. However, those policies where claims do occur will be incredibly unprofitable. What matters is striking a balance across all policies to provide a product that is competitive and appealing to customers, but also sufficiently profitable to make it a viable business proposition.

Note that this is a long-term view: in the short-term, avoiding the event is of course more profitable! But for multiple years or multiple risks, even if the insured event occurs, as long as on average the premiums charged are higher than the costs of the policies, there will be an underwriting profit. Of course, as it is uncertain when or if the insured event will occur, the cost of the policy will be unknown. Correctly determining the price is therefore a critical piece of actuarial work. It is also work that has to be re-visited regularly, as circumstances change (such as the cost of car parts or the **frequency** of natural catastrophes).

---

[1] This disregards the effect of investment returns, which are ignored in this chapter but covered in the next.

[2] These exact words were written by Benjamin Franklin on 13th November 1789, but a similar phrase was written even earlier, by Christopher Bullock in 1716: 'Tis impossible to be sure of anything but death and taxes'. And plagiarism, it seems! Physicists sometimes modify it: 'Nothing can be said to be certain, except death, taxes and an increase in entropy'.

[3] Again, disregarding the effect of investment returns.

> **Did you know?**
> Lloyd's of London has sold over 20,000 alien abduction policies.

In order to set the premium, first the cost of the policy must be estimated. The cost can be split into risk cost and non-risk cost. As suggested by the name, the risk cost is the expected risk-related outgoings, i.e., claims payments; and the non-risk cost is anything else, such as wages for insurance staff, rent for the office space, or broker commission. In this chapter, we'll focus on estimating the risk cost.

8.1 Why Estimate the Risk Cost?

The first step toward thinking like an actuary is to ask: what could possibly go wrong? Pricing risk is no different, as one policy is often exposed to multiple risks and **perils**. For example, a home could be burned down, it could sustain accidental damage, it could be flooded, burgled, or damaged in a windstorm. Perhaps something even stranger could occur. Some policies may have events that are specifically excluded in the policy terms and conditions, but anything else should be considered in pricing. Once the perils are listed, the expected future cost of each needs to be estimated.

In order to estimate the expected future cost, an actuary starts by looking at the past costs for that peril. Fortunately, established companies hold a wealth of claims data that can be used for pricing, so the total cost of all claims relating to that peril can be calculated. Ideally for this purpose, the data would consist of paid or settled claims, rather than claim estimates. Then, if it is known how many policies were in force, we can estimate the **burning cost** of each peril, that is, the claims cost per unit of exposure. This is calculated by dividing the total cost of claims by the amount of exposure over a defined time horizon.[4]

However, for those without huge quantities of claims data, burning costs are surprisingly difficult to estimate. For example, how much does a family spend on oranges every year? It is unlikely that most families keep such a record. But it could be estimated by considering how often they buy the item, and how much it costs every time they do. This is an intuitive way of estimating, as assumptions and calculations can be more easily checked. In this case, the estimate of the price of fruit could be checked against external

[4] See Chapter 3 for a refresher on exposure measures and units.

sources such as a supermarket website. Splitting the total cost in this way means that if there is a change in the behavior, such as inflationary pressure making the cost of oranges rise, it is easy to update a future expected total cost. Some permutation of this frequency-severity approach is often used in insurance pricing. Indeed, if the number or cost of fire claims increases by 20%, it is easy to increase your claims frequency or claims severity by the same amount. This gives an easy way to explain overall changes: if frequency increases by 20% but severity decreases by 4%, we can apply each change to the relevant term. This allows us to explain these changes more easily by attributing them to either claim frequency or severity.

Returning to claims data, if an actuary can access the claims data, they can calculate the **frequency** -how often it happens—and **severity**—the cost when it does happen. As well as being intuitive, this method is analogous to the concept of expectation in statistics; that is, the expected value of a random variable is equal to the probability of occurrence (the frequency) multiplied by the outcome (the severity).

More specifically, frequency can be calculated by dividing the number of claims by the exposure measure. The answer is often therefore expressed as a percentage. As most insurance policies allow more than one claim in a policy period (the obvious exception being life insurance!), and some policies may have claimed multiple times, the interpretation of frequency is simply the probability that a policy will make a claim for this event.

Severity—how severe a claim is—is expressed in monetary terms, and is calculated by dividing the total cost of claims by the number of claims. Severity can be thought of as an answer to the question: "If a claim happens, how much does it cost on average?", but crucially it assumes that a claim does occur.[5]

Providing we align how these measures are defined, the frequency-severity approach and the burning cost approach are equivalent. When multiplying the frequency and severity together, the number of claims rather satisfyingly cancel in the fractions, so it becomes exactly the same calculation!

$$\text{Frequency} \times \text{Severity} = \frac{\cancel{\text{Number of Claims}}}{\text{Exposure Units}} \times \frac{\text{Total Cost of Claims}}{\cancel{\text{Number of Claims}}}$$

$$= \frac{\text{Total Cost of Claims}}{\text{Exposure Units}}$$

$$= \text{Burning Cost}$$

[5] The alternative—the calculation for the average cost of a claim that never existed—is the actuarial equivalent of philosophy.

Example 8.1: A company had theft claims totalling £3m last year. Over the same period, it had 300,000 policies in force and 15,000 theft claims. Assuming policies in force are a reasonable exposure measure, what is the risk cost under a burning cost approach and under a frequency-severity approach?

The burning cost of theft claims $= £3\text{m} \div 300,000 =$ **£10 per policy**.

The frequency of theft claims $= 15,000 \div 300,000 = 0.05$

The severity of theft claims $= £3\text{m} \div 15,000 = £200$ per claim.

Multiplying frequency and severity together gives the same answer of $0.05 \times £200 =$ **£10 per year.**

Although the burning cost method is initially quicker, there is a further benefit to the frequency-severity approach. In order to calculate estimates of expected claim costs, actuaries often use actual claims data. However, this data is historic—it relates to past claims—and so it may need to be adjusted slightly to give a future looking view. There are lots of reasons why the cost of claims may change over time, including:

- Inflation (affects severity)

  Most people are familiar with the concept of inflation: the price of most items increases over time. This affects claims as the price of materials to rebuild a house or parts to repair a car makes up the cost of a claim. Even in injury claims, as minimum wage increases, the cost of carers or nurses to help the injured party increases. As inflation doesn't affect all things equally at the same rate, over time some claim types will become relatively more expensive than others.

- Technological advances (affects frequency and severity)

  Depending on the technology, this might affect the cost of claims by changing the likelihood of accidents, and therefore claims. For example, Autonomous Braking Systems on cars aim to reduce collisions by sensing distance and braking appropriately. However, they have a counteracting effect on severity as they make the vehicle more expensive to repair. The ABS system may need to be repaired as well as the basic bodywork of the car, and the ABS computer may need to be recalibrated by experts. Electric cars may also be more expensive to repair, as they may have batteries in the front and/or back of the car which are the most common collision points. Other advancements may only increase severity, such as expensive new medical treatments, or decrease it, such as advancements in making car parts that make them cheaper to produce.

- Court decisions (affects frequency and severity)
  Again, depending on the specific court case, this could affect the cost of claims in lots of different ways. Court decisions are unlikely to change the frequency of the underlying event, but may change the likelihood of a customer to make an insurance claim. For example, after the infamous *Stella Liebeck v. McDonald's Restaurants* case in 1994 (where a 79-year-old sued McDonalds for their coffee causing third degree burns), a host of other claims and litigation were made by customers to Starbucks, Dunkin' Donuts, and Burger King for similar reasons. Court cases can also set precedence for the level of court award or compensation received.

- Economic environment (affects frequency)
  The economic environment often affects how likely customers are to claim. In a recession, customers who might not previously have claimed for a small loss—perhaps due to the effort of claiming, or fear of losing a No Claims Bonus—may now need to do so for financial reasons. This is often more pronounced in smaller claims. For the same reason, recessions often also see more fraudulent or exaggerated claims.[6]

Did you know?
Insurers pay out over £1 billion for fire claims in the UK every year. The National Fire Chief's Council estimates that half of fires are started deliberately!
nationalfirechiefs.org.uk

Using the frequency-severity approach, it is much easier to adjust past data to be applicable to the future, as frequency or severity can be changed explicitly and independently. These adjustments are usually made manually after data-driven estimates of frequency and severity have been calculated. Some adjustments, such as inflation, can be quantified using official government statistics. Other adjustments, such as adoption of new technology can be more difficult to estimate and may require expert judgment. In these cases, assumptions should be reviewed often, and particularly if any new information emerges. Modeling frequency and severity can identify features and

[6] Fraud in a Downturn (pwc.com). https://www.pwc.com/gr/en/publications/assets/fraud-in-downturn.pdf

trends that may ultimately give rise to varying risk costs with this granularity, and is therefore preferable to a burning cost approach if data and time allows.

8.2 Getting to Know (People Like) You

'*We hold this truth to be self-evident: that all men are created equal...*'.[7] But it doesn't take a genius to recognize that not all risks are created equal—everybody knows someone who is a truly terrible driver. It is an evolutionary strategy to be able to identify very dangerous situations, but can the less clear-cut cases be ranked? In our own lives, we could probably estimate how often people we know claim on their insurance by considering whether they are careful, reckless, clumsy, or particularly precious about their item. However, an actuary or **underwriter** doesn't have time to get to know the policyholder of every risk, although for very large risks, an underwriter may check that certain risk-reduction measures are in place (e.g., making working sprinklers and alarms a condition of cover). So how can actuaries estimate the risk cost, and therefore set the price for an individual? Well, an actuary may not know you, but they know hundreds of people *like* you.

But which policies are truly alike?

The first step to determining this is to collect data on the policies already held. The exact information required depends on the nature of the policy as it should affect the level of risk in some way. For example, for a life assurance policy, medical history, lifestyle choices, and the age of the policyholder would be most relevant; for gadget insurance it would be more important to collect the technical specifications of the gadget and the cost of replacement. These pieces of information are called the **risk factors**. The assumption is that if all risk factors could be collected accurately, policies that had identical risk factor data would be a similar level of risk. Once these homogeneous groups are created, a risk cost for each group can be estimated using past claims' data, and so policies in the same group can be priced the same.

However, this method leads to some issues. The first is that it isn't possible to capture direct information about the true drivers of risk. For example, although driving ability is clearly important to the level of risk of a car insurance policy, it is not an objective or measurable factor, and some people have an over-inflated sense of their ability.

The second is a practical one: customers dislike answering questions. In order to collect the data, the customer must give the information to the

[7] Preamble to the Declaration of Independence.

insurer or broker and, as most adults know, lengthy forms are particularly unappealing. An insurer trying to collect too many risk factors from potential customers risks them not completing the process and ultimately going elsewhere.

Another practical issue is that some customers won't know all the relevant risk information. A lot of people don't know the horsepower or acceleration speed for the car they drive (which is relevant information for car insurance) without checking, and would definitely not be able to answer questions about the incidence of carjacking in their district. This information, if deemed important enough, could be sourced from elsewhere. This may be from publicly available sources or commercial vendors. For instance, police data might give the levels of different crimes in every police borough, or census data might record the population density (and therefore might inform how busy the roads are). DVLA/DMV records may even have records of car registration plates against the type of vehicle. By using such sources, a customer may only need to put in their car registration number and their zip code, and the insurer will enrich and collate the relevant information.

Lack of interest is not the only reason why customers wouldn't be able to give the information: prospective life assurance customers may not have been told their family medical history, perhaps because they were adopted or have complicated family relationships, or due to family members wanting privacy of their medical information.

Did you know?
Over 30 million people globally have taken a DNA test according to global genealogy service Ancestry.com. That's a huge source of potential data for health insurers!

To avoid these issues, an insurer might ask questions that don't directly relate to the risk but are good enough approximations that they can be used to determine the premium for a customer. These are called **rating factors**. Hopefully there will be considerable overlap between risk and rating factors, but the issues above mean that it isn't always the case.

Example 8.2:

- Instead of asking whether someone is a good driver (not a measurable question, and a driver is likely to be biased), ask about the number of previous claims or number of years of driving experience.
- Instead of asking if someone is at risk of a genetic form of cancer (they might not know), ask whether any close relatives have or have had it.
- Instead of asking whether a product contains any harmful substances (if it did, it would likely not be for sale!), ask about the risk mitigation practices in place to ensure purity and safety of ingredients.

Once information in respect of features for rating has been collated, customers can be categorized into groups. Creating homogeneous groups is a balance between making policies in the group truly similar, and having enough policies in each group. If the number of policies is too small, the claims costs of the group may be too volatile and not truly representative of the average claim cost for that group. On the other hand, if the group is too large and policies within it are not similar enough, individual policies within the group will be over- or under-charged.

8.3 Anti-Selection

In all industries, selling your product at a price that doesn't cover costs means that no profit is generated from that product, which is usually problematic. Imagine that it costs a baker £1 to bake a loaf of bread, including all costs. One day, for whatever reason, she charges all customers only 50p for a loaf of bread. At the end of the day, she realizes her mistake and she records a loss. The next day, she decides to charge her costs plus a little extra, and sells each loaf for £1.20. That day she makes a profit again.

If an insurer mispriced a policy, their mistake is not so quickly fixed, since policy cover usually lasts at least a year. However, their costs are not the same for all policies either, so it is less clear what should be charged to each person. Even less clear is how to remedy a price that is too low. The ideal situation would be that the insurer could accurately assess the risk cost for all customers. The insurer may still choose to charge the same premium to all customers, which it calculates as the average of previous claim costs, resulting in the following situation (Fig. 8.1).

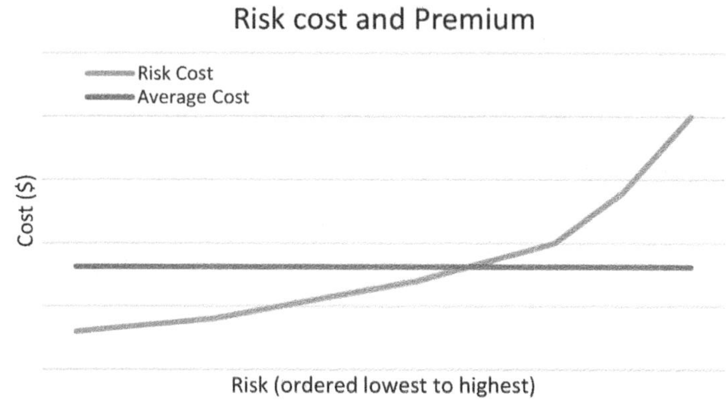

Fig. 8.1 A comparison between risk cost and the average cost over all customers

All customers to the left of the intersection are being over-charged, and they might recognize this, particularly if they shop around or use an aggregator site. If there was another company that charged their true risk cost, they would—in time—take their business to them.

At the same time, the customers to the right of the intersection are being under-charged. They also might realize this, particularly when they compare quotes from other insurers (although they may still feel like it is too expensive!). In addition, high-risk policyholders that used to be with other companies also notice they could get a lower price with this insurer, and so they move their business. Now the insurer is only holding policies where the premium charged is less than the cost, so it is loss-making. This is termed **anti-selection**: the underpriced risks are specifically selecting this insurer. This is not dissimilar to what happens in a bakery. If a baker was selling loaves at half their usual price, customers from all over town would flock to the bakery to buy one. The baker might sell more than the usual amount of bread, and so her loss would be even bigger. However, whether the customer is a usual or not is irrelevant to the production cost of bread. That is not so with insurance.

The next year, the insurer amends its prices to reflect the cost of claims. However, the cost of claims varies with the customer. The insurer has attracted bad risks all year and so their cost of claims has risen. If it charges the average now, the situation is as follows (Fig. 8.2).

Now, the cycle continues: charging a higher flat premium still attracts the worst risks and sheds the middling risks. In order to stay vigilant against anti-selection, insurers should keenly monitor business written after a new pricing structure is created. If the mix of business changes dramatically, or if a higher-than-expected number of policies is written in a certain customer

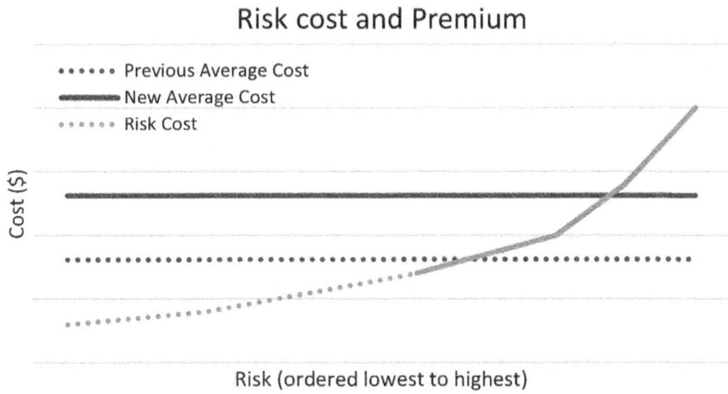

Fig. 8.2 The customers who were being over-charged take their business elsewhere, so on this chart those ex-customers are represented by a dashed orange line. The average claim cost of the current customers is now higher, shown by the solid blue line

segment it is likely that there is some mispricing. Therefore, pricing insurance is an arms race against other insurers to most accurately estimate the risk of an individual and price accordingly, so it is clearly in the interest of insurers to gain more information about customers.

Conversely, it is in the interests of customers who perceive themselves to be better risks to provide more information to insurers as well. By giving more information, they may be able to get a cheaper quote. Fortunately for those customers, and for pricing actuaries, the deluge of data that we all create daily is now able to be recorded and processed, such as using Smart Meters on home appliances or telematics systems in a vehicle. However, such effective risk segmentation also raises an interesting question for pricing actuaries: if everyone pays their own costs, is it really insurance?

In Chapter 2, **pooling** was described as a key part of insurance. By holding many similar risks, an insurer is able to lower the volatility of risk costs, and more effectively hold capital or reserves. However, if the full extent of telematics or smart meter data is used, there will never be another risk exactly the same as another. For example, telematics monitors can capture the length of the drive, the location of vehicle, and the time of day. There would never be another car who would do exactly all the same drives at exactly the same time as another vehicle: even if the data was slightly aggregated or simplified, there are so many combinations that two independent vehicles would not be classed as similar on all counts. But if no two risks are deemed similar enough, then there can be no grouping of risk, and so no pooling. What is the role of the insurer then?

Purists would say that at the point where all risks are estimated to that level of granularity and no risks can be deemed heterogeneous, it is no longer insurance. However, there does still seem to be a role for insurers, not as aggregators of risk but as holders of capital. In this role, they would collect premiums, invest on a large scale to obtain an investment profit, and then use the original premium amount to pay claims.

Another role for insurers here would be to calculate the risk in the first instance: a customer would be unlikely to be able to accurately compute and analyze all their own data to estimate a risk cost. However, this situation is still relatively far off: customers are still relatively squeamish about sharing that level of data, and techniques for organizing and processing the unstructured data the devices produce are still in their infancy. Telematics devices have been available since 1996, when President Bill Clinton signed a directive making GPS an international utility. Telematics products initially proved unpopular, and so were relatively unheard of for the following decade. The advent of smartphones opened up new possibilities and new attitudes to data collection, and in 2010 the first telematics-linked app was produced. However, telematics devices are still far from mainstream. Fewer than 1.5% of drivers in the UK have a telematics device in their personal car (although for commercial vehicles the take-up is significantly higher), so full adoption is still a long way off. Nevertheless, an applicant for insurance has to provide some data and this data leads to an assessment of risk. The applicant might alter their behavior after receiving a quote, e.g., by driving in a less aggressive way.

More accurate estimates of risk can also lead to higher premiums for those customers deemed riskier than previously. What happens to those customers? If they are unable to pay the higher premium, they may go without insurance and bear the risk themselves. Unfortunately, this doesn't just affect that customer, but anyone who has cause to claim against them such as a pedestrian that gets hit by their car. When those at highest risk don't have insurance, the benefit of insurance to society as a whole is compromised, and the state may be required to intervene, for example, by setting up pricing caps, putting limits on the amount of business that can be declined by an insurer, or even by not allowing insurers to decline business. The latter may require the government itself to set up a scheme to allow insurers to transfer any risk they deem excessive, while still writing it. Examples of schemes like these, as well as their benefits and drawbacks, will be discussed in Chapter 13. Alternatively, highly regulated pricing structures (perhaps even allowing only certain rating factors) may reduce the price sensitivity in the market. Price sensitivity can be good for customers as it forces insurers to compete on price and creates more affordable insurance. On the other hand, too much price sensitivity can mean that insurers only compete on price, and so other considerations such as the level of cover or good customer service are ignored in pursuit of the lowest premium. This can lead to **underinsurance**, where policyholders have insurance, but not enough to adequately transfer the risk. Over-arching these considerations is the recognition that insurance is a social good, and that while individual responsibility must be encouraged (e.g., by higher premiums for poor drivers, or encouraging risk mitigation), it is in everyone's interests that they—and those around them—are protected by a robust and efficient system of insurance.

Advocates of increased data and technology adoption say that using devices such as black boxes or smart meters reduces risk, as policyholders change their behavior. This can be true: if it is obvious that a black box is recording the speed of travel with a view to affecting premium, a customer may be more consistent at sticking to the speed limit. As this technology is still relatively new, there is continual refinement and research into the extent of behavior change. Initially, some black boxes measured the strength of braking—the hypothesis being that drivers who have the need to brake suddenly are more likely to be bad drivers. Explicitly collecting this data caused drivers to brake more gently even if an emergency stop was required, leading to an increase in the number of claims. With this feedback, providers have now generally switched to measuring the strength of acceleration, which effectively captures the aggressiveness of drivers and is not usually a reaction to the external circumstances. Similarly, driver cameras ('dash cams') are double-edged. They might protect a driver from losing money if their car is damaged by an obviously reckless driver in another car, but they will also record the driver's own driving behavior, including their speeding, so could serve to increase premiums for some customers. It is likely that adoption will hinge on public acceptance of this level of data collection, rather than any other factor.

Who are the better drivers: men or women?

The European Union have decided the answer: it doesn't matter. In the EU Gender directive, general insurers cannot ask the gender of the policyholder or use that information to determine motor premiums, nor can they use protected characteristics such as race, sexuality, or disability.

Undeterred, some motor insurers started analyzing characteristics that correlated strongly with gender instead, such as title (Mr., Mrs. or Miss) or even how 'masculine' the vehicle driven was. In order to stop this behavior, it was ruled that factors that proxy a protected characteristic should not be used for pricing.

Pricing actuaries aim to group risks into homogeneous groups based on certain characteristics. Even with huge datasets, armed with a host of statistical techniques and machine learning, this should be a very straightforward task; after all, even chimpanzees can group objects based on multiple factors. But while technology is really efficient at finding correlations, it is an actuary's job to decide what is causation.

Deciding what characteristics affect the risk is not always clear, as if hundreds of characteristics are collected then it can be difficult to separate out the effects of each, let alone determine what is random noise in your data. No matter what method is used to create the model, each rating factor must be carefully examined to identify correlation and causation.

To investigate correlation, a host of statistical techniques can be used. However, to investigate causation, an actuary must truly understand the risks they are considering. This is because statistical methods cannot shed light on what is not there: they can't consider the effect of a factor that was not captured or not filled out. This might seem obvious, but it is easy to get involved in complex modeling and lose sight of this qualification. Because of this, actuaries are increasingly intermediaries, not only between the model and the underwriters but between the data and the model. Actuaries need strong statistical skills, but also a sharp understanding of the wider context, purpose of the model, relevant regulation, and market practice. Inevitably, this creates specialization, even within actuaries themselves.

8.4 An Introduction to Generalized Linear Models for Insurance Pricing

So far, we have talked in general terms about the process of finding the risk cost. While that may be sufficient to gain a holistic view of the role of actuaries, this section aims to give some of the technical detail that traditionally sits behind those calculations. It is not for the faint hearted and assumes a basic understanding of linear regression. If this is your first introduction to statistical modeling, you may prefer to skip ahead to the worked example, which steps through all the stages of the modeling process without the complexities.

Before we start on **Generalized Linear Models**,[8] first, a recap of linear models. A linear model is of the form below, where y is the thing that we are estimating (called the response variable), x_i are the different variables, of

[8] Abbreviated to GLMs. This can be pronounced G-L-Ms or sometimes 'Glims'. Note (and avoid!) the common but infuriating tautology, 'GLM model'.

which there are n, and m and c are parameters. Linear models are always linear in their parameters.

$$y = \sum_{i=0}^{n} m_i x_i + c$$

Exercise 8.1: Which models are linear?[9]

$$y = mx^2 + c \quad y = mx + c^2 \quad y = \alpha^2 x_1 + \alpha x_2 \quad y = \alpha x_1 x_2 + c$$

Part of the attraction of linear models is that they are easy to work with and there are several easy statistical methods of parameter estimation for them, including Ordinary Least Squares and Maximum Likelihood Estimation. However, applying linear models to some cases can quickly generate absurd results. Most natural relationships aren't strictly linear, and require a more flexible model. We will return to this shortly but for now we consider an example where the use of (non-generalized) linear models is not appropriate.

Example 8.3: A linear model has been trained to model the salary of athletes in a particular sport. An athlete suffers an injury, so her salary is recalculated. If she has played 20 national games, what is her salary before and after injury?

$$\text{Salary} = \$480,000 + \$20,000 \times \text{Number of games played} - \$1\text{m} \times \text{Number of Injuries}$$

If she has played 20 national games, before the injury her salary is \$880,000. After the injury, the model predicts that she should have to pay the team \$120,000 a year!

To extend a linear model to any data that fits a distribution in the exponential family, GLMs simply apply a function, called the link function, to

[9] All but the third model are linear. Note that in the second model, c is squared (not linear), but can be rewritten as a new parameter, c', without affecting any other terms in the model. Whereas in the third model, because α is used in both terms, there is no way of rewriting α without one of them being non-linear.

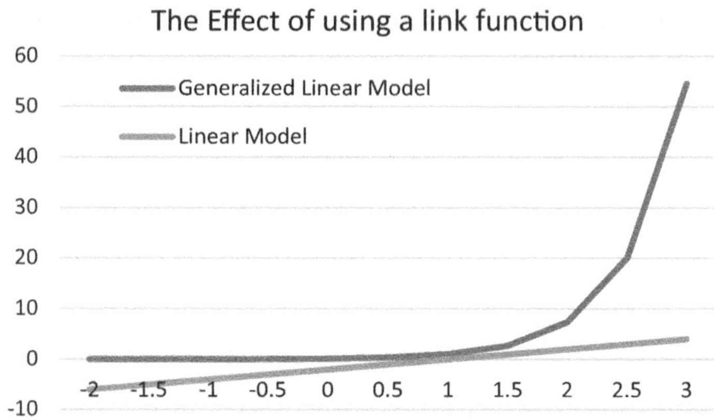

Fig. 8.3 A comparison of response variables in a linear and generalized linear model

the linear combination. This means that GLMs have the following form:

$$g(y) = \sum_{i=0}^{n} m_i x_i + c$$

This allows us to use linear regression techniques again. In the graph below, you can see how applying a link function (in this case, natural log) to the linear model (here, $y = 2x - 2$) allows the response to take on a very different shape than a straight line. This will help us model a wider variety of responses and situations (Fig. 8.3).

The choice of link function depends on the distribution of data you wish to model. The number of claims is usually modeled using a Poisson process. The prediction of the model is then divided by the time or exposure measure to give a **frequency**. The natural log function is the canonical[10] link function for Poisson distributions, as it never gives a negative number of claims. The severity of claims usually fits a gamma, beta, or Weibull distribution best, although all three can be parametrized to look remarkably similar. However, gamma is usually chosen as it is the simplest. The canonical link function for Gamma distributions is the inverse function.

But actuaries are not pure mathematicians, or even statisticians. Actuaries are doing calculations for a practical purpose: here, to explain and predict claim behavior. As the intent is to multiply the frequency and severity GLMs together to derive a risk cost, when using the canonical link functions, the

[10] Canonical here in the mathematical context, meaning 'the one that people use as standard'.

burning cost model becomes:

$$\text{Burning Cost} = y_{\text{freq}} \times y_{\text{sev}} = \exp\left(\sum_{i=0}^{n} \alpha_i x_i\right) \times \frac{1}{\sum_{i=0}^{n} \pm\beta_i x_i}$$

Although statistically correct, this is not a good model, because it is incredibly difficult to use this to explain the effect of any factor to a non-technical audience, such as an underwriter who has not studied advanced mathematics. For example, if x_3 increases by 10%, how much does the risk cost increase by? It's not obvious how the inverse linear combination and the exponential function of the linear combination will combine. So instead, it is common for pricing actuaries to disregard the suggested canonical link and instead use the log link function for both Poisson and Gamma distributions. This means that the model has a neat, multiplicative explanation for each factor.

$$\text{Burning Cost} = y_{\text{freq}} \times y_{\text{sev}} = \exp\left(\sum_{i=0}^{n} \alpha_i x_i + \sum_{i=0}^{n} \beta_i x_i\right)$$

The output of the GLMs will be a list of parameters for each variable, or a table of parameters if it is a categorical variable. These are then multiplied together to give a relativity for a risk, which is then multiplied by a calculated base rate to give a risk cost.

However, all this assumes that claims severity can be modeled by a Gamma distribution. Thinking about the shape of a Gamma probability density function (which is continuous, always non-negative, and positively skewed), making this assumption requires data preparation to remove any negative or nil claims. Nil claims often occur if the insurer pays out initially, but then claims the money back from another insurer or third party, or through invalid claims. Depending on the particular idiosyncrasies of an insurer's claim system, a substantial number of these claims may need to be removed. However, if there are administrative costs of dealing with nil claims, then an actuary may need to do some additional modeling to capture the frequency of nil claims.

Alternatively, a Tweedie distribution could be assumed. Tweedie distributions are not in closed form and so are quite difficult to perform calculations with, but they do allow for a point mass in a distribution that is otherwise non-negative, continuous, and right-skewed. A Tweedie distribution could therefore be used if a predictive burning cost model is required without explanation.

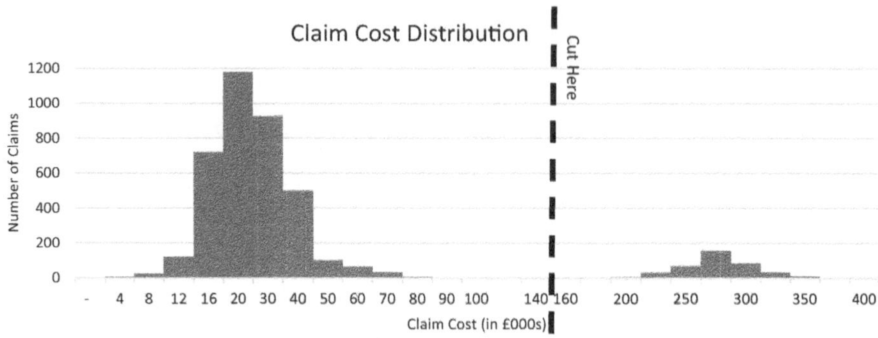

Fig. 8.4 A simple solution for bimodal data

The other statistical frustration is that claims severity distributions can sometimes be bimodal. It is very easy to spot this: by plotting claim costs against number of claims of that value, any graphs that have two 'humps', even if they are not of equal size, should be considered bimodal and problematic. There is no simple distribution that deals with bimodal data well, and so actuaries have devised an ingenious solution: cut the distribution into two (Fig. 8.4).

By cutting the distribution at a point between the two modes, claims costs below the cut can be modeled by a straightforward Gamma distribution. The cutting point is usually referred to as a cap. The interpretation here is that we firstly need to model if a claim happens or not, and if and only if it does, will it go above the cap? And if the answer is yes, then by how much?

To model this binary outcome—is this claim more than our cap?—a binomial model is used. Again, a GLM is suitable if there is sufficient data, but needs a suitable link function so that the response is limited to values between 0 and 1, to align to probabilities. The logit link function is canonical here: $ln\left(\frac{y}{1-y}\right)$. This model is referred to as the propensity model, and interpreted as "the propensity for the claim to be over the capped amount". Then, two severity models are created using Gamma distributions or otherwise: one for the costs below the cap, and one for the portion of the claim cost above the cap.

Our risk cost formula then becomes:

$$\text{Risk Cost} = \text{Frequency} \times \left[\text{Severity}_{\leq \text{Cap}} + \text{Severity}_{> \text{Cap}} \times \text{Propensity}\right]$$

This is a common approach, but there are valid alternatives such as splitting the two varieties of claims but not capping, modeling frequencies separately and using a shifted Gamma for the claims around the higher mode.

As with all modeling decisions, it depends on the context and quantity of the data.

Binomial data is also found when modeling lapses of policies. In this chapter, retention models have not been included as part of the risk cost, but for lines of business with multi-year policies and fixed premium increases, they are much more important.

Example 8.4: A health insurance product offers an annual policy. However, if a customer renews their policy, the increase in their yearly premium will be no more than 5%, and all conditions diagnosed in the previous year will be automatically covered by the policy.

An insurer would want to calculate upfront what the lifetime risk cost for this customer would be, rather than focusing on the risk cost for the first year. As customers who become ill are more likely to renew their policies in this offer, it would be crucial for an insurer to build a retention model to price the initial premium properly.

8.5 Worked Example: Pricing Pet Insurance

Imagine that we have 7 pets on cover. Last year, their claim costs were as follows:

Flora—$0
Rufus—$120
Skye—$135
Alfie—$130
Clara—$180
Fenton—$285
Jack—$55.

If the premium was set as the average cost, all pets (and all new pets) would pay $129. But the group of pets is very varied, so we may want to split by type of animal.

Flora, Skye, and Jack are cats, and the average of all of their claim costs is $63.

Rufus, Alfie, Clara, and Fenton are dogs, and they would be charged $179.

Let's split further: older animals are more likely to claim than younger animals. Splitting by whether the animal is over 8 years old (remember, it's important that rating factors are objective and specific), we get the following matrix, where the premium is the average cost of the animals that fall in that category.

	Cat	Dog
Old	$135 (Skye)	$232.5 (Clara and Fenton)
Young	$27.5 (Flora and Jack)	$125 (Rufus and Alfie)

Finally, we might want to assess how well-trained they are, as well-trained animals may be less likely to get into accidents, and so ask if the animal has completed a suitably endorsed training course. The answers might classify the pets as follows:

Cats	Trained	Not trained
Old	$135 (Skye)	-
Young	$0 (Flora)	$55 (Jack)

Dogs	Trained	Not trained
Old	$180 (Clara)	$285 (Fenton)
Young	$125 (Rufus and Alfie)	-

This data can then be used to form a pricing structure. If a new enquiry comes to an insurer with an old, trained cat, the risk cost will be estimated to be $135, the same as the average in that group (which currently only contains Skye).

Note that there are some groups where we have no information, such as the Not Trained Young Dog category. How would a prospective policy in that group be priced? There are many possible answers:

- It could be priced the same as Young Trained Dogs or Not Trained Old dogs (most likely whichever costs the most, to be prudent!)
- It could be priced using relativities to estimate it. In the above example, the relative difference in price between Trained and Un-Trained Old dogs was $285 \div 180 = 158\%$. Assuming the relativity would be the same for young dogs, Not Trained Young dogs should cost $\$125 \times 158\% = \$198$.
- It could be priced by a dog expert, rather than an actuary. Alternatively, an actuary could make an estimate using the other groups, and the experts adjust the actuarial figure using their own subjective assessment.
- It may not be important to gain an estimate for the group if it is unlikely a new policy will be in that group, or if it is not a target market for the insurer. In this case, policies in that group may be declined immediately.
- More data could be sought. This could be industry data or bought from a private company, or in another context, another source such as government statistics.

The other problem with this as a tariff is that the average cost for Trained, Young Cats is $0. That is clearly an unreasonable risk cost: if that were true, kitten insurance would be free (other than admin and insurer's expenses). An actuary would want to seek an alternative estimate for this category. It could do so using the methods above, or just increase the claim cost for this group a little, perhaps to $10. It may seem ridiculous that an insurer would never have had any claims at all in a category, but it isn't so far-fetched, particularly if the product is relatively new, or if claims tend to occur together, for example natural catastrophes. Looking at 3 years of past storm data may suggest that storms never happen, but extending your data to 5 or 10 years may reveal that they only tend to occur every few years, but are severe if they do happen.

This pricing structure uses historic data and is therefore backward looking to some extent. An actuary may know that due to the high cost of feline veterinarian equipment, all claims related to cats will increase by 8% next year. Instead of waiting for this increase to come through in claims, actuaries can be proactive and increase all the prices for cats by this amount. The risk cost for old, trained cats like Skye is now $145.80, and so on.

A simpler way to make a consistent change to all the prices without changing the ranking of risk would be to create tables based on relative risk, rather than monetary amounts. This means that in order to change all the risk costs by 8%, only the base rate needs to be changed. In this example, the Dog relativity table would be as follows, where $125 is the base rate:

Dogs ($125)	Trained	Not trained
Old	1.44	2.28
Young	1.00	–

Most pricing actuaries will do a version of this method but with millions of policies and hundreds of pieces of information for each policy. They will likely do this at frequency and severity level, thereby creating a table of frequencies for each group and a table of claim severities for each group and then multiply the two together to calculate a risk cost. This is usually performed for every peril. Because of the amount of data, it is often done using statistical techniques such as Generalized Linear Models. It is a time-consuming activity, not least because of the data preparation required, but also because a product with only 5 perils may require at least 10 models (Peril A Frequency, Peril A Severity, Peril B Frequency….). These models are created separately and then combined to create the final estimated risk cost to be used in the calculation of premium, which is the topic of the next chapter. But first, we will consider how to price big risks.

8.6 Pricing for Large Risks

The previous section described the process for pricing insurance when there is plenty of claims data to analyze. However, not all lines of business have that luxury. In addition, some lines of business are written based on certain limits or deductibles, where the insurer will cover the claim costs up to or above a certain point only. Commercial property and casualty policies are often written in this way. The previous pricing methods don't deal well with this awkward truncation of our claim distribution. Instead, Original Loss Curves are used.

Original Loss Curves use the idea that an insurer will usually experience many lower cost claims, but (hopefully!) not that many expensive claims. So, using the data for the cheaper claims as a guide, a curve can be drawn to estimate the higher claims. This is a general methodology that has been used for years by certain lines of business and so, like Darwin's finches, each curve methodology and name is now slightly different: you may hear about them as Increased Limit Factors, First Loss Scales, Exposure Curves, or Excess of Loss Scales. The derivation of them relies on differentiation of truncated cumulative distribution functions, but this section will focus predominantly on the method used in casualty business.

Casualty business has a big pricing problem: unlike property damage, there really isn't a cap on losses. In property damage, the value of a total loss is pretty much the limit for the worst possible event, whereas in casualty, court awards and medical inflation can mean that the worst-case scenario is very hard to judge ahead of time. Insurers, like most people, dislike the idea of taking on unlimited risk, and so casualty is usually written with limited cover, for example, $5m of cover. This also avoids levels of uncertainty and risk exposure that would make business untenable or premiums unaffordable.

The level of cover the customer chooses usually has little to no connection with their risk—it is usually determined by their risk appetite and their financial situation, as well as market standards or 'what they bought last time'. The effect of this is to protect the insurer from paying any claim cost above the limit, so the risk cost of interest to the insurer is therefore the risk cost capped at this limit.

So, what are the insurer's capped risk costs? They can easily be calculated in a table form using the Original Loss (i.e., the loss that the customer experiences) and the known limits, as below (Table 8.1).

But what if a customer wants a higher limit?

Firstly, notice that the limit is only important if a claim is over it, so increasing the limit won't affect the cost to the insurer for every claim. In

Table 8.1 An Increased Limit Factor example

Original Loss ($)	Loss with $500 limit	Loss with $1000 limit	Loss with $2000 limit
100	100	100	100
500	500	500	500
2000	500	1000	2000
5000	500	1000	2000

fact, the lower the limit, the more claims will be affected by the change. We can quantify this effect by using Increased Limit Factors. To do so, sum the capped claim costs for the old and new limits, and divide the new by the old.

Example 8.5: Using the table above, the Increased Limit Factor taking the limit from $500 to $2000 is:

$$\mathrm{ILF} = \frac{100 + 500 + 2000 + 2000}{100 + 500 + 500 + 500} = 2.875$$

After using a table of capped claim costs to calculate ILFs, a table of ILFs can be created. To save calculating the ILF for every pair of limits, it is usually calculated with respect to a very low (and arbitrary!) limit, reflecting a very basic cover. The choice of this basic limit may be a regulatory minimum, or it might be driven by the data that is available. While a table is the traditional way to show ILFs, graphing the ILFs is also interesting (Fig. 8.5).

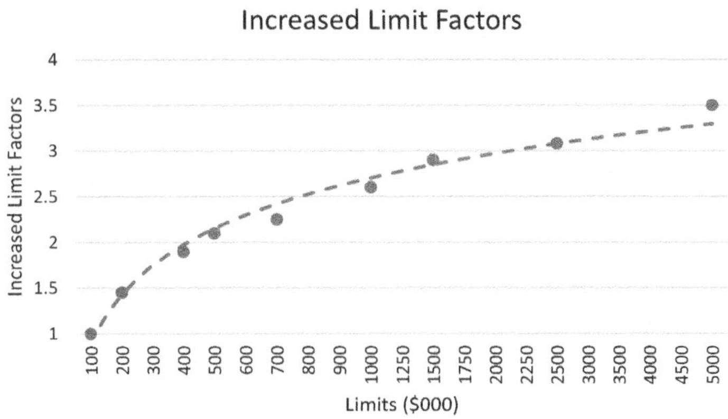

Fig. 8.5 Example Increased Limit Factor graph

This shape is typical for an ILF curve. This graph makes the origin of the name 'Original Loss Curve' clear: the method relates the uncapped (original) loss to the value that an insurer pays, and uses a curve to extrapolate data. Using this graph, it is simple to change the risk cost to account for new limits. This is done by reading the ILFs for the current and new limits of the graph and using the equivalence:

$$\frac{\text{Current Risk Cost}}{\text{ILF at current limit}} = \frac{\text{New Risk Cost}}{\text{ILF at new limit}}$$

Example 8.6: An insurer currently has a limit of \$1,000,000 and are proposing to change the limit to \$5,000,000. They currently estimate their risk cost to be \$22,000 per policy. Using Fig. 8.5 above, what is their new risk cost?

First, we read the ILFs of the old and new limits of the graph using the curve, not the data points. For \$1,000,000, the ILF is 2.7, and for \$5,000,000, it is 3.35. Now, we plug these numbers into the above formula.

$$\frac{\$22,000}{2.7} \times 3.35 = \text{New risk cost} = \mathbf{\$27,296}$$

This chapter has outlined the methods and considerations for determining the risk cost of a policy. You will have noticed that the methods for pricing lines of business with lots of data (that is, GLMs or even advanced machine learning techniques) are very different to the approaches used on heterogeneous large risks! Regardless of how it is determined, the risk cost is not the premium that should be charged to the customer. To get the premium, *all* the costs of administering a policy need to be considered, which we will do in the next chapter.

9

Pricing: The Art of Estimating Everything Else

Chapter Objectives:

- Estimate non-risk costs for a policy, including fraud costs
- Understand the types of tax applicable to insurance products
- Combine the risk cost and non-risk costs to find the total premium
- Explain the insurance cycle and how environmental factors affect insurance premiums.

You may depend on it that I will not draw the King into any great expenses.
—Marie Antoinette, in a letter to her sister Maria Theresa, dated July 30th, 1774

If you've ever been on an all-inclusive trip, you will know that the cost is rarely 'all-inclusive'. The headline price includes your flights and accommodation, yes, but not the local tax payable at the hotel. You have to get a cab to take you from the airport to the hotel—that's not included. You need to tip the service staff—not included. The headline cost might be the largest single cost, but there are plenty of hidden extras that can quickly add up.

Similarly, the risk cost is not the only expense for insurers: they also have a host of non-risk costs which need to be covered by the premium. Non-risk costs are costs associated with trading and servicing the policy and any relevant claims. However, insurers have the added complication of how to

© The Author(s), under exclusive license to Springer Nature Switzerland AG 2022
C. Townsend, *A Risky Business*,
https://doi.org/10.1007/978-3-031-11673-5_9

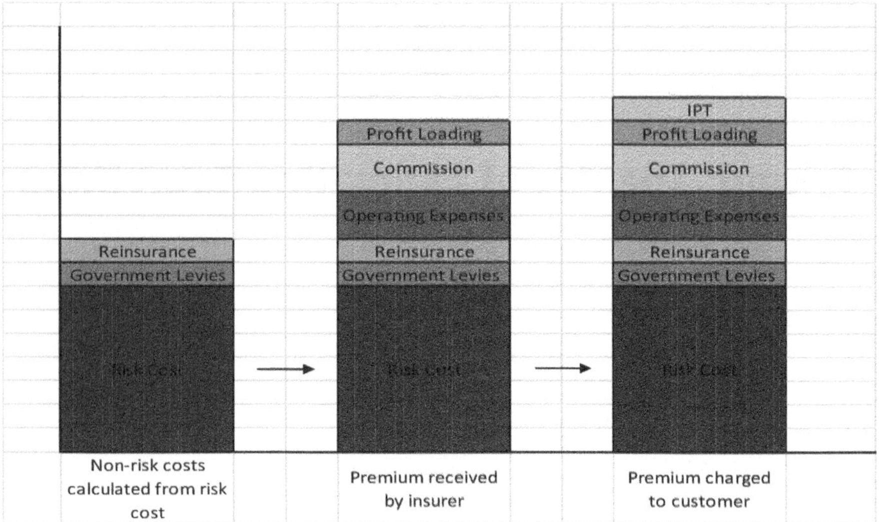

Fig. 9.1 From risk cost to final price

add this cost into the premium, as the risk costs for each policy differ. There are two main methods for including non-risk costs: a flat amount (e.g., £5 on top of the risk cost), or a proportional amount (e.g., 5% of the risk cost). In order to decide how best to allow for non-risk costs, each source of expense must be understood.

We'll now look at each of these non-risk components in turn. They include:

- Operating expenses
- **Reinsurance** and levies
- Commission or other acquisition expenses
- Profit (allowing for corporate tax rates)
- Investment Income
- Insurance Premium Tax (IPT) (Fig. 9.1).

9.1 Types of Non-risk Costs

9.1.1 Wages, Rent, and Other Operating Expenses

As with other businesses, the cost of staffing needs to be covered. Insurers also have general business expenses such as IT systems, HR services, stationery, and marketing. IT costs for insurers may be higher than other businesses as they store huge amounts of data and, as it is of a personal nature, needs to be particularly secure. Processing data also has a cost: as well as the computers themselves, specialist software might need to be purchased. There is specialist software for just about anything of course, but an insurer would almost certainly want software for data processing, software for policy and claims administration, software for building pricing models, software for **reserving**, software for underwriting and monitoring or catastrophe models, and software that calculates capital requirements. It may need software or a platform for sales as well. For even one of these, the licensing cost for a few computers is likely to be six figures (although insurers are increasingly looking to open-source software such as Python because of the flexibility and cost-saving it provides). These software programs are often produced by specialist actuarial consultancies, which in themselves are a very fast-paced and interesting area for actuaries to work. An insurer may choose to outsource work to actuarial consultancy firms, for example, a specific pricing project or a reserving exercise, so working in a consultancy often provides the opportunity to work on a variety of actuarial work.

Insurers, like all businesses, will also need to pay for rent or building costs. Many insurers have offices in the City of London, near the Lloyd's of London building (which is itself near the original Lloyd's coffee shop) so brokers can visit multiple **underwriters** relatively easily. However, their central locations on Lime Street, Lombard Street or Gracechurch Street give rise to particularly high rents. For this reason, many insurers will also have offices elsewhere. In particular, global insurers may have offices in any country they write business in to avoid tariffs or foreign trade regulations.

Like many companies, insurers often outsource work to less developed countries where rent and wages are cheaper. Outsourcing has traditionally been associated with call centers, but now includes data and analytics capabilities. For example, in the past 20 years India has increased its university provision and now produces the third highest number of graduates in the world (approximately 9 million every year, behind only China and the US). This increase in technical skills makes it an ideal place for a large insurer to base an IT or data team.

To allow for the above expenses, an insurer would usually add a proportional (percentage) loading onto the risk cost, but the percentage amount may vary. For a particularly staff-intensive product, such as something that requires a lot of underwriter involvement or is written exclusively from a central London location, the percentage may be set at a higher level. Some insurers may calculate how much of the floor space is 'used' by each product or how much IT resource is required for each product, and charge accordingly. However, this can be time-consuming and it can be difficult to allocate the time of cross-product staff such as HR.

There are two concerns with adding a proportional loading onto the product: from a customer fairness perspective it means that bad risks are shouldering a lot more of the expenses than a good risk, even though there may be no difference in the expense cost to the business from the two policies. This can significantly decrease the affordability of insurance for the worst risks, which may mean that they will go without insurance. The other concern is that writing less risky business may lead to not having enough money to cover fixed expenses. This means that attracting lower risk customers perversely creates another risk for the business! In fact, any change in the mix of policies that an insurer takes on may lead to the risk that expenses may be higher than expected, or that attracting lower risk policies leads to not enough money being raised to cover expenses.

Example 9.1: An insurer writes home insurance, predominantly for homes with a high risk of flooding. The pricing team calculates an average risk cost of $1,000 per policy, and a 20% expense load is applied to this. On average, this means that $200 per policy can be put toward expenses.

The business wants to reduce its exposure to flood events so refocuses on home insurance for homes inland. After a heavy marketing campaign, the type of policies on cover change to being predominantly lower risk, with an estimated average risk cost of $600 per policy. The pricing structure is not changed, so now each policy only generates $120 toward business expenses. Now, selling 100,000 policies only raises $12 million instead of $20 million to cover expenses. To counteract this, the pricing team will need to recalculate the percentage loading applied, for example increasing it from 20 to 33%, or with the help of the business, try to reduce expenses.

9.1.2 Insurance and Reinsurance

Insurers have to have insurance themselves for their day-to-day business operations, just like other businesses do. On top of that, they also have their **reinsurance** costs. Reinsurance is effectively insurance against claim costs being larger than expected, and is explained in more detail in Chapter 13. Some types of reinsurance effectively set a maximum for the insurer's liability for claims costs, ceding the cost above this threshold elsewhere. All types seek to transfer some risk from insurers to reinsurers. For this, reinsurers will charge the insurer a premium just as insurers charge customers a premium to take on their risk.

The reinsurance premium might be considered a risk cost, particularly in the case of large commercial properties where reinsurance is bought specifically to cover one policy. For example, if a high-profile city skyscraper was insured, an insurer might be comfortable to pay rebuild costs up to $100 million but no more. In this case it would buy reinsurance specifically to cover the possibility of a claim event costing more than that amount. For pricing purposes, it would assess the premium based on risk up to the threshold and load on the cost of ceding risk above that point. It is not important whether reinsurance costs specific to a policy are included in the risk cost or not, only that the cost of reinsurance is included once and once only in the premium.

For smaller policies there is usually no specific reinsurance bought per policy, but overall reinsurance that covers the risk of the cumulative cost of claims exceeding a certain level. Because the risk of an individual policy contributing to breaching this limit can't be calculated without considering the other policies, this reinsurance cost is considered separately to the risk cost. This is usually allowed for in the form of a percentage load, as the bigger the risk cost the bigger the risk, so the more that the individual policy would be expected to contribute to the cumulative claims cost.

Some markets and products may be subject to additional levies, paid to the government or local services relating to the risk. For example, a government may have a risk sharing scheme for areas with a high risk of flooding, in order that homeowners are able to buy affordable flood insurance. More detail about how schemes like this work is given in Chapter 13, but the cost to insurers to be involved in this kind of arrangement is usually a percentage or flat cost applied to the exposure measure.

9.1.3 Commission and Acquisition Expenses

The cost of finding the business depends on how the policyholder came to buy a policy from the insurer. For non-Lloyd's business, there are three ways:

- Directly
- Attracting through media channels
- Through intermediaries such as insurance brokers

If a policyholder goes straight to an insurer's website or sales line and buys a policy, the cost of acquiring that policy is low—only the cost of maintaining the sales center, which would be a general business expense and not attributable to any specific policy. This is obviously therefore preferable, but it requires a strong public brand and perhaps a high marketing spend (marketing is not usually counted as an acquisition cost but as a cross-business expense). Renewals also fall into this category, as there is no additional cost: a letter is usually sent to advise a customer their policy is ending, and the same letter usually offers a renewal price. Insurers therefore want a high renewal rate: like other industries, it is cheaper to retain customers than find new ones.

But not many people browse insurer websites for fun, and so attracting customers through multimedia channels is a more realistic prospect of obtaining new business. This used to be telesales or postal flyers, but now this is primarily online. Customers search online for insurance, click through to an insurer's website, and buy a policy from one of the companies they find. This also may have a relatively low cost to attracting the business—the only cost here is the cost of marketing, or buying a top spot in a search engine's responses.

Increasingly customers turn to comparison websites to do this for them: they put their details into one website which then scans multiple insurers' websites to get quotes. The customer can then compare quotes from hundreds of insurers, including smaller companies that they hadn't heard of, and choose the one they would like to buy. It costs insurers money to be included in this list which is usually charged as a percentage commission for each policy sold. This percentage needs to be added on to the risk cost. Insurers often refer to comparison websites as aggregators, as they aggregate quotes from multiple insurers.

The use of intermediaries to buy financial products is particularly common for commercial insurance. This is traditionally an insurance broker who acts in a very similar way to an aggregator website, but through their relationship with underwriters or companies may be able to access cheaper premiums or personalized policy wording on behalf of their client. As well as acting in an advisory role, brokers also sometimes do the policy administration on behalf of the insurer. Brokers charge an insurer commission for doing this, commonly a percentage of the policy, say, 15–30%. Some brokers have agreements with certain insurers that they will even do the administration of claims on the insurers' behalf. These brokers, known as delegated authorities, will charge an even higher commission.

Although commission can be a large part of the premium, brokers give access to a wider range of clients and niche business. This can help an insurer diversify. Commission is not the only way brokers are remunerated: insurers may incentivize careful risk selection by brokers by entering into agreements to share any profit from certain policies. The expected cost of these agreements is usually considered in the profit loading. There is increasing regulatory pressure on brokers to prove that they are adding value to insurance-buying process as part of wider fair-pricing regulation, so the next few years will be pivotal for brokers as they refocus their offering and services.

9.1.4 Profit Loading

An insurer doesn't just want to break even on a policy however: it wishes to make a profit! Only a small percentage of the premium is expected to be profit - perhaps between 5 and 10%, depending on the market and line of business. It is only *expected* to be profit as it is impossible to know the exact cost of the policy until any claims are settled, which could be years in the future. Profit loading is usually based on strategic aims rather than any actuarial science, so it is often dictated by a company's targets and shareholder expectations. This target will often be expressed in the form of a **Combined Operating Ratio**: the sum of claims and expenses divided by premiums (note that this ignores the result of any profit due to investment). Therefore, if claims and expenses are estimated, an insurer adds a little extra to its premiums to reach its required COR—this little extra is the profit loading.

However, the insurer may not receive all of this expected profit, even if the profit is made! It may have profit sharing agreements with brokers, and will also be subject to corporation tax. The percentage profit required by the shareholders should therefore be increased to take account of these considerations.

Example 9.2: An insurer wishes to achieve a COR of 92%. It therefore wants 8% of the premium to be profit for the business. If the corporate tax rate is 25%, the company should use a profit loading of 8% × 1.25% = 10% in order to pay its tax and hit its profitability targets.

Some segments of the market may consciously make losses for strategic reasons. UK motor insurers tend to record a COR of more than 100% as they fight to establish brand and market share in a very price-competitive environment. The hope is that customers will then take out additional, profitable policies (such as home or gadget insurance) with their motor insurer, rather than having to deal with multiple companies. The UK motor market 2019 COR is estimated to be 105.7%, and 2020's COR is 107%, so insurers don't expect any profit in this product for these years. The US auto insurance market shows similar trends, with a 2019 COR of 109.4%: a loss of $4 billion[1]!

9.1.5 Investment Income

Insurers hold huge sums of money. They need some assets that are easily accessible in order to pay the majority of claims, but also some assets to pay for extreme events and larger than expected claims. All of these assets can be invested in order to yield an investment return, in the same way you receive interest on the money in your bank account. The majority of assets need to be in fairly safe, low-yield vehicles such as government bonds, but some capital could be invested in ways that are riskier but give a higher return.

To see how this works, consider just one policy. The insurer expects that there will be a claim of $100 to be payable in exactly one year's time. Ignoring all non-risk costs, would charge $100 to cover this risk. The insurer collects the premium on the inception date and invests it in an easily accessible way. If it was invested for the entire year, the $100 might earn the insurer $3 dollars. In reality, the insurer has received $100 when the policy is written and then $3 in a year's time—extra profit! In a particularly price-sensitive market, the insurer may feel able to reduce the premium to $97.09. Investing this amount in a way that yields 3% per year would give the insurer exactly $100 to pay the claim as expected in one year's time.

Although listed as a non-risk 'cost' here, investment income will typically reduce the price charged. However, investment returns are not certain, and may be lower than expected (this is called investment risk). As investment returns are largely determined by wider economic conditions, all competitors will benefit or suffer similarly, and so investments aren't a competitive lever: they cannot be controlled by an insurer in order to create competitive advantage. Actuaries may choose to specialize in investments: it is an industry all by itself, as well as a tool used by the traditional insurance and pension companies, so there are lots of opportunities in this field. The main role of actuaries here is to quantify investment risk, and use the estimates of the amounts and timings of an insurer's or pension provider's costs in order to find suitable matching investments (that is, investments that have the same characteristics as the company's liabilities, so that overall risk is minimized). This increasingly encompasses **ESG** concerns too, as companies want to invest ethically

[1] fitchratings.com.

and in environmentally friendly assets, and investment actuaries are enabling that cultural shift.

9.1.6 Insurance Premium Tax

Finally, in addition to the premium, a UK policyholder must pay Insurance Premium Tax. Like its more famous sibling, Value Added Tax, IPT is included in the premium advertised and charged as a percentage of the premium. There are two levels of IPT: the standard rate of 12% for **general insurance** products such as car or home insurance, pet insurance, and health insurance; and a higher rate of 20% for gadget insurance and travel insurance, as well as car insurance for brand new cars. Life insurance or assurance are exempt from IPT altogether. One of the reasons for having a specific insurance tax is that VAT doesn't apply to insurers, as they are selling neither a good or a service, but a promise. IPT was introduced in 1994 at the level of 2.5% of premiums, but has been subject to multiple increases since (Fig. 9.2).

In the US, states levy their own State Premium Tax, which varies in amount between a minimum of 2% and a maximum of 4% (although 2.5% is the most common).

Politically, taxes on insurance premiums are an excellent way to fill government coffers. Firstly, it is collected by insurance companies rather than the government itself, and as it is not as well-known as other forms of tax, some customers may not realize they are being taxed at all, which is why many call it a 'stealth tax'. It also means that insurance companies are the focus of ire about prices and, as IPT is included in the price advertised, customers often feel it is insurers charging too much for car insurance, rather than viewing it as a price plus tax. Like VAT for retailers, insurers don't include IPT in their

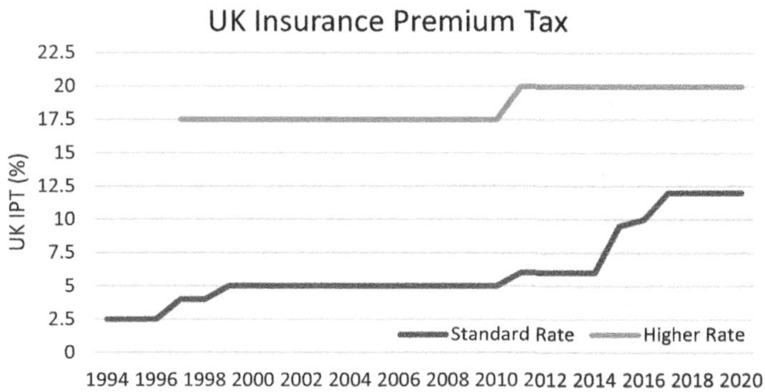

Fig. 9.2 UK Insurance Premium Tax since 1994

financial reporting, so any financial ratios (such as the **Combined Operating Ratio**, **Expense Ratio** or **Loss Ratio**) will not include the effect of IPT.

Secondly, IPT is advantageous because it is a tax on an often-mandatory product. Car insurance that covers the cost of third-party injury is a legal requirement for all UK drivers, and home insurance is often necessary in order to secure a mortgage. Therefore, it is a tax on a product that for most is a necessity. This is unlike other taxable products, such as luxury goods, biscuits, alcohol, or gambling. Because of this, as spending wanes in a recession so do VAT receipts, but spending on insurance products may be less subject to the same contractions, and IPT may be a more stable form of income for governments. It also raises a lot of money: in the 2018/2019 tax year, UK IPT raised £6.3 billion—triple the amount from gambling, and double the amount raised from tax on wine! With other forms of taxable spending, it is often argued that a reduction in consumption is good for the customer due to health benefits or the financial benefit of not buying an unnecessary item. Perversely for those needing to reduce their spending on insurance products, reducing the level of cover they receive leaves them more exposed to risk and potential costs. IPT is effectively a tax on doing the responsible thing!

Further, IPT is largely a regressive tax: it is disproportionally paid by those who can least afford it. Taking the example of car insurance (as the insurance that most people will have—about 20 million households in the UK have at least one policy): **risk factors** might include:

- Where you live, e.g., less salubrious areas often having higher level of theft or arson claims
- Level of voluntary **excess**—the higher the excess, the less an insurer has to pay in the event of a claim
- Number of previous claims, which may penalize those who don't have the disposable funds to pay for minor car repairs so need to claim on their insurance.

The Association of British Insurers has estimated that the least affluent households spend over 4% of their income on insurance products, whereas for the most affluent it is less than 2%. Insurers need to price risks accurately due to **anti-selection,** but because IPT is calculated as a percentage of the premium, those with the highest insurance costs are hit again with the highest absolute amounts of tax. This is also true for young drivers, particularly those under 25, who on average pay £110 in IPT alone on their car insurance. Some argue that it is a progressive tax as the most affluent have the most to insure, but the risk of the higher sum insured is often offset against the lowering of risk that being able to pay higher excesses or living in an affluent area affords.

While there is little evidence that increasing IPT increases the number of uninsured drivers (although due to the difficulty in measuring the number of uninsured drivers, that is still a concern of many insurance industry bodies), in health insurance the effect is more easily observed. Between 2015 and 2019, the level of IPT doubled and the number of British people covered by health insurance dropped by 10%. Of course, correlation doesn't mean causation, but common sense suggests that in optional products, price elasticity is much lower. The number of people covered privately directly impacts the number of people needing to access state-provided NHS care, so there are wider considerations and wider costs to raising this tax.

9.2 Insurance Fraud

The cost of fraudulent claims is covered in the premium as part of the risk cost. This is because if it is known that a claim is fraudulent, the claim will be rendered invalid and the claim won't need to be paid by the insurer, so the claim cost will become zero. However, the cost of undiscovered fraudulent claims, and the cost of fraud detection and prevention, is paid by every customer, so it is in the interests of companies and customers to improve fraud spotting measures. It is surprisingly big business: it is estimated to cost around $80 billion in the US, and almost £1.5 billion in the UK. In some European markets, fraud is estimated to be 10% of the total claim cost!

Fraudulent claims can occur in many guises and sizes: a smaller example is an exaggerated claim, when the claim event occurred but the amount being claimed is greater than the cost to the customer. This is often done by individual opportunists, as is policy fraud, when a policyholder lies on an application form to obtain a cheaper premium. One type of policy fraud is called fronting, and is common in motor insurance: a younger driver is listed on a policy as an additional driver, when in fact they are the main driver and the older driver (usually a parent) is only listed to reduce the premium. Fortunately, this is relatively easily spotted by taking the age difference between main and youngest drivers, and can be checked before a policy is underwritten.

The Horse Murders Scandal

James Druck couldn't find a buyer for his daughter's show-jumping horse. In 1981, he wanted to sell it for $150,000, but the highest bid was only $125,000. Instead of chalking it up to a loss, he realized that if the horse died, he would receive the full $150,000 that it was insured for. He hired a guy, Tommy Burns, to electrocute the horse in a way that would be indistinguishable from a death from colic, and collected the money.

This fraud was soon turned into a reliable money-maker by an organized group of criminals. The group targeted wealthy widows to buy shares in what were supposedly prize-winning horses. The widow would pay the 'trainer' for a share (the trainer being part of the group), and the group of conmen would live off this money for a while. If she got suspicious, the horse would die from 'colic', and she'd be encouraged to buy a share in another champion. One widow, Helen Brach, got a little too suspicious, and mysteriously disappeared, although no conviction was ever made.

This scam continued for almost 20 years, when the FBI convicted those responsible. During that time, at least 50 horses (and perhaps one wealthy widow) were killed.

Organized criminals often go further, and either fabricate claims entirely or create an event themselves. One motor example is what is known as 'crash for cash' claims, where a criminal drives in front of another vehicle and brakes suddenly, causing the other driver to crash into the back of them. Often, the criminal would claim for damage to their car as well as soft tissue injuries, and perhaps soft tissue injury claims for any (including fictitious!) passengers. This type of fraud was so common in the early 2010s that it was costing £340m a year in the UK alone,[2] the cost of which was being passed on to all policyholders and driving up the price of compulsory motor insurance. At the same time, law firms were being set up specifically to represent people who had 'been in an accident that wasn't your fault', in order to claim for small injuries, so the fraudulent cases were being drowned in a sea of legitimate claims.

In order to reduce the cost of these small injury claims, the UK introduced the Legal Aid, Sentencing and Punishment Act of 2012 which reduced the amount of legal fees a firm could claim for. While this act has meant that many people no longer have access to legal aid, it was successful in reducing the cost of small injury claims: the legal costs of motor-related soft tissue

[2] Insurance Fraud Bureau | Crash for Cash. https://www.insurancefraudbureau.org/insurance-fraud/crash-for-cash/

claims reduced by 65%, and car insurance premiums reduced by 15% within the year.[3]

Because of these costs, it is unsurprising that insurers are always looking for new ways to spot fraudulent claims. Traditionally, insurers have a fraud team who comb claim submissions to spot and investigate suspicious claims. Nowadays, these teams are increasingly augmented by data scientists and artificial intelligence (AI) algorithms. Given a set of historic claims that have been marked as fraudulent or not fraudulent, AI can be used to create a decision tree of rules.[4] By examining known fraudulent claims for similarities, both humans and machines can use claim features to spot suspicious claims. For example, long notification delays for property damage are unusual, so may be flagged for investigation. So too might policies that appear **underinsured** or have a particularly small premium, claim descriptions that contain certain words or phrases, or claims payments requested to be paid to a different accountholder's name. Once the AI is trained on this dataset, it can be used to predict the likelihood that a new claim is suspicious. A member of the fraud team will investigate the claims with the highest likelihoods, as well as a small sample of other claims. This gives the AI feedback on how predictive it was, which it can then use to create a new decision tree. This **Insurtech** is already being used by large insurers and can help fraud teams use their time more wisely and save on paying invalid claims. This helps insurers control and reduce their risk cost, and therefore their premiums.

Show Me the Monet In 1994, Steven G Cooperman claimed for the burglary of two paintings: Monet's 'Customs Officer's Cabin at Pourville', and Picasso's 'Nude Before a Mirror'. Mr. Cooperman, an optometrist from Los Angeles, had been on holiday in New Jersey when the paintings, worth $2.5 million, were stolen. The police reported that the burglary seemed unusual, but the case went cold. Despite their value, Mr. Cooperman had insured the paintings for $12.5 million! Lloyd's of London alleged policy fraud because of this over-insurance and on behalf of its syndicates, initially refused to pay out. However, it eventually agreed to settle at $17.5 million to stop the case going to court.

Three years later in Ohio, Pamela Davis and her ex-boyfriend James Little were having a legal battle of their own by filing complaints against each other for violence and stalking. In one of the police interviews, as testament to his character, Miss Davis mentioned Mr. Little had a storage locker containing

[3] Figures from the Institute of Actuaries.

[4] When data used to train the algorithm is marked in this way, it is described as a supervised learning or classification problem.

stolen goods from California. Lo and behold, in that storage container, the Ohio police found a Picasso wrapped in a blanket and a Monet in a cardboard box. Mr. Little later confessed that he received them in 1994 from the man he believed was the owner (who was in fact Mr. Cooperman's lawyer). Mr. Cooperman was sentenced to 3 years in jail.

Unfortunately, fraudulent claims are not the only way that the insurance industry is affected by crime. The industry is also a target of money laundering, and life insurance products are particularly susceptible due to the high premiums upfront and the wide variety of ways money can be recovered from the product.[5] Some of the techniques well-known to the industry are below, but more creative ways will inevitably be discovered.

- **Annuity** products are often desirable for money launderers, as no unusual activity is required to withdraw money from them: as annuities are simply an initial payment in exchange for a series of regular payments, once the premium is paid, a criminal only needs to use the product as anyone else would, perhaps arranging for the payments to be paid to a different account.
- Single premium products of any type can be used in conjunction with consumer-focused cancelation rights. Many regulatory regimes require a short initial cancelation period, where the policy can be canceled (and the premium returned) with no penalty. Life insurance policies often have surrender terms, where the policy can be canceled at any time but at a loss to the customer. Criminals use both of these procedures, although note that the former is more lucrative but more easily spotted.
- Transferring ownership of policies can also be used for money laundering, as a criminal can opt to transfer the policy to a different individual, who then withdraws the money or cancels the policy.
- There are also more complicated ways of using life insurance policies that involve using the policy as capital or collateral against a loan. Taking out the loan against the policy (paid for with dirty money) may be the extent of it, but different loan terms could be leveraged such as not having to repay the loan, but using the policy to repay the loan and interest when the policyholder dies.

[5] It is left as an exercise to re-read Chapter 3 with a view to spotting (theoretical!) methods of laundering money and committing claims fraud.

Because of the detrimental effects of criminal gangs on society, governments in almost every country have put regulations in place to help insurers spot money-laundering activity. For example, in the US, the Bank Secrecy Act lists certain transactions that must be monitored for criminal activity. If a transaction is over $5,000 and involves the purchase of permanent life assurance products, annuity contracts, or insurance products with investment or cash-value withdrawal features, the insurance company must check the transaction and flag any suspicious activity to the Financial Crimes Enforcement Network. Insurance is big business, so it's a big target for financial crime.

There are plenty of creative, fascinating, and often gruesome insurance fraud cases! Nowadays, digital claims databases and Insurtech solutions (as well as improved post-mortem methods) would allow insurers to easily spot these, but criminals continue to think up new ways to defraud companies.

9.3 The Insurance Cycle

If an insurer correctly estimates the risk and non-risk costs, then it should generally make a profit (apart from years in which there are extreme events). However, in reality, profits from non-life insurance are somewhat cyclical. Most people are aware of boom-or-bust cycles in other industries, but insurance has its own cycle which doesn't necessarily reflect wider economic circumstances. Because of this, the insurance cycle affects profits due to underwriting activity only, i.e., it doesn't directly affect investment returns.

Let's start by assuming the insurance market is breaking even, and companies have an average **COR** of 100%. Recall that this means their claims costs and their expenses are on average the same as the amount of premium they receive. Some companies will have a COR of less than 100% and will be making a profit. This profit means that companies will have more capital, so they are able to take on more business, or write a new type of business. Companies' capacity to write insurance increases. At the same time, the profit in the market is attractive to new entrants. These new companies start writing business, and so the supply of insurance further increases (Fig. 9.3).

Insurance isn't typically an exciting new purchase: customers usually buy what they need consistently. Demand for insurance fluctuates a little with economic cycles (although the exact nature of this is difficult to quantify: in hard times some customers will realize they are less well-off and can bear less risk so will buy more insurance whereas some will be less able to afford

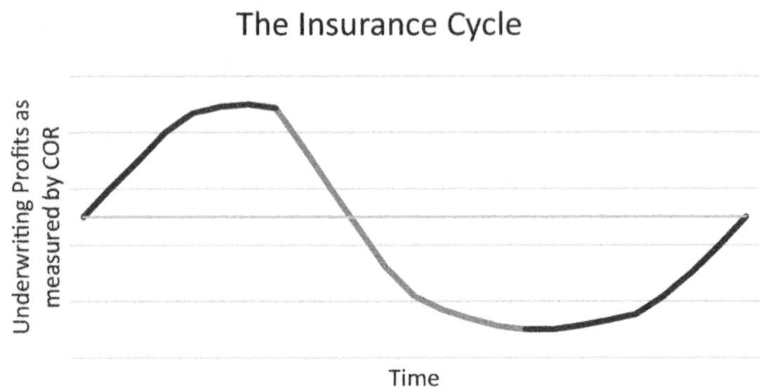

Fig. 9.3 Underwriting Profits and the Insurance Cycle: the orange line denotes 100% COR (breakeven). No figures for COR are given (other than 100%) as the extremity of the cycle varies. However, the higher the y-axis the higher the COR, and therefore the lower the profitability. The pale blue line denotes the hardening market, whereas the darker line denotes the softening market

the premiums and buy less cover),[6] but if lots of companies have increased capacity, supply is likely to exceed demand.

Competitors will naturally want to retain their market share, so will often reduce their premiums in reaction to this. Once one competitor does this, the others will quickly follow suit, especially if business is often bought through price comparison websites or if the product is particularly price sensitive. As this happens, prices reduce and the market softens—with insurers competing for customers' business, brokers will usually be open to negotiating terms. In order to make their product more attractive, insurers may also add additional covers to their product, or loosen their underwriting criteria (that is, they become less strict on who they sell policies to) in order to drum up more business. This is called a soft market, characterized by low average premiums but high total premiums driven by volume.

After a while, insurers will be unable to drop their prices any lower without eroding their profits, and underwriting profit will level off. However, in fighting for more policies, an insurer may have accepted too many poor risks, or loosened the terms and conditions a little too much, and suddenly be exposed to several claims that are higher than expected. Increasing the **loss ratio** means the **COR** increases. As COR increases, profits decrease, and so insurers may need to dip into their assets to pay claims. This reduced level

[6] It is much more affected by the occurrence and awareness of events and disasters: people tend to buy coverage after the fact. Insurance for luxury items such as gadgets and travel does see much more variation in demand, but again, this tends to be in line with economic cycles and the news cycle, not the insurance cycle.

of capital means that they have less capacity to write business. Some insurers may even find themselves insolvent and unable to write any business at all!

As the supply of insurance reduces throughout the market, insurers can be more choosy about the risks they take on. They might tighten their underwriting criteria again, or add more exclusions to their policy. They will increase their premiums to try and keep revenue steady, and increase their profitability. This market contraction can be sudden, especially if there is an extreme event or catastrophe that suddenly increases loss ratios. In this hard market, average premiums are high and number of policies sold are low. Customers may find it difficult to get insurance cover at an affordable price. The regulator or government may step in to increase the supply of insurance or impose price caps, or the market may regulate itself if new entrants are able to join. As profits increase again, premiums are driven down, and so the cycle continues…

The insurance cycle doesn't affect the methodology outlined in this chapter, but it does impact the strategic decisions an insurer makes, including their target profitability. As profitability rises and falls in the market, an insurer needs to price in a realistic expectation of profit. In a softening market, insurers should take care to not use historic acquisition costs, but calculate 'per policy' costs rather than 'per $100 premium' in order that their expenses include the administrative cost of writing many policies, rather than fewer high-premium policies.

We conclude this chapter with a worked example. If this seems very simple, remember that the devil is in the detail. Calculating the assumptions is part of a pricing actuary's role and is usually anything but straightforward in real life!

Example 9.3: A pricing actuary has calculated the risk cost of property damage policy to be £500. The sum insured is £500,000. The following non-risk costs are added:

- Reinsurance for this policy is 0.005% of the sum insured = 0.001% x £500,000 = £5
- The London Fire Brigade Levy is charged at £35 per million of sum insured = £35 × 500,000/1m = £17.50
- The policy was sold via an aggregator website who charge the insurer 10% of pre-tax premium.
- The insurer has operating expenses of £24 million for this product. It expects to sell 75,000 policies this year with an average premium of £1000 (excluding tax). Therefore, it expects to receive £75 million in premium and spend £24m on expenses. It decides that charging expenses as a percentage of pre-tax premium is easiest, so calculates its **expense ratio** as £24 million ÷£75 million = 32%.
- The insurer's target profit is 7%, determined by its corporate plan. The corporate tax on profits is 19%, so it needs to target 7% × 1.19 = 8.33% of pre-tax premium.

- Investment returns are expected to be 1% of pre-tax premium.

Therefore, the insurer needs to receive a premium of:

$$(£500 + £5 + £17.50) + 0.1P + 0.32P + 0.0833P - 0.01P = P$$

where P is the pre-tax premium. Rearranging gives:

$$\frac{(£500 + £5 + £17.50)}{(1 - 0.1 - 0.32 - 0.0833 + 0.01)} = P = £1031.18$$

Insurance Premium Tax is charged at 12%, and local regulation says that tax must be included in the premium shown to the customer. Therefore, the premium charged to the customer is £1031.18 × 1.12 =**£1,154.92**.

10

Reserving: First Principles

Chapter Objectives

- Understand why reserves must be held, and the role of actuaries in reserving
- Understand the benefits and challenges with different types of claim data [aggregate data; incremental and cumulative; Incurred But Not Reported claims; Incurred But Not Enough Reported claims; paid claims.
- Create claims data triangles for different types of data
- Understand why some claims are settled as Periodic Payment Orders, and the benefits and challenges of PPOs.

Forewarned, forearmed; to be prepared is half the victory.

—Miguel de Cervantes

10.1 Principles of Reserving

In Chapter 1, we explored the life cycle of a policy and noticed how in actuarial work, the cost of providing a service (in the form of claims costs or pensions) will be incurred in the future. A key part of actuarial work is estimating these future **cashflows**. If these estimations are used to determine the

© The Author(s), under exclusive license to Springer Nature Switzerland AG 2022
C. Townsend, *A Risky Business*,
https://doi.org/10.1007/978-3-031-11673-5_10

amount of money a business needs to hold to meet future obligations, then this work is called *setting the reserves*, or simply **reserving**.

To recap the claims journey from Chapter 1: an insurer is made aware of a claim at some notification date, on or following the date of loss. Based on the information the policyholder gives, the claims handler will give an initial estimate of how much the insurer would be likely to pay. As time goes on and more information becomes known about the claim event and the damage, the estimate of the claims cost will change. It may increase or, hopefully, decrease. If a claim is found to be invalid—perhaps due to the event being excluded from cover, or due to fraud or a breach in the terms and conditions of the policy—, the estimate would become zero. However, eventually the claim will be settled: both the insurer and the policyholder agree on the amount that should be paid. The payment is made and the claim is said to be closed.

A similar story is found in **defined benefit** pensions: the provider is informed that a person has retired and the claim is opened. Here the monthly cost is known, but the uncertainty is how long the pension will need to be paid for, and therefore the final total claim cost is uncertain.

Both insurers and pension providers wish to know how much this ultimate claims cost will be, long before it is paid in full. It is the role of actuaries to quantify this future uncertainty.

To get a feel for the importance of this process, imagine that you know you will receive an electricity bill in three months' time, but not the exact cost, as that depends on your energy use. What would you do? Well, you could do nothing until the bill arrives, that is, until there is no longer any uncertainty in the situation. However, that may leave you scrambling for the money, or at the very least, needing to re-budget the next month. This could leave you in financial difficulty and is clearly not a good outcome if you want to ensure the long-term stability and security of your budget. Finding the money last minute may mean expensive borrowing agreements, late fees charged by the power company and a hit to your credit rating.

Alternatively, you could estimate what the bill might be: you might use last quarter's bill or your bill at this time last year. You could also use your meter readings and a cost per kilowatt-hour, and set aside that amount of money now. Unless you inadvertently underestimated the cost, by doing this you can feel reassured that this bill won't cause financial trouble in three months' time. In exactly the same way, businesses want to ensure their long-term financial security. Realize the importance of *uberrima fides* here: customers buy insurance or pension products in good faith, trusting that they will receive money when they need. If it is paid late, or not paid at all, this not only means that the customer will be unlikely to return, but that the poor reputation of the company may deter other customers. Industry regulators reinforce this with financial penalties or strict restrictions on writing future business (analogous to being moved to a pay-as-you-go meter after repeatedly not paying your bill). This may mean that a company is limited in the amount or type of business it can write, or extra reporting requirements in order for the regulator to check you will be able to meet future claims payments. Putting money aside for future costs is clearly a sensible thing for companies to do: the question is how much.

10.2 Types of Claim Data

In order to quantify these future claim costs, actuaries need data. Fortunately, established products have historical claims data which can be used[1] to estimate. First, let's look at where this data is coming from.

As all statisticians know, one data point is not much use. In fact, we are so used to '*data*' meaning many data points, that the singular '*datum*' is now only used by pedants. Looking at the total historic claims cost for a product is therefore not very helpful as it only gives us one data point. Instead, actuaries will split their claims data up into multiple time periods, and get a claims cost for that period. Then, they will have multiple data points with which to apply statistical techniques. But how to split the data?

If you were estimating your future quarterly electricity bill, you might want multiple quarterly electricity bills to base your estimate on. It almost goes without saying that you wouldn't need your water bill or your shopping bill. In the same way, actuaries want to split their data by product type, so that the claims costs in the historic data relate to the costs that may be incurred in future. Ideally, each product would be reserved for separately. However, that may not be possible in practice.

If the product doesn't have a lot of claims data (perhaps because there are very few claims, or very few policies sold), then reserving for it alone may be tricky. Remember, if risks aren't **pooled**, then claims costs may be very unpredictable, with extreme values (either no or very low, or very high claims totals). For this reason, actuaries may group several similar products together. As a corollary, erratic (or volatile) claims totals lead to a high degree of uncertainty in the estimate for future claims, as well as reducing the ability to confidently spot errors in the data. So, wherever possible, reserving actuaries

---

[1] New products don't have this history, so reserving will require more assumptions and data from other sources. Different methods for reserving products old and new will be covered in the next chapter.

try to use aggregated claims data—claims data grouped together, rather than considering individual claims—as long as the claims in the group are suitably similar in their characteristics. However, there's a tricky balance here between having enough data (including more years, or grouping several similar products) and having relevant data. There's not a cut-and-dry rule here, so the actuary must use their judgment.

Insurers commonly group claims not only by product, but also by when the claim happened—the accident date. At Lloyd's of London, claims are grouped by underwriting date—the date the policy was incepted. This is because profits from Lloyd's need to be divided between the members of the **syndicate**, and so profit needs to be attributable to a certain year of insurance business. It isn't necessary to split the profits in this way at an insurance company, and so accident date (or for liability lines of business, sometimes notification date), is more commonly used.

Did you know?
The largest UK motor insurance claim to date is the Selby Railway Disaster of 2001, where a Land Rover derailed a train of passengers, sending it into the path of a freight train. 13 people died and 70 more were injured, and the insured cost was over £40 million.

Regardless of how claims are grouped, the claims estimates are not static: they will change over time as more information is relayed to the claim handler. Even for a single claim, the claim estimate could change multiple times. These changes could be shown in two ways: either as the incremental change to the claim cost, or the cumulative claim cost. Incremental claims data only shows the increase or decrease to the previous claims estimate, whereas cumulative data shows the new total cost. Both are useful in different contexts: for claims that are paid regularly (such as annuities), incremental costs may be more relevant, as the company doesn't need to hold reserves for costs that have already been paid. However, cumulative claim costs are easier if an overall claims total is needed, for example, for estimating the profitability of a pool of claims.

Example 10.1 Tom retired in March this year. He received £1200 a month from his pension provider, starting from the month he retired. He died in November, after the pension for the month had been paid. The claims data could be shown either incrementally or cumulatively, like so:

	Incremental	Cumulative
January	0	0
February	0	0
March	1,200	1,200
April	1,200	2,400
May	1,200	3,600
June	1,200	4,800
July	1,200	6,000
August	1,200	7,200
September	1,200	8,400
October	1,200	9,600
November	1,200	10,800
December	0	10,800

The conversion from incremental data to cumulative data is simple: sum all the increments together to find the cumulative claims cost. Don't be surprised if increments are negative, particularly in **general insurance**. The damage may have been overstated to the insurer initially, or the **FNOL** team may routinely round up in their initial claim estimates out of prudence.

Try it for yourself!

To understand how volatility makes data analysis difficult, try to answer the following questions for these two sets of data:

Set 1: 2, 4, 5, 8, 110, 12, 14, 16, 19, 21....

Set 2: 2, 2, 2, 2, 8, 8, 18, 21, 21, 21...

- What comes next in each sequence?
- What will the 20th entry in each sequence be?
- Which entry or entries are likely to be an error?

Despite having the same start and end point, the sequences are very different. The first is more like pooled data costs: a gentle incline to an end point. The data input error of the extra digit on the number '110' is easily spotted and queried (we'd want to check whether the true value would be 10 or 11), and future entries can be estimated without too much uncertainty. The sequence isn't always exactly adding two each time, but it's not far off. We could reasonably expect the next value in this sequence to be 23, or give a range of 22–24. Therefore, the 20th entry would likely be 41 or thereabouts. It would also be relatively easy to monitor when it was not going as expected. All of these questions, while seemingly trivial, are part of standard actuarial work, albeit for slightly more complicated data.

The second sequence is much more like the claims total of a single or small number of risks. It jumps unexpectedly and is hard to predict. Previous 'jumps' have been between 0 and 10, so the next entry could be 21, or 31 or anything in between. The 20th entry is even harder to predict. The last three terms are the same, so perhaps it has leveled off and won't increase further at all. Perhaps it might have a few more 'jumps' of 10? Using this logic, it could still be at 21 or it could be 50 + .

It is also harder to spot errors. The '1' which makes an '8' into an'18' was erroneous (and exactly the same error as the first sequence) but the volatility in the sequence generates a cloud of doubt: a jump from 8 to 18 might be plausible if a jump from 2 to 8 has been seen.

Tables of claims data (either incremental or cumulative) can be created for any claim or group of claims. However, estimating the ultimate claims cost can be tricky if the claim is very undeveloped. Reserving actuaries try to identify patterns in how claims develop in order to apply those to new claim estimates. These development patterns can be quantified in lots of different ways which will be covered in the next chapter, but in order to do this, the claim estimates of lots of groups of claims needs to be easily compared.

To make a table in order to easily compare claims development, let's start by making the axes. The rows of the table will be different cohorts of claims (whether that is grouped by accident year, notification year, or underwriting year). The x-axis will show how that cohort develops over time, so the first data column should relate to the first development of that group of claims. For products that are settled quickly, this data may be collected monthly. Alternatively, this could be quarterly or yearly. The length of time between the collection of claims data is called the development period. Development periods are referred to as a number: period 1, period 2, etc. For each cohort, this relates to a different calendar date: the first point at which the reserving actuary receives data about a 2018 claim may be December 2018. So, December 2018 would be development year 1 for the 2018 claim. However, for a 2017 claim, December 2018 would be development year 2 (with December 2017 being the first year).

Once the axes are constructed, the table can be populated with data from the incremental or cumulative claims data tables for each cohort (for example, a row of the table in Table 10.1). The most striking thing about these tables is that they are half empty! The most recent cohorts have lots of blank entries as they haven't had time to reach the later development periods. For this reason, these tables are commonly known as claims triangles.

Table 10.1 A Cumulative Claims Triangle showing the position as of December 2021 of losses occurring in 2016–2021

Cumulative Claims	Development Year					
Accident Year	1	2	3	4	5	6
2016	9,657	16,955	30,344	40,599	44,235	44,383
2017	16,995	40,188	59,140	71,777	70,289	
2018	28,670	47,432	70,322	70,656		
2019	27,066	46,782	54,802			
2020	21,577	33,741				
2021	23,632					

10.3 Using Patterns in Data

The claims triangle can be visualized as a graph. When looking at this graph of claims development, the development patterns are much clearer. In this example, the claims costs increase steeply for 3–4 years and then mostly level off. For some reason, 2016 has much lower claims costs, but the more recent years look more similar to each other. As the end goal is to predict where each year will flatten off, years that are similar to the one you are trying to predict can give clues as to the development might continue (Fig. 10.1).

Exercise 10.1 Where do you think the 2019 year will flatten off? How confident are you in that?

So far, this chapter has focused on historic claims data, with very little discussion of future costs at all! But the historic claims experience informs

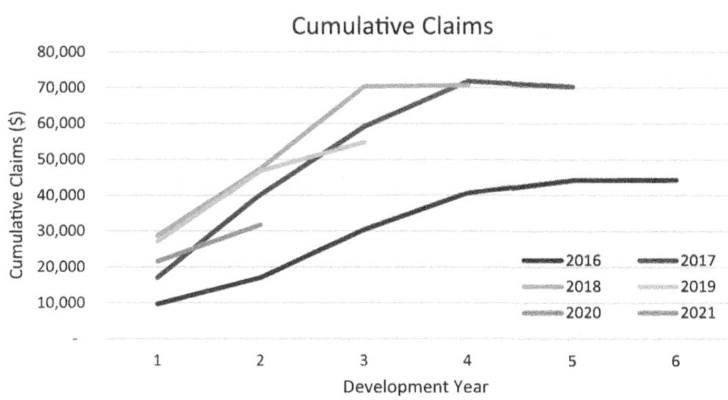

Fig. 10.1 Graph of Development Patterns

the estimate of the ultimate claim cost, and the ultimate claim cost informs the level of reserves, so understanding the historic data is the crucial first step of reserving.

Reserving is all about detecting patterns in historic claims data and using a combination of historic patterns and current data to adjust your expectations about the future. To see why this gets complicated, let's consider a simple scenario. Imagine you are traveling to work. Your journey usually takes an hour: you drive 20 minutes to the train station, take the train for 35 minutes, then walk 5 minutes to your office. Before you leave home, you estimate your travel time to be 1 hour.

However, when you arrive at the station on a particular morning, the train is delayed by 3 minutes. How would you estimate your travel time now? Depending on your level of optimism (and your previous experiences with rail travel!), you might adjust your estimate in any of the following ways:

- You might expect the train to arrive 3 minutes late, but otherwise travel as normal. Your travel time estimate would be 1 hour 3 minutes.
- Perhaps the train driver would be very conscious of the tardiness and drive faster to make up the lost time. Your estimate would be 1 hour.
- You might think that the 3-minutes delay is only a sign of things to come, and assume more delays are most likely: your estimate might be at least 65 minutes.
- The delayed train might mean you have to walk on busier streets, increasing your walking time by 1 minute as well as the effect on your train journey. Your estimate might be 1 hour 4 minutes.

All of these estimates are valid, and show that combining past trends with current information can be done in a variety of ways. The next chapter will discuss ways of quantifying patterns and applying various methodologies, but let's remain with our delayed commuter a little longer. What if it was their car journey that was delayed? Of course, the travel time estimate could be adjusted in the same way as the train delay, but there may be a traffic jam and a train delay. Or, there could be a traffic jam and the train could be running ahead of schedule so you still arrive on time. The earlier in the journey, the harder it is to estimate travel time with certainty. Similarly, the more undeveloped the claim cohort, the more uncertainty in the ultimate claim cost.

Reserving actuaries do not rigidly apply development patterns to the current data: a large part of their role is to understand the wider context and drivers of claims and how claims are being handled in the company. In Table

10.1, the 2016 cohort of claims had lower claims costs and developed more slowly than the more recent years. Because it developed so differently to the other years, an actuary might want to disregard its pattern entirely. Similarly, if a new trend emerges, the development pattern should change too. In our commuting scenario, if some engineering work on the track means that from next week, all train journeys will be twice as long, a commuter should adjust their travel time estimates now, rather than waiting until Monday morning at the train station to do so. For actuaries, this could be a court case that changes the payout for a type of claim, or it could be a new procedure that means that claims are settled more quickly.

Claims development patterns often follow the same shape: they increase sharply at first, then increase more slowly until they level off. This shape is explained by claims moving through the claims journey. On notification, the claims cost for the policy moves from 0 to the initial estimate. Then, as more information is gained by the insurer, the estimate might move upward or downward to take account of this. Once the claim is settled, the cost doesn't change.

Even though the ultimate claims may not change for a fully developed cohort, the reserves will. This is because as claims are paid, the pot of money required to pay future claims for that cohort doesn't need to be so big. For this reason, reserving will predominately focus on more recent cohorts; the older years will likely be stable, if not settled and paid.

However, just because very few claims have been reported in undeveloped cohorts, that doesn't mean that ultimate claim costs are low. A reserving actuary expects that more claims will be reported. So, as well as setting money aside for the cost of claims that the insurer is already aware of, a reserving actuary should also set aside money for claims that are expected to be reported soon. This is called the Incurred But Not Reported (IBNR) reserve: the claim has happened, but the insurer hasn't been notified yet. In the first few development periods, IBNR will be a large portion of the total reserve, but as the insurer is made aware of more claims, the IBNR gets converted into estimates that relate to a specific claim—known as case estimates.

Example 10.2 At the end of 2021 there are estimated to be 100 unreported claims for the 2020 year, each one with an estimated average claim cost of \$1,500. The IBNR for 2020 is therefore estimated to be \$150,000. During 2022, 50 more claims are reported relating to loss events occurring in 2020, so data relating to them is added to the 2020 cohort in the claims triangle. If the estimate of the ultimate number of claims and cost per claim are unchanged, then the IBNR for 2020 is now reduced to 50 x \$1,500 = \$75,000.

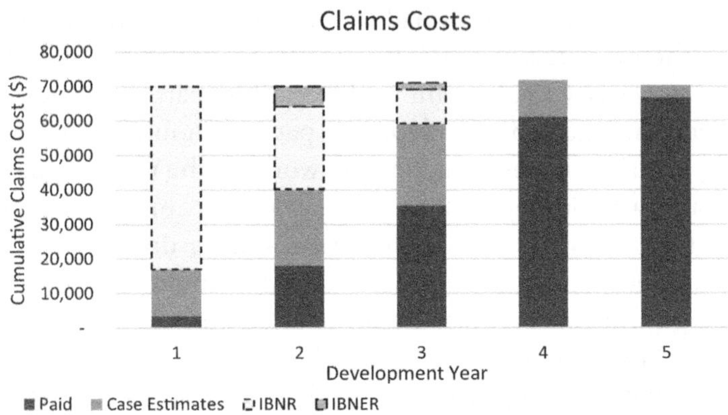

Fig. 10.2 Claims cost, split into paid, case estimates, IBNR, and IBNER (*this data is a more granular view of the 2016 cohort of Table 10.1*)

The ultimate claims cost also should take into account the fact that case estimates are likely to change over time. Specifically, if an actuary thinks that the case estimates are too low, they may want to set aside a little bit more money to allow for the possibility that these estimates are not adequate. This pot of money is unimaginatively called the Incurred But Not Enough Reported (IBNER) reserve.[2] Initially, this is very small as very few claims have been reported and therefore have their own case estimates. However, particularly in the middle of development when claims have been notified but no further information has been provided, the IBNER might increase. As claims are settled, the IBNER reduces as there are fewer estimates that have the potential to change. IBNER can be negative if it is believed that the current estimate is higher than the ultimate cost.

Figure 10.2 shows how paid claims, case estimates, IBNR and IBNER might make up the proportion of the claims costs as time goes on—the exact proportions and development periods are affected by the line of business, but this is the kind of pattern that a general insurer could expect. Reinsurers and pension providers use a similar method to determine their ultimate claims cost: determine the exact sources of costs and, using historic claim patterns and understanding of current context, estimate future costs.

[2] Reinsurers and Lloyd's of London tend to report their IBNER reserve as part of their IBNR. Sometimes insurers will do the same, but call the combined IBNR + IBNER, the IBN(E)R reserve.

10.4 Injury Claims and PPOs

Usually in **general insurance**, full payment follows swiftly after a settlement: the policyholder wants their money and the insurer would usually rather pay and be done with it! However, some claims are not paid in this way.

Unfortunately, sometimes a third party will sustain a life-limiting injury.[3] They will require a high level of care for the rest of their life, and the insurer of the person responsible will be liable for this payment—one of the reasons that third-party injury insurance for drivers is mandatory many countries. However, for the injured party, a lump sum can be another thing to worry about: will it last for the whole lifetime? What happens if care costs increase? How should it be wisely invested? To reduce these risks for victims in the UK, the Courts Act 2003 introduced a different way for claims to be paid.

> **Did you know?**
> Prior to 2019, instead of buying car insurance, UK drivers could deposit £500,000 with the Department of Transport.

A **Periodical Payment Order** (**PPO**) is a court order that requires an injured person to be paid compensation in the form of a series of regular payments, rather than a lump sum. Although payments have been structured in this way before, previously both parties had to agree to it. Lump sum payments limit the risk that an insurer is holding—once it is paid, there are no uncertain future payments to be made—so naturally insurers would be reluctant to voluntarily pay in alternative ways. When PPOs were introduced in 2005, it was novel because insurers couldn't opt out once a court had ordered it.

PPOs are awarded in cases of severe injury, either because of an accident or because of malpractice. So, they occur most often in motor insurance and liability products—in fact, most of the clinical negligence claims settled against the National Health Service are settled as PPOs. Periodical payments work in a similar way to an **annuity**: the payer must pay a specified amount to the injured party at regular intervals. The first payment is usually a bit larger to account for the upfront costs of being suddenly injured, such as making accommodation accessible. The rest of the payments are usually indexed to a relevant inflationary factor, such as the UK-based Annual Survey

[3] If the insured is injured in an accident that is their fault, their medical bills may be paid by their insurer if they have Personal Injury Protection cover.

of Hours and Earnings, which acts as an index for care costs. The payments continue for life, and are set at a level that not only indemnifies them against the cost of medical care, but also makes up for a lifetime of lost salary. PPOs were designed to increase security for the injured party by transferring risks to the insurer, specifically:

- Longevity risk—the risk that the victim will live longer than expected (and their lump sum runs out)
- Inflation risk—the risk that inflation is higher than expected (and therefore that the lump sum doesn't cover as much as originally planned)
- Investment risk—that the vehicle the reserve is invested in doesn't generate as much investment return as expected (not only does a PPO transfer this risk to the insurer but it also transfers the related investment administration that can be a significant source of stress)

Administering a periodical payment is much more complicated for the insurer than a lump sum claim. Reserving for a PPO is also complicated, as an insurer usually won't have many PPOs to administer at a time. Further, some of the cost of the claim can usually be recovered from a reinsurer (recall reinsurers are insurers for insurers, protecting them from the financial impact of very large losses from a claim or event. A PPO will usually cost millions, which may trigger a **reinsurance** payout).

So, without much historical claims data, claims triangles aren't appropriate for reserving PPOs. Instead, techniques from life-dependent annuities valuation—see Chapter 6 for a recap on these—can be used to estimate the ultimate claims cost. The reserving actuary will need to make some assumptions, such as the rate of earnings inflation, long-term investment returns, and crucially the life expectancy of the injured party. The standard life tables that we used in Chapter 6 could be used, adjusted for the fact that people with injuries may not live as long, or specific tables for those with illness and disability could be used. These are traditionally called impaired life tables.

Exercise 10.2 Estimate the present value of the total cost of a PPO, paid annually, in which the first payment is £150,000, paid immediately and second payment (payable in exact 1 years' time) is £50,000, paid to a man who is 52 at the time of the accident and has standard life expectancy (so use the standard life tables). Subsequent payments increase at a rate of 5%, to match the cost of wage inflation, and regular inflation is 3%.

Other assumptions, such as the interest rate that must be used in the calculation of these annuities, may be prescribed by the regulator or government.

In the UK, this is called the Ogden discount rate. In March 2017, the government announced a sudden change to the Ogden rate from 2.5% to −0.75%. This meant that insurers needed to increase their level of reserves for PPOs. As an example, AXA reported that an 18-year-old plasterer with a spinal cord injury would previously receive a lump sum of £7.6m would receive £19.3m under the new rate. After much industry consultation, in 2019 the rate was increased to −0.25%. The next review is scheduled for 2024 at the latest.

Some claimants may prefer a lump sum, particularly if they no longer wish to deal with an insurer to receive their annual payment (payments may be payable only if the claimant is still injured to a certain extent, and the claimant may find this evaluation of disability undignified or intrusive). They may also prefer a lump sum if they think their life expectancy has been overstated: a lump sum could be passed onto heirs (although PPOs are most commonly awarded to young claimants), whereas a PPO is often only payable to the victim.

Exercise 10.3 Are there any benefits or drawbacks of PPOs to any other parties?

Ultimate claims costs are not only useful to the reserving team. They are a core part of business planning, determining the risk cost, portfolio monitoring, and regulatory capital requirements. Reserving teams also synthesize information from lots of parts of the business, for example understanding the change in the mix of business for a portfolio, anticipating the effect of a change in claims management processes, or analyzing the impact of a recent catastrophe. Because of the importance of the ultimate claims costs, signing off to ensure that the reserves are correct and appropriate is a role that only qualified actuaries can do.

Now you are familiar with the concepts of reserving, it's time to put numbers to the process. There's a seat reserved for you (*ahem*) in Chapter 11, where we'll work through some quantitative methods. Grab your calculator and let's begin!

11

Reserving: Methods

Chapter Objectives

- Use the Chain Ladder method to calculate a reserve
- Use the Expected Loss Ratio method to calculate a reserve
- Use the Bornhuetter-Ferguson method to calculate a reserve
- Be familiar with other reserving methods, and understand when each method would be appropriate

Chapter 10 outlined the aims and principles behind reserving. This chapter builds on that foundation by detailing some of the quantitative methods used to do this. While the math isn't particularly complicated, this chapter does involve multi-step calculations. The broad outline of each method might be sufficient for your purposes but to get the most out of this chapter, I highly recommend that it is deferred until you have a pencil, paper, and calculator ready.

> *There is no greater joy than to see that this curious activity we call calculation—the depositing of ink on paper, followed by throwing away the paper and depositing new ink on more paper—can actually tell us something about reality.*
>
> —Alan Guth, theoretical physicist

To recap Chapter 10, the main processes in reserving are:

© The Author(s), under exclusive license to Springer Nature Switzerland AG 2022
C. Townsend, *A Risky Business*,
https://doi.org/10.1007/978-3-031-11673-5_11

1. Split claims into homogeneous groups. This is usually by product or **peril** and a time factor—often accident year, or underwriting year at Lloyd's of London. Key considerations here are the balance between having enough data in each group and the group being sufficiently similar (homogeneous).
2. Create claims triangles (these could be incremental or cumulative).
3. Use the claims triangle to predict the cost of claims next period.
4. Extrapolate to calculate the ultimate cost of claims.

Steps 3 and 4 cover a multitude of techniques including stochastic reserving, statistical methods, and more classical actuarial methods, but the purpose remains the same. Most techniques rely on the assumption that future claims develop in roughly the same pattern as previous years, so we are assuming that the past is good guide to the future. Of course, actuaries should always question whether or not this is true!

The first method we will consider is a very intuitive classical actuarial technique called the Chain Ladder method.

11.1 Chain Ladder Method

When first faced with a data sequence, it is intuitive to consider the 'jumps' between data. In the 'Try It Yourself' sequence from Chapter 10, we determined that the 'jumps' between data points were additive; specifically, adding around 2 each time. Jumps can also be multiplicative, such as a number doubling.

The Chain Ladder method calculates these jumps for each cohort and development period, then assumes that all cohorts will follow the same development pattern, that is, that the 'jumps' between development periods will be the same for all cohorts. In actuarial work, these 'jumps' are assumed to be multiplicative, and are called development factors.

Chain Ladder requires a cumulative claims triangle. If you have incremental data, you will have to do an additional step to convert it into cumulative data. The steps of the Chain Ladder method are as follows:

1. Create a cumulative claims triangle.
2. Calculate the multiplicative 'jump' between each data point for each cohort.
3. Decide which development factor you think is most representative for each development period.

4. Apply the selected development factors to the current cost of claims to calculate an ultimate cost of claims.

Therefore, the ultimate cost of claims as estimated by the Chain Ladder method is:

$$\text{Ultimate Cost}_{k,n} = \text{Cost}_{k,n} \times \prod_{t=n,n\geq 1}^{\infty} \text{Selected Development Factor}_{t,t+1}$$

where k is the cohort and n is the time of the latest known position.

This equation reads as 'the ultimate cost is the current cost multiplied by all the selected development factors relating to future periods'. It would also be accurate to summarize it as 'start where are you, and jump all the way to the end'. Figure 11.1 shows this visually: we have claims data for development periods 1 and 2. Using historic claims data, we select development factors for periods 2–3, 3–4, 4–5, and 5–6. We apply this multiplicatively to the cumulative claims cost at the most recent development period (here, development period 2) to find our claims cost at period 6.

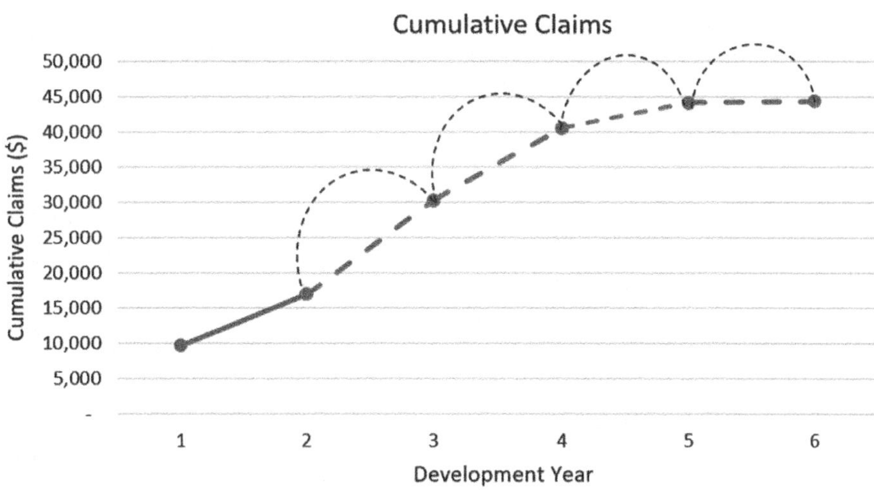

Fig. 11.1 A visual representation of the Chain Ladder method

Exercise 11.1 At the end of development period 2, the cumulative claims cost for a product is $23,632. The selected development factor for period 1–2 is 1.8; for 2–3 it is 1.6; 3–4 is 1.33; 4–5 is 1.12; 5–6 is 1.08; 6–7 is 1.01. What is the expected cost of claims at the end of period 7?

Chain Ladder is easy when the selected development factors are given! Choosing the development factors to apply is where actuarial judgment comes in. First, let's calculate the development factors for each cohort and each development period. As you'd expect, the development factor between period t and t + 1 is simply:

$$\text{Development Factor}_{t,t+1} = \frac{\text{Cost at time } t + 1}{\text{Cost at time } t}$$

From a claims triangle, we can calculate a development factors triangle! The triangle below is calculated using Table 10.1 from the previous chapter. There are two things to notice: firstly, that this triangle is one column and one row smaller than the claims triangle. This is because every entry here is calculated from two values in the claims triangle. The second thing to notice is that the development factors tend toward 1 as time goes on and claim amounts are settled. If it is believed that the claim amount will never change beyond some point in time or development, the future development factors will all be 1, and the ultimate position will have been reached (Table 11.1).

Reserving actuaries can use this table to help them select the development factor to apply for the current year. They might want to calculate a selected development factor, for example by calculating the average development factor for each period. Some examples of calculation methods are:

- Simple average

Table 11.1 A triangle of development factors. Notice how the bottom right-hand corner can't be completed until more development has occurred

	Development Factors				
Accident Year	1–2	2–3	3–4	4–5	5–6
2016	1.76	1.79	1.34	1.09	1.00
2017	2.36	1.47	1.21	0.98	
2018	1.65	1.48	1.00		
2019	1.73	1.17			
2020	1.47				

- Weighted average (weighted by exposure measure. This is very important if a product's sales volumes have grown or shrunk significantly in recent years, as it means that more importance is given to the development pattern when there was more data).
- Average excluding high/low (a simple or weighted average, but where the highest and lowest development factors are excluded from the calculation. This means that the most unusual years are not used to project ultimate costs).

Deciding on a calculation method is not the only input an actuary has into this process: they also need to decide what cohorts they will use in this calculation. With a development factor triangle, it is much easy to spot the cohorts where the development pattern is atypical. For example, in Table 11.1 the development factor for the 2019 accident year between periods 2–3 seems unusually low. Actuaries need to decide what to do with this information, and this is where actuarial judgment comes in. There is no one-size-fits-all method but a range of considerations:

- Are there any obvious data errors that can be removed?
- Are there any years that were unusual, and the development pattern is not likely to be repeated this year? For example, a one-off weather event where a lot of claims came in all at once, or a legislative change that impacted insurance claims.
- Which years are most relevant to the year being extrapolated? This is usually the most recent years, unless there's been an unusual circumstance or event. For example, a change to the internal claims handling process may change the development pattern, but would be likely to continue.
- Are there any trends or patterns within the development factors? For example, are they increasing or decreasing over more recent cohorts?

Proprietary reserving software can be used to make the process of creating claims triangles, calculating development factors, investigating and selecting development patterns, and selecting development factors as simple and quick as possible. Algorithms are increasingly being used to automate the Chain Ladder method, identify development factors that deviate sufficiently from the average, and then check whether the underlying claims was anomalous or influenced by a catastrophe event. Using this, it might then calculate the weighted average development factor for the most recent 8 cohorts, excluding the highest and lowest factors (and any that have been significantly affected by an unusual event). An actuary of course needs to sense check the selections, and apply judgment based on any qualitative considerations, or trends not yet seen in the data, such as known future changes to regulation.

Another shortcut is to calculate the development factors to the ultimate position, so instead of having to jump the data forward by one time period multiple times, the Factor to Ultimate (abbreviated as FTU) takes data from the current position to the ultimate position. The FTU at time t is calculated as the multiple of all development factors from time t onwards, as follows:

$$\text{FTU}|_t = \prod_{i=t}^{\infty} \text{Selected Development Factor}_{i,i+1}$$

A FTU that is greater than 1 means that the ultimate claim cost is expected to be higher than the claim cost is currently; a FTU that is less than 1 means the current claim cost is higher than the ultimate cost. The inverse of the FTU also acts as a rough quantification of how developed a cohort is: the higher the FTU, the more development to ultimate there is to go.[1]

The Chain Ladder method is simple, but serves as the foundation for many of the other reserving methods. It can also be used just as it is but with some adjustments to the initial claims triangle:

- Removing or capping large claims from the triangle may reduce the variation in development factors
- Removing catastrophe or event losses from the triangle again helps reduce variation in the development factors.
- The effects of inflation can be stripped out from the claims triangle before development factors are calculated.

One flaw with the Chain Ladder is that the data needs to be prepared properly, taking care to remove anomalous data entries or claim events. For instance, under this method, if a cohort has that slightly more claims in the first quarter (perhaps due to an event), that extra claims cost will be assumed to continue in future periods. This effectively magnifies the data issue, and can lead to holding an unnecessarily high level of reserves.

The Chain Ladder method is simple and intuitive, but it is only useful when there are years of relevant historic claims data available. For products without this history, what can be done?

[1] When the FTU is less than 1, using the inverse to calculate how developed the cohort is gives an unhelpful and meaningless answer.

11.2 Expected Loss Ratio method

In order to price a product, the **underwriter** must have some expectation of the cost of claims that will arise from it. Explicitly or not, there is an expected loss ratio that is baked into the pricing (calculated by dividing the risk cost component of the price by the premium). The Expected Loss Ratio method for reserving uses that assumption alone and doesn't need any development patterns, so it is very useful if the product is new.

Specifically, the expected loss ratio of a product refers to the ultimate expected loss ratio. Using this method, calculating the ultimate cost of claims is very simple: it is the expected (ultimate) loss ratio multiplied by the premium. This is a quick rearrangement of the definition of the **loss ratio**.

To calculate the reserves using this method, take the paid claims from the ultimate claims cost.

$$\text{Reserves} = \text{Earned Premiums} \times \text{Expected Loss Ratio} - \text{Paid Claims}$$

This method is incredibly simple, with only 1 assumption and basic total premium and total claims data required. If data or time are in short supply, then this can give a ballpark figure. It can also be calculated using **written premiums** if required. However, it is too simple for most purposes as it is just too simple—it doesn't allow for any more information such as the current reported claim amounts. On the plus side, it isn't affected if there are anomalies in the data, for example if a large claim is included in the claims triangle. Crucially, it relies on the expected loss ratio being correct, and is very sensitive to changes to that estimate.

There are lots of ways to come up with such a loss ratio including industry benchmarks, using the loss ratio assumption used for pricing the business, historical loss ratios for the product, or even just underwriter opinion. But this method is usually not preferred, as actuaries wouldn't want to rely so heavily on a number that is so subjective if possible. However, if historic or target loss ratios are used, there is a risk that this method will replicate bias. This is a classic 'error-carried-forward' situation: if the target loss ratio was wildly optimistic and the same loss ratio is used for reserving in this method, the reserves will also be wildly optimistic. A feedback loop into the pricing or underwriting team is one way to reduce the potential for self-validating errors like these.

That's not to say that this method is useless—it is used in reserving at global insurers, but in a limited way. This method is usually consigned to reserving the first few development periods for a product, and then as more claims data comes in and the claims are more developed, the Chain Ladder can be used. Also remember that methods can be used concurrently, and then the actuary will weigh up the relative merits of each method in the particular circumstances. The final estimate might use just one method, or it may be somewhere in between the two.

Example 11.1 An established but small product has a historic loss ratio of 64%, as estimated by experienced underwriters. At the end of June 2022, $125,000 of premium has been written, which is typical for this product, but only $3,000 of claims have been paid and only $20,000 reported—this is lower than historically, but after investigation this appears to be just luck.

The reserving actuary decides to reserve based on the Expected Loss Ratio method. This gives an ultimate cost of $64\% \times \$125,000 = \$80,000$. The reserve should be $\$80,000 - \$3,000 = \$77,000$. The Chain Ladder method gave a lower estimate of $62,000.

At the end of the year, the reserves need to be updated. $10,000 of claims have been paid and a further $40,000 of claims have been reported but not yet settled. This claims experience is more typical compared to previous years. The Expected Loss Ratio method suggests the reserve should be $\$125,000 \times 64\% - \$10,000 = \$70,000$. The Factor to Ultimate is 1.7, so the Chain Ladder method gives $\$40,000 \times 1.7 = \$68,000$.

Now there is more data, the actuary feels more confident that using the Chain Ladder method is sensible, as the claims development pattern now appears to be similar to previous years—in June this was not the case. The actuary might decide to use a pure Chain Ladder estimate of $68,000, or they could take the simple average of both methods and use that value, $69,000, as the reserve. In the next period they will likely rely solely on the Chain Ladder methodology.

The Expected Loss Ratio method is very simple, but almost too simple to use for most purposes. Notice how the ultimate cost of claims never changes over time when using it; the reserves will only change as the amount paid increases. This is another problem with the method: even if the cohort is fully developed (in that there are no more claims being reported and all claims are settled), this method will always give the same ultimate claim cost unless the loss ratio estimate is amended. More generally, at what point do we believe the actual claims costs?

Despite these issues, the idea of combining an expected loss ratio with more recent claims data was too potent to ignore. To address this, another reserving method was devised: the Bornhuetter-Ferguson method.

11.3 The Bornhuetter-Ferguson Method

Created by American actuaries Ronald Bornhuetter and Ronald Ferguson in 1972, the (unimaginatively titled) Bornhuetter-Ferguson (B-F) method has since become a classic actuarial method. It combines the stability of the Expected Loss Ratio method, particularly when a cohort is not well-developed, with the concept of development patterns.

To use the B-F method, you need three things:

- An Expected Ultimate Loss (use the expected loss ratio and the premium: ask an underwriter, consider relevant benchmarks or make a guess)
- A development pattern for the product
- Current claims data (usually incurred claims, but you could use paid claims data if needed)

Once you have sourced these ingredients, you can get started!

Ultimate Cost for Ferguson
In 2008, Ronald Ferguson, of Bornhuetter-Ferguson fame, was found guilty of conspiracy and various types of fraud. As the CEO of Gen Re., along with other board members, he created artificial reinsurance transactions that made it seem like reserves were higher than they really were. This boosted the insurer's stock price. He was sentenced to 2 years in prison and had to pay a fine of $200,000, although this was overturned in 2011 and a retrial was demanded on the basis that Gen Re.'s stock price should not be allowed into evidence.

In 2013, the federal government declined a retrial. The case was dismissed and Ronald Ferguson was judicially exonerated.

Step 1: Assume you are exactly where you need to be.

Use the development pattern to determine how developed claims data usually is at this point in time. You can do this by dividing 1 by the Factor To Ultimate (FTU) as given by the Chain Ladder method for this time period. The Bornhuetter-Ferguson method assumes the claims data you hold is exactly as developed as the development pattern suggests. For example, if the development pattern suggests that at the end of the first year your claims data should be 20% developed, assume your actual claim data is 20% developed.

Step 2: Assume the future will go as expected.

Once you know where you are in your development pattern, you can work out how much development is left to come. Now assume that this portion of claims will follow the expected loss ratio. Multiplying the proportion undeveloped and the expected ultimate loss gives an estimated future claims amount (the IBNR[2]).

Step 3: Assume the best of each other.

Add together the two parts: the actual claims data with the estimate of future IBNR, to give the ultimate expected claim cost. This method therefore has the advantages of simplicity and stability from the Expected Loss Ratio method, but the benefits of Chain Ladder is that it uses the actual experience as well. B-F is particularly good for early to mid-stages of development and for small lines of business. In an extreme situation where you have no reported claims to data, the Chain Ladder method will project no ultimate claims, whereas the B-F method will at least give a non-zero ultimate!

To do the B-F method in one step, the formula is:

$$\text{BF Ultimate Claim Cost} = C + (1 - p) \times \text{Expected Ultimate Loss}$$

where C is the actual claims cost and p is the percentage developed.

Example 11.2 A product has incurred \$1.2m of claims. Using the Chain Ladder method on historic data suggests that the FTU at this time is 4.5. This cohort wrote \$10m of GWP and was priced using an expected loss ratio of 68%.

First, let's prepare our ingredients. The incurred claims data can be used as-is, but we need to calculate how developed our actuals are: % Developed $= \frac{1}{4.5} = 22.2\%$. We are therefore 77.8% undeveloped. We also need to calculate our expected ultimate loss using the expected loss ratio and premium. Plugging these numbers into the equation:

$$\text{BF Ultimate Claim Cost} = \$1.2m + 77.8\% \times (68\% \times \$10m) = \$6.489m$$

However, the B-F method also combines the disadvantages of the two methods. It requires both an expected loss ratio (which might be subjective or hard to determine) and enough historic data to create a development pattern. It also assumes that claims never track downward, which can occur in reality, especially if the **FNOL** team tend to put cautious estimates on claims. In practice, actuaries tend to do all three methods listed here and compare the results. They might choose one result or take an average of, say, B-F and Chain Ladder results. Reserving reports therefore usually state the projected ultimate loss and the method used.

[2] Strictly speaking, this might be referred to as the IBN(E)R, as this amount will include reserves for claims not reported and for claims that were under-reported. For simplicity, this will be referred to as the IBNR, as is the convention for reinsurers and London markets.

11.4 Alternative Reserving Methods

We have so far seen three classical reserving techniques: the Chain Ladder, the Expected Loss Ratio, and the Bornhuetter-Ferguson methods. Although these are the three most famous methods, there is plenty of innovation in the reserving space. This section isn't intended to cover all the detail for every method, but will give you an outline of how and when these might be used. Some of the methods are very statistics-heavy and can be omitted on first reading—it is enough to know that they exist.

11.4.1 The ACPC Method

The ACPC, or Average Cost Per Claim, method can be used alongside other reserving techniques for a more granular estimate of ultimate claim costs. Similar to pricing methodology where the **frequency** and **severity** are modeled separately, this method creates two claims triangles: one for total claim cost and one for the number of reported claims. Then, the actuary can divide the former by the latter to create a triangle of the average claim cost for every cohort and development period. Chain Ladder or B-F could be used to project the ultimate number of claims, and then that can be multiplied by the ACPC. This method therefore requires more granular data than the previous methods as two claims triangles need to be created, although average cost per claim could be taken from industry benchmarks, then combined with an insurer's own claim numbers to do a projection if data was lacking.

This method is particularly good if claim costs are fixed, as it means that you are directly projecting the uncertainty (for fixed cost claims, this is the number of claims) rather than a semi-certain value (a number of claims multiplied by a fixed cost). Inflation, or any other trends, can be an explicit assumption. However, if the product has a lot of nil claims, that is, claims that have a cost to the insurer of zero, these can skew the data, particularly if they are more likely to occur at a specific point in development. Because of this, nil claims are usually excluded from these claim triangles—although knowing how many nil claims there were at each development period requires yet more data!

11.4.2 Curve Fitting

Instead of calculating development factors to calculate the IBNR, a line of development could be drawn. Whereas development factors are based on a

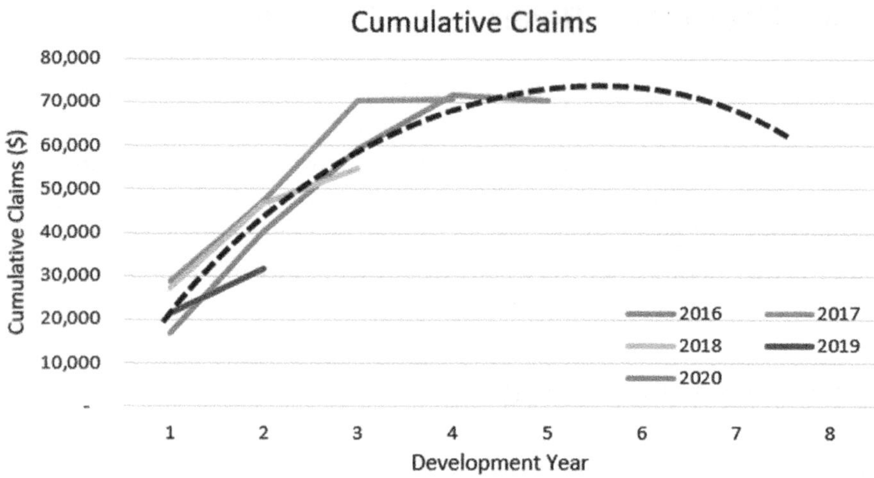

Fig. 11.2 Curve fitting. Here, the dotted line shows the development curve and the solid lines are the actual historic data

single point in time, a development curve allows reserves to be recalculated in between development periods. This method is not only visually appealing, but also is particularly good at allowing for development beyond the oldest year (remember that the Chain Ladder method assumes that your oldest year is fully developed).

To perform this method, plot the development of multiple years of data. Choose the shape of curve to fit,[3] then use regression to estimate the best parameters for this line. As you are estimating parameters for the curve, the line can continue past the development period of your oldest data indefinitely (if you think that is a reasonable assumption)! The other benefit of this method is that trends in your actual data can easily be examined: simply plot a curve for each cohort of actual data that you have. By comparing the parameters over time, you can spot trends in the development curve which can then be extrapolated into the future. It's a graphical method, so the visual appeal might mean it is more easily explained to non-technical board members (and other interested stakeholders). For **short-tailed** products it is unnecessary to go to the trouble of extending the curve beyond the actual data, and so not commonly used in these lines of business (Fig. 11.2).

[3] Usually, incremental development is plotted using a Weibull or exponential distribution, but in Fig. 11.2 I have fitted a polynomial curve to the cumulative development instead.

11.4.3 Stochastic Methods

One drawback of the previously mentioned methods is that the result is a single number—the ultimate expected loss cost. Of course, the actual ultimate loss is unlikely to be exactly as expected, and so it is sometimes preferable to have a range of the likely ultimate cost. Stochastic methods provide us with a **best estimate** and a confidence interval for that estimate. Confidence intervals are widely used in statistics as a way of communicating the uncertainty in an estimate: the bigger the interval, the less certain we are about our figure. Quantifying the uncertainty in our reserves is the main benefit of using a stochastic model, but it does require considerably more assumptions, statistical knowledge, and time.

In a stochastic model, the claims cost is assumed to be a random variable, with actual claims to date being just samples of this random variable. It is possible to use purely statistical formulae to calculate the uncertainty, the most well-known of these being the Mack method. The Mack method is relatively simple to implement but it doesn't predict extreme values of the distribution well.

If a purely analytical calculation is not your style, another way to determine the underlying distribution of claims costs is to use the bootstrapping method. Bootstrapping is a general statistical technique that tries to answer the statistician's constant demand for more data by making some new data up! Specifically, it randomly selects a certain number of your actual claim costs, takes the average of that sample, and records it. Then, it takes another random selection from your actual claim costs, takes the average of that sample, and records it. It repeats this, creating at least 10,000 random selections from your data in order to calculate the mean of each sample. Graphing these results give a histogram of sample means, which can be used to estimate an underlying claims distribution. Once you have this distribution, you can use it in other stochastic methods, including the Mack method.

The Mack Method

This is a stochastic but purely analytical method, created by Thomas Mack in 1993. Using your data to calculate the mean and variance of the underlying distribution, the Mack method then uses these to derive estimates for ultimate claims for each claims cohort. By taking the square root of these estimates, the standard error for each estimate can be calculated, giving a quantification of uncertainty. It makes a lot of assumptions similar to those in Chain Ladder, so is often considered a stochastic extension to that method. These calculations are easily done in a spreadsheet.

To use bootstrapping in the Mack method, first use bootstrapping to create an underlying claim distribution. Then, apply the rest of the Mack method (using the distribution to derive estimates and standard errors for each cohort) as usual.

There are lots of other stochastic reserving methods. As computer power has increased, simulation methods have become possible (remember, reserving is usually completed to tight regulatory deadlines). In these methods, an underlying distribution is required. This can be derived from the mean and variance of actual data, or bootstrapped, or just from prior expectation. Then, samples are taken from that distribution, and an average ultimate loss cost for that sample is calculated. This is repeated many times—usually at least 10,000 times—to give a distribution of average ultimate costs.

The graph below shows an example of ultimate claim costs from 10,000 simulations. From this, we can calculate all kinds of statistics. For example, the 50% percentile is somewhere between \$9m and \$10m (we could interpolate the exact value if we wanted). Or we could say that the most likely occurrence is to have an ultimate cost of \$10m. But the real benefit of simulation methods is in the tails: here, 149 of our simulations had ultimate costs of \$20m or more, so if we wanted to hold enough money to pay all claims in 99% of cases, we'd need to hold \$21m. As we'll look at in more detail in Chapter 12, the regulator would be very interested in the tails, as an equivalent interpretation is to say that if the company held \$20m it would have a 1.5% probability of being under-reserved. Equivalently, it would be expected to fail every 67 years! (Fig. 11.3).

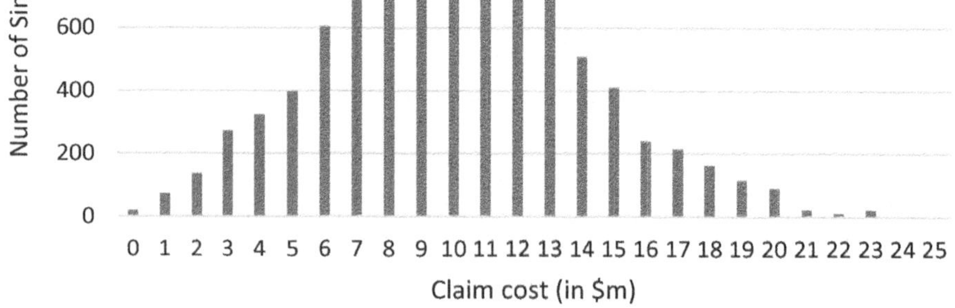

Fig. 11.3 Simulated Ultimate Claim Costs after 10,000 simulations

11.4.4 Machine Learning

Reserving methods are starting to become increasingly automated, and some companies are even starting to trial machine learning methods in the reserving context. One area of particular focus is the choice of method and hyperparameters for that method (the parameters that are set before even training the model). For example, when trying to decide upon a claim distribution for use in a stochastic method, we might select a certain distribution, such as lognormal. Then, the stochastic method will determine the parameters for that lognormal distribution using our data. But what if a lognormal distribution is not the best choice?

Machine learning algorithms can test hyperparameters quickly, but they need training data and test data. Training data is the data that the algorithm learns from, and test data is the dataset on which it tests that set of choices. Given we have limited actual claims data, these algorithms use a method called k-fold cross validation. This method is well-established in data science, and it works in the following way:

1. Split the data into *k* groups.
2. Make one of those groups (say, the first group) your test dataset, and all the others together form the training dataset. Test on this dataset and give a quantitative measure of how well the model fits.
3. Now, make a different one of the *k* groups (say, the second group) your test dataset, and all others together are your training dataset. Repeat the process of creating, testing, and scoring the model.
4. Repeat so that all *k* groups have been the testing dataset. You have *k* scores for this model, so take an average to give an average score for that combination of hyperparameters.
5. Repeat steps 2-4 for different model approaches (or hyperparameters within an approach).
6. Compare the scores for each model. Using these scores and actuarial judgment, select the model that is most appropriate and well-fitting for your data.

Although this method is common in data science, it is only starting to be turned to actuarial problems. But given how many reserving methods there are to choose from, an algorithm that picks the couple that are best statistically could not only improve estimates, but also save actuaries a considerable amount of time.

This chapter concludes with a longer worked example that will demonstrate the Chain Ladder, Expected Loss Ratio method, and Bornhuetter-Ferguson on one set of claims data. The next chapter explores what happens when the reserves held aren't enough to pay claims—and unlike these calculations, it isn't pretty!

Example 11.3 A product has an expected loss ratio of 75%. The cumulative claims triangle, as well as each year's GWP, is given below. Find the ultimate loss cost for each year using the Chain Ladder, Expected Loss Ratio, and Bornhuetter-Ferguson methods.

		Development Year					
Accident Year	GWP	1	2	3	4	5	6
2015	60,000	9,657	16,955	30,344	40,599	44,235	44,383
2016	82,000	16,995	40,188	59,140	71,777	70,289	
2017	94,000	28,670	47,432	70,322	70,656		
2018	90,000	27,066	46,782	54,802			
2019	72,500	21,577	31,741				
2020	87,000	23,632					

Let's start with the Chain Ladder method. First, we calculate the development factors for each year, between each development periods (recall, these are the multiplicative 'jumps' between each data point). For example, to calculate the development factor for 2017 between periods 2 and 3, divide 70,322 (the claims cost at period 3) by 47,432 (the development at period 2) to give 1.48.

Do this for each year and consecutive pair of development periods to get a development factor triangle. Now we need to choose a development factor to use for each period.

	Development Factors (to 3 decimal places)				
Accident Year	1–2	2–3	3–4	4–5	5–6
2015	1.756	1.790	1.338	1.090	1.003
2016	2.365	1.472	1.214	0.979	
2017	1.654	1.483	1.005		
2018	1.728	1.171			
2019	1.471				
Selection	1.652	1.375	1.185	1.034	1.003
FTU	2.796	1.692	1.230	1.038	1.003

(continued)

(continued)

	Development Factors (to 3 decimal places)				
Accident Year	1–2	2–3	3–4	4–5	5–6
% Developed	0.358	0.591	0.813	0.964	0.997

Looking at the triangle, 2016's 1–2 development factor (2.36) looks much higher than the other 1–2 factors. After making a call to a friend in the claims department, there was a large claim that was reported in the second development period for this cohort. This should have been excluded from our data and allowed for separately for this cohort, but the next best thing is to exclude it now, so let's exclude that development factor. The other factors for 2016 won't be affected.

For the development period 2–3, the development factors seem to be trending down. Because of this, we choose to exclude the oldest (and highest) factor. All other factors look reasonable, so excluding the two factors we know about, we calculate (simple, not weighted) averages to get our selection.

Now we can calculate the Factor To Ultimate. The oldest cohort is assumed to be fully developed, so the FTU for period 6 is set equal to 1. The FTU for all other periods is the product of all subsequently selected development factors. It's not required for the Chain Ladder method, but we can use the FTU to calculate the % Developed at each development period by dividing 1 by the FTU—we'll need it for the B-F method later.

Now we have all our ingredients for all the methods, let's start calculating some ultimates!

For the Chain Ladder ultimates, we need to multiply each year's current position, taken from the diagonal of the claims triangle, with the corresponding FTU.

The Expected Loss Ratio ultimate cost for each year is also simple to calculate. It's the GWP multiplied by the Expected Loss Ratio given in the question. Here it was 75% for all years, so has not been included in the table below.

Now for the B-F ultimate costs. Remember, this method combines the current position with the expected loss ratio ultimate, weighted by how developed we think a year is. To calculate the B-F ultimate, multiply the Expected Loss Ratio Ultimate Cost by how much development is still left to go (mathematically, 1—% Developed). Add this to the current position to give the B-F ultimate cost.

Accident Year	Current Position	FTU	Chain Ladder Ultimate	GWP	Expected Loss Ratio Ultimate Cost	% Developed	B-F Ultimate Cost
2015	44,383	1.000	**44,383**	60,000	**45,000**	1.000	**44,383**
2016	70,289	1.003	**70,524**	82,000	**61,500**	0.997	**70,494**
2017	70,656	1.038	**73,332**	94,000	**70,500**	0.964	**73,229**
2018	54,802	1.230	**67,426**	90,000	**67,500**	0.813	**67,440**
2019	31,741	1.692	**53,706**	72,500	**54,375**	0.591	**53,979**
2020	23,632	2.796	**66,072**	87,000	**65,250**	0.358	**65,544**

Exercise 11.2 Recalculate the Chain Ladder, Expected Loss Ratio, and B-F ultimate costs without excluding any historic development factors in your selection.

As we have seen, reserving is important not only for cashflow modeling, but also for business planning, estimating risk costs for pricing and calculating regulatory capital requirements. As such, having up-to-date projections of the ultimate claim costs is vital. Therefore, reserve reviews will be performed at regular intervals, usually quarterly or annually, depending on the line of business. Although the techniques detailed in this chapter are considered traditional actuarial methods, new applications of machine learning, stochastic models, and AI are increasing reserving automation and sophistication.

12

In Case of Emergency: Capital Requirements

Chapter Objectives

- Understand why insurers should hold capital, and the pros and cons of doing so
- Understand the broad principles of regulatory capital requirements (including which risks are covered in the capital requirement)
- Understand the potential consequences of holding inadequate capital

I always keep at least three different kinds of Brie in the kitchen for emergency situations.

—Jonathan Coe, Mr. Wilder & Me

I don't like holding money. I have two reasons for this: firstly, I don't like the dirty, metallic smell it imparts to my hands. Secondly, I once watched an episode of *The Simpsons* where Homer squeezes a penny so hard it moves under his skin and circulates around his body. While I admit this is irrational, I still don't hold coins too tightly.

Businesses don't like holding money either, albeit for (presumably) very different reasons. Money sitting in a bank account is money that could be better used for other things—investing in the business, or perhaps investing in other people's businesses—to receive a higher return. In this chapter we will be considering how much money insurers and pension providers need to

© The Author(s), under exclusive license to Springer Nature Switzerland AG 2022
C. Townsend, *A Risky Business*,
https://doi.org/10.1007/978-3-031-11673-5_12

hold. Like most actuarial science, it involves balancing risk and practicalities. And like most *Simpsons* episodes, it involves a touch of reality, a dash of the absurd, and a talent for predicting the future.

12.1 That's Capital!

Businesses don't tend to talk about money. Not because it is gauche to do so, but because it is capital that concerns them far more than money. The difference between capital and money is mainly time: capital is money invested for longer periods of time. Money is for short-term use, usually up to about 3 months. There are a few corollaries from this that are important to note. Money, only being invested for very short terms, is easily accessible (economists call it highly liquid), whereas capital investments are hard to withdraw if you urgently need them. Further, money is very low risk. There just isn't a lot of time for it to increase or decrease a lot in value. On the other hand, capital is higher risk than money. This means that you can earn a higher return on capital investments, both in terms of a higher return and also earning it over a longer time (and note that we use 'return on investment' for capital rather than the money term 'interest earned'). While 'money' conjures images of coins or notes, for businesses these very liquid funds are only partially held in this way: government bills or bonds (known as T-bills or T-bonds in the USA, or gilts in the UK), mutual funds or stocks are more common. In this chapter, I will use money and capital relatively interchangeably, in order to explain the basics of capital requirements, and to draw parallels with how we as individuals use our own money. Now we have a basic understanding of *what* capital is, let's talk about *why* it is.

Financial service companies don't require much in the way of specialist equipment to get going: proprietary software is often helpful but non-essential, computers of reasonable speed and security can be bought off-the-shelf, and data storage is increasingly cheap.[1] But insurers and pension providers[2] are the owners of huge funds. So, what do they need all that capital for?

Well, it's not all capital; a small amount will be plain old money ready to smooth over any timing mismatches in **cashflow**, to allow the business to pay

[1] I bought my first memory stick, holding 128 MB, for about $10 in 2008. Last month I bought 50 GB of cloud storage for under a dollar.

[2] For brevity, in this chapter I will refer to general insurers, pension providers, and life and health insurers under the umbrella term of insurers. Pension providers in particular hold huge amount of capital—in 2018 the Investment Company Institute estimated the assets held by US pension providers alone at $8.6 trillion!

suppliers and claims in a timely manner without rejigging their long-term investments too regularly. If you've ever run a tombola or lemonade stand, you will know that you start the day with a few coins in the register from your own piggy bank, so you can give change readily to your first few customers. This pot of cash is only a small proportion of insurers' funds however.

In Chapter 10, we discussed the principles behind **reserving**. It's a very important actuarial function, but the summary is that the insurer expects some of the collected premium to be required for future claims. Therefore, the company holds that money aside ready to pay the expected future claims, in a similar way to keeping part of your paycheck aside to pay your gas bill at the end of the month. However, most reserving is performed in order to give a **best estimate** of claims cost (although as in Chapter 11, stochastic reserving methods give confidence intervals and ranges of estimates). This is the most likely cost of claims, according to the actuary. Remember that different actuaries will make slightly different selections when applying their own actuarial judgment, and so each actuary might have a different best estimate.[3] So, when setting funds aside, is that best estimate of future claims cost the right amount of capital that an insurer should hold?

If we assumed that the total cost of claims will follow a gamma distribution,[4] the most likely estimate (the mode) is less than the mean, as the distribution has a positive skew. Holding the most likely estimate of claims cost is less than the statistical expectation of claims cost! To add further confusion, the median of a gamma distribution is not easily calculable. Even if it was, holding the median amount would mean that the likelihood of being above that value is the same as the likelihood of being below. In this scenario, there would be a 50:50 chance that holding the best estimate would be insufficient. In other words, holding this amount would be like tossing a coin: heads, your business is safe; tails, you are ruined. But what then is the correct amount to hold? If you held a little buffer over and above the best estimate, what then would be the probability of bankruptcy? Is an 80% chance of survival enough? 90%? 99.99%?[5] To frame this in another way, an 80% chance of survival suggests that the company would go bankrupt once every five years. That is clearly not a safe way to run a business!

[3] In a reserving team, in practice it is usually the opinion of the most senior actuary that wins out. However, auditors might get several senior actuaries to independently reserve and then compare best estimates.

[4] The sum of independent gamma distributions (as the cost of claims are) is itself a gamma distribution.

[5] An insurer can't ever be 100% sure they are holding enough if they have written any amount of unlimited cover.

And if survival isn't a compelling reason enough to hold sufficient capital, recall that insurance is based on the principle of *uberrima fides*—utmost good faith that claims can and will be paid. A company that is known to be low on funds is not one that customers will want to buy from, in case they go bust and can't pay claims. A lack of faith in one company can spread like wildfire through the system of risk transfer and financial services. This wouldn't necessarily just be limited to insurers or pension providers: if enough people thought that their pension fund was at risk, they might decide to withdraw it from the provider and keep it under their mattress. To pay customers their funds back, the provider might have to withdraw the assets from investments. Even a small contraction in available assets could trigger a global panic, as global pension fund assets have been valued at over $32 trillion[6]! And that is quite aside from the possibility that the public and companies would no longer feel able to perform everyday activities due to the level of financial risk to themselves.

Equally, an insurer that is overly cautious and holds too much capital, or invests it in very low-risk bonds, won't make much of an investment return and therefore will be less profitable. This is particularly true if this cautious behavior is anomalous in the market: meaning other companies are receiving higher investment returns, therefore able to offer a lower premium and therefore increase their market share. This in its own way is a risk to the survival of the company!

In order to instill confidence in the system, an independent regulatory body needs to check the level of capital insurers and pension providers are holding.[7] With this information, customers can compare the financial strength of these companies. In other industries, financial strength can be measured based on assets or profits alone. For risk transfer companies, these measures aren't as useful. This is because of the inversion of the product cycle: as discussed in Chapter 1, for insurers and pension providers, the product is sold long before the costs are certain.

[6] OECD Global Pension Statistics project, June 2020.

[7] Regulators also do a host of other important functions, such as reducing financial crime, applying and enforcing the rules (of which capital requirements are only a small part), and vetting senior leaders of financial companies as fit and proper.

Did you know?
Allianz, the world's largest insurer, was founded in 1890. It has assets of around $1,200 billion.

Example 12.1 Supermarket A made a profit of $18 million in 2021, while Supermarket B made a profit of $23 million. Supermarket B has therefore increased its absolute value more in 2021 than Supermarket A. In contrast, Insurer X wrote $10m of premium in 2021, and at the end of the year only had $3m of claims reported. The true profitability of their 2021 business won't be known until perhaps 2025 or later.

Assets are not directly the best assessment of financial strength for insurers either, as the funds they currently have will in some part be depleted by future liabilities: namely, future claim costs for business that has already been sold. Therefore, directly comparing the absolute value of assets doesn't give a reliable indicator of insurers' financial strength.

Example 12.2 Supermarket A has assets of $150 million. Supermarket B has assets of $145 million. Therefore, supermarket A has more assets which could be sold off if the business wound down. In contrast, Insurer X has assets of $32 million and Insurer Y has assets of $54 million. However, insurer Y writes a lot more business and experiences higher claim costs than insurer X. Therefore, more of their assets are required to pay claim costs, and so they are not free to be invested, paid as dividends, or sold off.

It falls to industry regulators to assess if the capital that insurance and pension providers are holding is adequate. From this, they can judge the financial security of the company. As we have seen, we can't use standard accounting metrics such as assets or profitability, but require measures that take into account assets and the level of risk and future liabilities that the company is exposed to. Capital requirements ensure sufficient assets are held and robustly managed against the company's liabilities, the business written, and the traditional conditions. In normal circumstances, insurers are likely to be holding enough money to pay claims, as it is a requirement of their company's survival to do so! But it is in case of emergency that capital requirements really come into play: when unexpected or unlikely claims arise and blow standard estimates of claim costs out of the water. Akin to a piggy bank that can be raided on a rainy day, having an emergency buffer is vital for companies, particularly those who routinely take on large and unpredictable risks.

12.2 How Much Capital Should be Held?

The regulator requires a certain amount of capital to be held, dependent on the level of risk that a company is exposed to. Of course, the level of risk first needs to be quantified. Hopefully by now you recognize that this sounds like a job for an actuary! There are two main components to this risk: the insurance risk (the risk attributable to writing insurance business) and then all the other risks that a business is exposed to.

The level of insurance risk is a function of several factors, such as:

- The amount of business written (number of policies or **GWP**)
- The product or type of business
- The risk mitigation strategies in place such as **reinsurance** arrangements, or setting maximum claim costs in the terms and conditions of policies
- Whether the risks are homogeneous or diverse

Other sources of risk include:

- Operational risk—the risk of direct or indirect losses resulting from inadequate or failed processes, people or systems, or from external events. Considerations might include how the business functions operationally: is business written through brokers or direct? Are there strict controls over processes and security?
- Market risk—the risk that financial markets won't act in the expected way, affecting investments, inflation and potentially claims as well. For instance, recessions often trigger a higher number of claims, or over-stated claims, as unscrupulous individuals try to make some extra cash.
- Credit risk—the risk that a creditor will not repay what is owed to the company at the time agreed.

For some of the above factors, the reasoning is obvious: if you write fewer policies, you are taking on fewer risks. However, the effect on the level of risk is not necessarily linear, or even obvious. If you write very fewer policies, that might be because the **underwriters** are carefully selecting only the low-risk policies to take on. Alternatively, writing fewer policies might leave you exposed to mis-pricing, as with less data to train your pricing model it is less likely to be accurate. Fewer policies also might give rise to a lack of cashflow or funds to cover fixed expenses.

Some products are higher risk than others. Fixed benefit insurance is lower risk than car insurance because it is a fixed claim cost and therefore limited liability for the insurer. It is a very predictable line of business: the only uncertainty is the number of claims, not the amount per claim. But car insurance is lower risk than public liability cover, which may be particularly subject to very large claims or even **latent claims**, and is more sensitive to legal rulings. The less predictable the claims, the more capital required to support it.

The type of business written is usually linked closely with the risk mitigation strategies in place: if the insurer is writing large, unlimited liability policies it is likely to want extensive reinsurance cover to manage or limit their own exposure to risk. Companies with reinsurance (or other risk mitigation in place, such as strong terms and conditions) won't require as much capital as they would have without reinsurance because they are reducing their own risk. Unfortunately, when working with any external supplier, reinsurer or company, there is a small chance that they might default. This would mean that the goods or services that have already been paid for (or any money owed to the company) would be lost—and this is a risk in itself! Credit risk doesn't have to be a complete collapse of a company; even a partial or late payment is a risk to the cashflow of the insurer. In an extreme scenario, lots of companies in an industry would simultaneously default on credit or go bankrupt (see the LMX spiral in Chapter 13 for a dramatic example of this!). This means that overall, buying reinsurance reduces the insurance risk-related capital requirements a lot, but increases those of credit risk, overall reducing the capital required.

Capital requirements are there in case of emergencies. This could be an extreme external market event like a severe recession or a catastrophe such as a natural disaster or terrorist strike, or anything else that would cause a sudden increase in claims and/or a sudden contraction in available credit or liquid assets. As we have mentioned, lines of business that are exposed to these types of claims are higher risk, and therefore more capital intensive. But it also depends what else an insurer writes. Insurers who have a large and homogenous portfolio of risks, say, only writing one line of business, or writing only in one country, are more vulnerable if there is a shock to that product or location. Ideally, insurers would write products that cover opposite risks—insurers writing both life insurance (that pays out on death) and also life-dependent annuities (that pay out only while the recipient is alive). Then, in the case of an extreme event, perhaps a pandemic where there were thousands of excess deaths compared to usual levels, the insurer makes a loss on life insurance but might recoup some of that back from lower-than-expected **annuity** costs.

Example 12.3 Miles runs two ice-cream vans. The vans make a profit of \$500 per van, but only on sunny days. There are usually 200 sunny days a year where Miles lives, so he expects a profit of \$500 × 2 x \$200 = \$200,000 each year.

Henrietta lives in the same town as Miles, and she runs two food vans; one ice-cream van and one that sells grilled cheese. Each van can make a profit of \$500 per day, if it is operating. Grilled cheese vans only make a profit on non-sunny days. Henrietta expects \$500 × 365 = \$182,500.

However, if the weather is unexpectedly poor and there are only 50 sunny days a year, Miles will only earn \$50,000 in a year. In contrast, Henrietta will earn her usual \$182,500. Miles has not diversified his food truck offering, and so his income is more unpredictable and more variable than Henrietta's. He should hold a bigger emergency fund than Henrietta to live off in case of a year of adverse weather.

We have so far thought of factors that will impact the amount of capital required to be held, but the effect of each factor on the amount of capital is complicated. One role for regulators is to create a standard formula that uses the above factors (and many more!) to quantify the level of risk that the company is exposed to, and therefore the amount of capital that should be held. We'll see more of the technical methods used in Sect. 12.4. For now, let's digress and discover what else regulators can do.

12.3 Solvency

It is not just the size of the capital held that matters, although clearly that is important. It is essential that some of the capital is secure but accessible in an emergency. This is where things start getting complicated: it is essential that enough capital is held as an emergency buffer, and therefore is not likely to lose all value in a catastrophe. Assets such as property that frequently see volatile returns, are more likely to do so—imagine how galling it would be if a hurricane not only ruined your home but also instantly emptied your bank account! Regulators will therefore want to impose some other rules about capital—not just how much, but when and where it should be held.

Regulators also need powers to audit insurers and check that they are in fact keeping to the regulator's rules. This usually means that mandatory reporting at specified time intervals should be submitted, in a standardized format. A regulator might also do more detailed spot-checks on insurers to ensure that the submitted reporting is backed up by a rigorous process. If an insurer is found to be in violation of the rules, the regulator needs to have powers to enforce sanctions. These are usually fines, but in extreme cases they could stop the insurer writing new business entirely. As well as looking at assets held, mandatory reporting can include standard measures of profitability such as **COR**, information about the products sold, and risk exposure. It can also cover non-financial information, such as the number of complaints received, any operational events or near-misses (such as security breaches), and the company's internal governance structure.

These rules, and the sanctions that accompany them, may seem like overkill. However, the regulator is trying to maintain confidence in a necessary (but privatized) public service. That is a big task, and requires a big rulebook! In practice, meeting capital requirements isn't binary: there are different levels of capital requirements. Think of these like a traffic light system. To start with, there is the bare minimum level. Holding below this level means there is a serious concern that the business wouldn't be able to pay claims in adverse circumstances. This is called the Minimum Capital Requirement (MCR) and it is a red light: if capital is below this level the regulator will intervene significantly, i.e., stop new business being written.

Your Local Regulator

Wherever you work, it is essential to know the regulator and regulatory requirements that apply. Here are just a few:

- Prudential Regulation Authority (PRA)—UK
- Australian Prudential Regulation Authority (APRA)—Australia
- Bundesanstalt für Finanzdienstleistungsaufsicht (BaFin)—Germany
- Autorité de Contrôle Prudentiel et de Résolution (ACPR)—France
- EIOPA, the European Insurance and Occupational Pensions Authority supports local regulators.
- Insurance Regulatory and Development Authority of India (IRDAI)—India
- Financial Services Board (FSB)—South Africa

In the USA, insurance is regulated at state level, but the National Association of Insurance Commissioners (NAIC) supports these organizations. This is similar to the Canadian system, where provincial Superintendents of Insurance are supported by the federal Office of the Superintendent of Financial Institutions.

Of course, the regulator doesn't wait until that point to act. If a company got close to their MCR, the regulator might require more frequent reporting.

But at what point are the regulators concerned? Well, the regulator has another minimum in mind, called the Solvency Capital Requirement (SCR). This is the amber light: below this level, the regulator thinks you are holding too little for your exposure to risk, but still enough to function. Increased supervision and strong recommendations to hold more capital ensue. In practice, most insurers will want to hold at least 20% more than their SCR, to allow for fluctuations in asset value. One metric is the **Solvency Ratio**: the amount of funds the insurer has, divided by their net **written premiums**. There is a recommended level set by the regulator. For global insurers, solvency ratios of 140–160% are common; that is, they are holding 40–60% more capital than their regulator would suggest as a reasonable minimum.

There are some issues with these methods. Firstly, any reporting of capital is only a snapshot in time. If taken on a good day, the value of the assets might be slightly inflated as assets natural fluctuate in value. If a market was particularly volatile or a large claim was reported, sufficient assets yesterday may not mean that the company is holding enough today. Because of this, the regulator's 'green light' situation is when insurers are holding more than the SCR, to allow for these fluctuations.

Each region will have its own regulatory body, requirement, reporting scheme and much more. Now we look in more detail at just one regime, which fundamentally reformed capital requirements throughout Europe.

12.4 Solvency II

So far, we have talked in general terms about capital requirements. However, there are very specific rules for each region. Solvency II is the set of rules used in Europe. Using this regulation as an example, we'll see how some of the aforementioned ideas work in practice.

12.4.1 What is Solvency II?

Solvency II (shortened to SII[8]) is an approach that bases the amount of capital that should be held on the level of risks the company is exposed to. It applies to all insurers and reinsurers in the European Economic Area who have premiums over €5m, or who hold over €25m in reserves (including a risk margin).

[8] This short form is only used in writing, not speech, so read SII as "Solvency Two".

Solvency II has a standard formula for calculating the Solvency Capital Requirement. The formula is very complex, but it says that the SCR should be equal to the 99.5% confidence interval—i.e., the insurer should only expect to have insufficient funds once every 200 years. The MCR is set at a confidence interval of 85%; an expected failure rate of once every 7 years! The MCR is between 25–45% of the SCR.

As its name suggests, Solvency II is the second iteration of a European capital regime. Solvency I was not one piece of legislation but a series of European Union directives that were passed between 1973 and 2003. Rather than a considered regime, it was regulatory firefighting: urgent shortcomings in **general insurance** were tackled first, then six years later, life insurance regulation was created. Then, some risks were found to be not adequately covered and so more piecemeal legislation was introduced. However, this was the first attempt at unifying capital requirements across all insurers in the European Union, so perhaps it can be forgiven for being a little slow to market. In contrast, the entire Solvency II regime came into force on January 1st, 2016.

12.4.2 How Much is the Capital Requirement Under Solvency II?

The beauty of Solvency II is that all European insurers[9] must abide by the same rules, including Lloyd's of London.[10]

So, what does the SII SCR include? It takes account of all the insured risks, whether that's non-life, life, or health (or all three!), including catastrophes. It also considers the risk of default: of reinsurers, suppliers, or other asset-managers defaulting; market risk (the standard economic factors such as inflation, spread, interest rate…) and operational risk such as the risk of security breaches or processes failing. It also includes intangible asset risks—risks that arise from intangible assets being lost, damaged, or misused. In insurance, this is largely data and models—if these fell into the wrong hands, not only would there be data protection regulation fines (such as GDPR or equivalent), but a competitor could effectively recreate the insurer's models, and then undercut them! Given the cost of data storage and the salaries required to pay actuaries to build such assets, the risk of losing them requires capital to be held explicitly. In addition to all of these risks, the SCR under Solvency

[9] There are a couple of exceptions, such as very small insurers and death benefit funds who instead have to abide by Solvency I.

[10] Although the UK is no longer part of the EU, at present all UK-based insurers (including London markets) must still abide by their rules. However, the view of UK insurers is no longer represented in the annual review of Solvency II.

II also includes an adjustment for deferred taxes, but we won't go into that level of detail here.

But the ground-breaking part of SII is that it also covers group risk. Group risk is the risk that arises if an insurer is part of a wider group. This can be a parent company with operational entities set up in different countries, or it could be several insurance companies forming a group to work more efficiently (whether that's operationally, or for capital efficiency, or tax efficiency). But despite the benefits, being part of a group comes with some risk as well. For example, if another member of the group is loss-making, or isn't holding enough capital, other group members might be required to cover the shortfall. Solvency II tries to capture this risk and feed it into the capital requirement.

To apply the Solvency II standard formula, insurers need to calculate the SCR for each line of business. Then, the formula accounts for diversification benefits by giving a correlation matrix between different lines of business. If there is positive correlation between two lines (i.e., higher losses in one would tend to occur alongside higher losses in another), more capital should be held. It also allows for geographic diversification between risks, although that is assumed to be included as part of the SCR for each line of business. The formula for the SCR under Solvency II is:

$$\mathrm{SCR_{overall}} = \left(\sqrt{\sum_{i,j} \mathrm{Corr}_{i,j} \times \mathrm{SCR}_i \times \mathrm{SCR}_j}\right) + \mathrm{SCR_{intangible}} + \mathrm{SCR_{operational}} + \mathrm{Adjustment_{def.\ taxes}}$$

The square root term denotes the SCR directly related to each line of business. It is calculated in this way, rather than a sum of SCRs, to take account of the pairwise correlations between each line of business. In the above equation, the summation is done between every possible pair of lines of business. The correlation matrix is defined as part of the SII rules.[11] Other parts of SII specify a calculation method for the SCR for intangible assets risk, SCR for operational risks, and an adjustment for deferred taxes. These three terms are simply added on as there are no correlations with other risks to be considered.

[11] Directive 2009/138/EC, for the legislation aficionados.

Example 12.4 If an insurer who writes life, health, and general insurance (GI) will have 6 pairs of (i,j) for this equation: life x health; life x GI, health x GI, life x life, health x health, and GI x GI.

In short, we must calculate the SCR for every line of business, add a little to allow for any correlations between lines, add an amount to cover losses to intangible assets and operational risks,[12] and make an adjustment for deferred taxes. Simple, right?![13]

12.4.3 What Kinds of Assets are Allowed Under Solvency II?

Solvency II also has rules around what types of assets can be invested in, so that the insurer is able to access funds in case of emergency. To do this, it groups assets into one of three tiers. Tier 1 is the most accessible and most suitable for the liabilities that the insurer has. It includes shares and retained profit, and can be liquidated whenever the insurer chooses to do so. Tier 2 is slightly less easy to access, and includes letters of credit or 'called up' capital—capital can that be called upon when required, but relies on that party having the capital they promised! There are also often limits on when the capital can be called, which is why this capital is less liquid than Tier 1. Tier 3 is the least accessible in the case of loss, and includes deferred tax credit or subordinated debt. Solvency II uses these tiers to define what kind of assets count toward the MCR or SCR in the following way:

- 80% of the MCR must be Tier 1 capital.
- Tier 3 capital can't be used to cover the MCR.
- 50% of the SCR must be covered by Tier 1 capital.
- No more than 15% of the SCR can be covered by Tier 3 capital.

These are surprisingly straightforward rules, given the complexity of financial systems!

[12] The SCR for operational risks is calculated as the maximum of 3% of annual Gross Earned Premium or 3% of best estimate reserves + risk margin. This is subject to a further limit based on the overall SCR.

[13] This formula is not simplified at all, it genuinely is this simple! Calculation of the basic Solvency Capital Requirement | Eiopa (europa.eu) -the official SII rulebook—shows the same formula. The devil is in the detail; namely, the calculation of the SCR for each line of business and the adjustments.

Solvency II is not just about holding the right amount of capital in the right assets. It is based on three pillars: quantitative, risk-based capital requirements; qualitative assessment of governance, systems, and risk management; and reasonable and transparent disclosure. With these three things in place, regulators have the information they require to spot companies needing extra supervisory attention. Reporting is not just helpful for the regulator though; market pressure means that public reporting in itself encourages companies to clean up their act!

We are all individuals!

—Monty Python, Life of Brian

When a standardized formula for SCR was initially proposed, companies protested on the basis that they are all individual, and a standard formula wouldn't allow for the nuances of their particular business or circumstance. Most didn't want a standard formula as they feared it would force them to hold more assets, drive inefficient use of capital and thus incur the wrath of shareholders. To address these concerns, Solvency II has another option. Companies are free to use their own internal model for capital modeling, as long as the internal model and outputs are approved by the regulator.

Internal models can give less onerous capital requirements, and they are particularly good for companies who have a non-standard risk profile. If the company already has a risk management model, repurposing that rather than using the standard formula can also be a time-saver. However, to get regulatory approval of an internal model, models must pass the following tests:

- Is it statistically sound? Is the data accurate?
- Does the internal model cover all the risks that the standard model does?
- Does the internal model suggest an SCR equivalent to the 99.5% confidence interval from the standard model?
- Is the model regularly checked and validated, at least annually, to include all new information?
- Is the model well-documented, and does it demonstrate compliance with other regulatory standards?
- Is the model actually used in the business for something other than capital modeling, for example, for governance or risk management?

Perhaps surprisingly, one of the hardest tests is the final one here: the use test. This is because the company's decision-making process, and the role that the internal model plays in each decision, needs to be well-documented.

Reviewing the internal model and evidencing that you have done so needs to be embedded into every big decision. Like the ice-cream maker in my kitchen cupboard, it is easy to own one; to use it is an entirely different matter!

Most large insurers will use their own internal model. The costs of creating and documenting one are often more than offset by a slightly lower capital requirement. Lloyd's of London also has its own internal model that can be used by all Lloyd's **syndicates**. This calculation of the SCR at Lloyd's must be performed using two different bases: firstly, with a one-year view, consistent with Solvency II's standard model. Secondly, the calculations must be repeated using an ultimate view, looking at all future risks and any adverse development of claims (i.e., claim costs increasing) for all liabilities. This view is then used by Lloyd's to set their overall Lloyd's of London SCR, which determines the amount that should be in the Central Fund and central assets combined. The Lloyd's internal model for each syndicate is:

$$\text{Capital Required for Syndicate } i = (\text{Ultimate View SCR}_i + \text{Loading}) \times 1.35$$

The loading in the above formula is applied by Lloyd's, if it is felt that the managing agent of the syndicate has understated the risks for their syndicate. The 35% additional load is required by Lloyd's in order to maintain Lloyd's credit rating. Once the SCR for the syndicate is calculated, that view will be passed on to each member, for it is the members at Lloyd's that must be solvent, not the syndicate.

Calculating each member's requirement is a big task for Lloyd's syndicates. The Member's Capital Minimum is calculated based on the member's share of each syndicate, and the capacity of each syndicate. There are some diversification benefits if the member is part of multiple syndicates, but generally, a member's capital minimum is calculated as follows:

$$\text{Member's Capital Minimum} = 40\% \times \text{Syndicate Maximum Capacity} \times \text{Member's Share}$$

A member is said to be 'in line' if their **Funds at Lloyd's** exceed this minimum. Members must 'come into line' in November, and have until June to lodge assets with Lloyd's.

There is a slightly oddity here. In the Solvency II standard model, capital is based on the risks undertaken. In the above equation, capital is based on the maximum risk capacity, not the actual level of risk. Sometimes, syndicates have more capital than they know what to do with—they can't find profitable business in the market for which to use the capital as backing. But the member's capital minimum stays the same whether the syndicate writes 100% of its capacity or 10%!

In this situation, faced with a relatively high capital requirement for the amount of business written on their behalf, it is no wonder that members would start to complain to managing agents! The managing agents will therefore rush to write more business, sometimes of dubious quality, to placate their members. This is called overcapacity, and happens from time to time at Lloyd's. In 2018, Lloyd's ordered that syndicates that have been unprofitable for three consecutive years should be put onto remediation plans, or close, to reduce overcapacity. Profitable syndicates were told to improve or exit their worst performing segments of business. An increase in supply, such as a new syndicate or insurer can also trigger overcapacity fears, potentially forcing members to deposit Funds at Lloyd's for business never written.

Capital requirements can seem horribly complicated, and in the most part, they are: there are 312 articles in the Solvency II directive alone, excluding appendices! But the principles behind them—what to hold, how much to hold, and what risks to include—are actually very straightforward. The next section will cover some of the detail of these calculations but if you are keen to get into the gory detail of insolvencies, you can skip ahead to Sect. 12.6.

12.5 Capital Calculation Methods

Capital modeling is about managing the risk of adverse scenarios. There needs to be a balance between solvency and efficient use of capital under management, so quantifying this problem can allow a mathematically optimal solution to be found. But capital modeling isn't just for regulatory purposes! Businesses have limited funds at their disposal, so might want to understand how which lines of business or which geographic regions are more capital efficient (or which are more capital intensive) in order to use what they have to make the biggest return. This means that insurers and shareholders could then use the common metric, return on equity, to compare not only lines of insurance business but also investing in other industries.

Before we can define a method for capital allocation, we need to choose a measure for risk as we are using a risk-based approach. There are lots of possible choices for this, all requiring some prior statistical knowledge. This section will give a précis of some commonly used measures, their flaws, and their application to risk quantification specifically, but won't include a deep-dive into the underlying statistics. First, let's have a look at some options for a risk measure.

12.5.1 Risk Measures

Option 1: Variance (alternatives: standard deviation, or semi-variance)

Although it's commonly used in high-school math lessons and in the theory of statistical distributions, in practice variance calculations aren't particularly useful. Semi-variance, which considers only the variance on one side of the mean, tries to capture the idea of downside risk—that we are only interested in the worst scenarios, not the very best scenarios—but all forms of variance are flawed for capital allocation methods. This is because these measures assume a symmetrical distribution (which is why we can use the mean ± 2 standard deviations to calculate confidence intervals, rather than concern ourselves with two different deviations for above and below the mean). But in insurance, the loss distributions are rarely symmetrical, so this measure gives distorted results. The other problem with variance is that it is a measure of volatility, which is not directly what we wish to measure: we wish to measure financial loss. A volatile profit is better than a certain loss, but using the variance as a metric would suggest otherwise.

Option 2: Expected Policyholder Default

A really intuitive risk measure, this gives the answer to "Given I experience a loss that is more than the company's emergency capital, what is the further expected loss?". The name comes from the fact that if a company has used up all its buffer, the policyholder will not receive their full payment amount, and so policyholders will experience a loss. For fans of conditional probabilities, the full formula can be described as follows, where c is the level of company capital. It can be read as: the expected policyholder default is the probability that the loss is greater than the level of held capital, multiplied by the expected loss over that amount, given that the loss is over that amount.

$$\text{Expected Policyholder Default} = \text{Probability}(X \geq c) \times E[(X - c)|(X \geq c)]$$

This measure is very intuitive, which is a definite advantage when trying to communicate risk to non-technical directors or board members. But this measure doesn't translate the advantages of **pooling** into statistics. As we saw in example 2.1, if the insurer held \$995 of capital per car, it would fail with probability of 1/10.

Example 12.5 Using the numbers from Example 2.1, the expected policyholder default is $= 0.1 \times E[\$5|Loss \geq \$995]$. Given that the loss can only be equal to $1000 (or there is no loss), the expectation of the loss over the threshold is therefore $5. So, the expected policyholder default is $0.50.

Example 12.6 In the 10-car example from chapter 2 and assuming the amount of capital held is $9950, the expected policyholder default is $= 0.0000000001 \times E[\$50|Loss \geq \$9950]$. Again, the only possible loss values in this area of magnitude are $9000 or $10,000, so if the overall loss is over $9950, it must be $10,000. So, the expectation expression simplifies to $50, and the expected policyholder default is $0.000000005.

Where this measure falls over is when there are multiple risks, because this measure doesn't scale up in a linear way. For large portfolios, it does not give results consistent with those from other measures, and specifically it gives lower estimates for the capital required.[14]

Option 3: Value at Risk (VaR)[15]

This measure gives an amount of capital that we should hold, given an acceptable probability of financial ruin. Therefore, it measures the downside risk only (remember, best-case scenarios should be celebrated, not calculated!). It can be expressed statistically, with c denoting the capital and y is the acceptable probability of ruin, as follows:

$$\text{Value at Risk}_{y\%} = c, \text{ such that Probability}(Loss \geq c) = y$$

Example 12.7 Using the 10-car example, the Value at Risk with a 0.05% probability can be found by calculating the cumulative probability of the claims cost. We found that 4 claims ($4000 of claims) had cumulative probability of 0.9984, but that 3 claims ($3000) had a cumulative probability of 0.9872. Ideally, we would take the claims cost where the cumulative probability equals 0.995, so we could either interpolate or take the higher capital value. Interpolation is suitable when there is a lack of data for a continuous distribution, but here we have a 'lumpy' (discrete) claims distribution so interpolation is not appropriate. Therefore, the Value at Risk for 0.05% is $4000.

[14] The Casualty Actuarial Society showed this for large auto risks in 2007 in their Spring 2007 Forum. Consistent Measurement of Property-Casualty Risk-Based Capital Adequacy (casact.org). https://www.casact.org/sites/default/files/database/forum_07spforum_07sp113.pdf

[15] Pronounced 'v-ahr' (to rhyme with 'car'), not V-A-R.

Value at Risk is the measure used in Solvency II to calculate the overall SCR (specifically, with an acceptable probability of ruin of 0.05%, denoted $\text{VaR}_{0.05\%}$). However, statistically it is not ideal for a few reasons. Firstly, it doesn't make any reference to what happens after the selected loss threshold has been reached. For the business, there might be a difference in outcome depending on whether the capital was insufficient by \$5,000 or \$5 million. It could be argued that this doesn't matter: if you are insolvent then that's that, whether you are by an inch or a mile, but using VaR takes that information away from decision-makers. The good news is that it is widely understood, not just in insurance, but also in banking and wider financial sectors. Secondly, and more problematically for insurance, VaR doesn't recognize the benefits of pooling. A bigger portfolio increases the VaR, even if the growth would diversify the business and reduce the risk. Because of this, using VaR as an internal decision-making tool to assess capital efficacy is not a good idea, as it never suggests that the business should be grown!

To combat these weaknesses, an adjusted version of VaR—called Tail Value at Risk (shortened to Tail VaR or T-VaR) is more commonly used in insurance (apart from for SII calculations where VaR is demanded!). Tail VaR is slightly bigger than VaR as, like the expected policyholder default, it considers the expected loss given that the loss is greater than the capital threshold. It is calculated using the VaR like so:

$$\text{Tail VaR}_{y\%} = E[\text{Loss}|\text{Loss} > m] \text{where } m = \text{VaR}_{y\%}$$

Example 12.8 In the 10-car example, the Value at Risk for 0.05% was \$4000. What is the corresponding T-VaR?

First we need to calculate the conditional probabilities of 5, 6, 7, 8, 9, and 10 losses, given that one of these outcomes occurs. To do this, we calculate our conditional probabilities, $\frac{\text{Probability j claims}}{\sum_{i=5}^{10} \text{Probability of i claims}}$ for $j = 5, 6, 7, 8, 9,$ and 10. Now, plug them into the formula, using a good number of decimal places to try and capture the least likely outcomes:

$$\begin{aligned}\text{Tail VaR}_{0.05\%} = E[\text{Loss}|\text{Loss} > \$4000] &= \$5000 \times 0.910148 + \$6000 \times 0.084273 \\ &+ \$7000 \times 0.005351 + \$8000 \times 0.000223 \\ &+ \$9000 \times 0.000006 + \$10,000 \times 0.000,000,06 = \$5096\end{aligned}$$

Exercise 12.1 Sense check this calculation: is the calculated T-VaR higher or lower than the equivalent VaR? Is that what you'd expect?

There are many other risk measures available. The most important thing to note is that they are all arbitrary, and all have strengths and weaknesses. For simple distributions, like the 10-car example, they are relatively easy to calculate. For continuous distributions you may find yourself embroiled in tricky integrals!

12.5.2 Risk Allocation Methods

Risk measure selected; we move onto allocation methods. For now, we are specifically allocating the risk, but under a risk-based approach such as Solvency II, this can also be seen directly as capital allocation methods. Again, there are many choices here, some with exciting names such as game theory, option pricing framework, and co-measures. However, we will look at just two possibilities: the proportional method and the Shapley Method.

Option A: The Proportional Method

The proportional method is delightfully simple. Once a risk measure has been chosen, it is used to calculate the risk for each line of business, and then to calculate the risk of the overall business (so $n + 1$ calculations in total, where n is the number of lines of business)—don't be alarmed if the sum of the individual lines of business' risk measures doesn't equal the overall company-wide measure at this stage. Next, we scale the individual line of business' risk up or down so they sum to the overall company's risk measure, but stay in the same proportions to each other.

Example 12.9 Using T-VaR, a reinsurer determines its overall company capital requirement as \$100m. It has three lines of business, which have individual T-VaRs as follows: A has a T-VaR of \$55m, B \$30m, and C \$25m.

A should be allocated $\frac{\$55m}{\$55m+\$30m+\$25m} = 50\%$ of the \$100m, so \$50m. B and C should be allocated 27.2 and 22.7% respectively: \$27.2m and \$22.7m (to 1 decimal place).

This method is very simple, not just to calculate but also to communicate to non-technical board members. However, it doesn't make any allowances for lines of business that are correlated or diversified, which means that generally this is too simple and gives too high a capital requirement for well-diversified portfolios.

Option B: The Shapley Method

The Shapley method has an advantage over the proportional method because it takes into account correlation between different products. However, it requires more calculation than the previous method.

The Shapley method considers the effect on the risk measure of adding in each line of business one by one. By considering this marginal effect for all possible combinations and averaging across the combinations, the line of business is allocated capital. This means that for every combination of lines of business, we must use our risk measure to assess the marginal risk.

Lloyd Shapley (1923–2016)

Lloyd Shapley is most famously known as an American mathematician and game theorist, but his Nobel Prize for Economics in 2012 was awarded for his work on allocation methods.

Lloyd Shapley's first award was the Bronze Star medal, given to him during the Second World War. Drafted while a mathematician at Harvard in 1943, he served in China, where he broke the Soviet weather code, revealing crucial information for aircraft sorties.

Example 12.10 An insurer writes three lines of business: public liability, employer's liability, and home insurance. What is the Shapley allocation for public liability insurance?

To find the allocation for public liability, we need to add it in to every possible combination, in every possible order. First, we imagine we write nothing, and therefore our VaR = \$0. We add in public liability. Now the VaR is, say, \$10m.

Next, we imagine that we already write employer's liability insurance, and our VaR is \$8m. We add in public liability. Because of the high positive correlation between the two lines of business, our VaR for this new two-line portfolio is \$20m. To state the obvious, adding public liability increased our VaR by \$12m.

Now we imagine that we already write home insurance and our VaR is \$4m. We add in public liability, and due to low correlation between the lines, our VaR becomes \$12.5m; an increase of \$8.5m.

Finally, we imagine that we already write both home and employer's liability, and have a VaR of \$11m. Adding public liability gives a VaR of \$23m, an increase of \$12m.

The average of these four marginal contributions, and therefore the capital allocation for public liability is (\$10m + \$12m + \$8.5m + \$12m) /4 = **\$10.625m.**

This method draws on the game theory approach to collaboration and coalitions. It is computationally intensive, especially for insurers with lots of products! As well as this practical drawback, the Shapley method also gives different allocations if portfolios are split in a different way. For example, if home insurance was split into buildings and contents insurance, or into different geographic regions of home insurance. The Shapley method penalizes high correlation very harshly, and equally gives generous diversification benefits, so results should be taken with a pinch of salt.

In this section we have performed some capital modeling calculations. In real life, assessing the correlations between lines of business and geographic regions takes considerable effort. Assessing **accumulations** of risk is particularly important for property lines of business that are increasingly impacted by climate change related natural catastrophes. It's also an interesting area to work in if you are interested in investment, as there is considerable overlap between modeling capital required and deciding which assets to keep that capital in, while adhering to various regulatory requirements.

The calculations we performed in this section were simple, but at least the math was correct! The next section shows what happens when capital modeling goes wrong...

12.6 Inadequate Capital: A Warning from History

In Chapter 9, we introduced the insurance cycle. We noted how at certain times there would an influx of new entrants to the market, and then capacity would reduce later in the cycle. Clearly, the market cannot sustain an infinite number of companies all vying for business. Some might choose to reduce their capacity for whatever reason, but many will become insolvent. In this case, business failure means company is not able, or will not likely be able, to pay their future claims. I include the latter because historically, struggling insurers have been bailed out before their total collapse—not by governments but by other, larger insurers.

There are lots of reasons a large insurer would want to do this. Firstly, recognize that the cost of collapse of an insurer is not only borne by that company. The effects ripple into the wider market and society. For example, a loss of trust of insurers by the public would mean that existing insurers may lose business as a result of an insurer failure. The effects might be borne by wider society, especially if the failed company insured small or medium-sized businesses. If claims to those policyholders were not paid, those companies (in non-insurance industries) could become bankrupt as a result of the insurer failure. For homeowner insurance, governments might have to pick up the costs for the public if an insurer failed after a catastrophe. The cost to the public is not just theoretical: a report by the Subcommittee on Oversight and

Investigations of the US House of Representatives estimated that in 1990, insurer failures cost the US public $5 billion.

A large insurer buying out a smaller, failing company is one way to limit damage,[16] but there are less altruistic reasons as well: the insurer taking over might be taking on new lines of business, new regions, or just increasing its market share in existing segments. If you are well-capitalized and wish to grow quickly, buying another company is a relatively cheap and easy way to do so!

It is not necessarily a lack of capital that causes insurer insolvencies. According to one study by the Institute and Faculty of Actuaries, 1 in 3 insurer failures were due to insufficient capital,[17] and a further 9% due to over-stated asset valuations, another side of the capital requirement coin. Accurately modeling catastrophe losses, which make up another 8% of failures, has been a priority for capital modeling regulators since this data was analyzed, and you may recognize 'Affiliate Collapse' as another name for group risk, which Solvency II specifically tries to address. Fraud, although not necessarily a 'capital' offense, is the target of the qualitative and disclosure pillars of SII: by carefully vetting professionals and requiring extensive reporting and disclosure, as well as clear and efficient internal processes, fraud is (hopefully) more difficult to get away with (Fig. 12.1).

But enough of generalities: let's look at some specific examples of what happens when capital modeling goes wrong.

12.6.1 Transit: 'The Titanic of Insurance Insolvencies'

Founded in St Louis in 1945, Transit quickly became a big player in (unsurprisingly) transit risks, especially public transportation risks. These commercial policies were generally profitable but decreasingly so, and in 1964, the company was bought by Beneficial Standard, a life insurer based in Los Angeles. This buy-out led to a convoluted company structure, with multiple directors, a president, and a CEO. New management encouraged Transit to diversify into non-transportation coverages. In 1979, just as the insurance market was softening (becoming less profitable), Transit created a Risk Management division, to keep track of all the new risks. Transit underwriters had no experience of non-transportation risks, and so 17 different

[16] The Institute and Faculty of Actuaries' Insurance Company Failure Working Party notes that in the 1960s and 70 s, in the UK there was an unwritten agreement that large insurers would take it in turn to bail out smaller companies. While this 'charitable' practice was discontinued in the late 70 s, the practice of large, well-capitalized insurers buying smaller companies in financial difficulty (whether that is widely known information or not) continues to this day.

[17] Microsoft Word - Massey.doc (actuaries.org.uk). https://www.actuaries.org.uk/system/files/documents/pdf/insurance-company-failure.pdf.

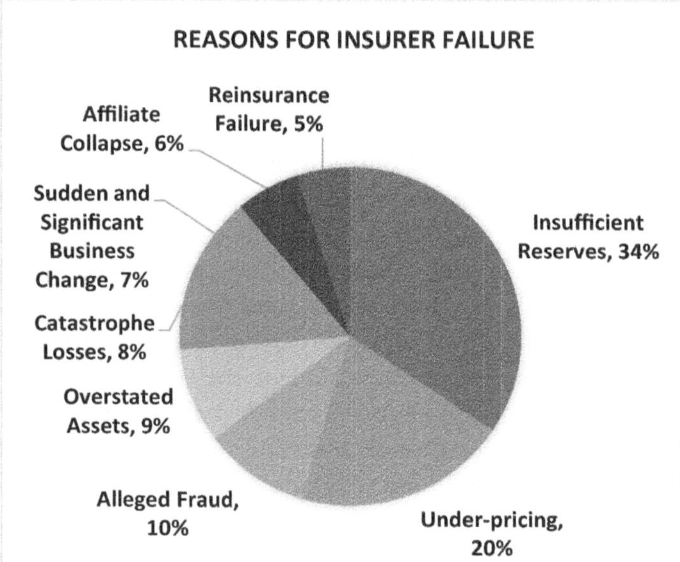

Fig. 12.1 Reasons for Insurer Failure. Based on 640 US insurer failures between 1969–1998 From the Insurance Company Failure Report, Institute of Actuaries' Working Party; Microsoft Word - Massey.doc (actuaries.org.uk)

agents were used to place business to achieve the company's dream of rapid growth. These agents would find and write business on Transit's behalf. This was considered low risk by Transit's directors, as the idea was that Transit would be used only as a fronting company—that the agents would reinsure all their risks, while paying a small amount to Transit for use of their brand and capital backing. This is akin to letting 17 strangers use your car. It might be fine; indeed, they might even get it cleaned and waxed for you, but... < spoiler alert! > it might lead to you being the subject of a police investigation.

Transit's agents were prolific, and expansion was rapid. Transit was used to writing 500–800 transportation policies a year and, as wasn't unusual in 1980, processed all policies and claims manually. Their reinsurer of choice was a single, well-known provider, based in the USA. The business plan for 1981 was ambitious, aiming for 37,000 policies to be written that year alone. Unsurprisingly, manual records of all of these policies were difficult to keep and store, and so 256 locations across the USA had to be used as storage.[18] Reinsurance was now up to the agents to source and, as they were unconcerned with the level of risk that Transit was holding, they placed

[18] Nowadays we can hold 26,000 documents in the palm of our hands, but after the company went into administration, the administrators recovered only 26,000 of the 37,000 policy documents for 1981. These documents alone were stacked 8ft high and covered 20,000 square feet!

their business with the cheapest reinsurer. This led to risks not being fully reinsured, and to Transit having reinsurance policies with over 1,400 reinsurers! These were mostly offshore or foreign companies and, in their haste to reinsure then write more business, agents had failed to ensure that all the reinsurers were even authorized to cover risks based in the USA. Received premiums skyrocketed from $93m in 1979 to $227m in 1984. As expected, their reinsurance costs increased too, from $23m to $217m, showing how much reliance Transit were placing on the reinsurers. And Transit's capital? That remained about constant over the same period, hovering around $44m.

Manually dealing with 37,000 policies a year was every bit as difficult as it sounds, and the newly acquired computer system was not ready for the task. There were no controls on the agents, no underwriting guidance or rules, and no limits on the lines of business that could be written. Even creating a record of total premium written was a near-impossible task, let alone the process of aligning claims with policies, or policies with premiums. Chaos reigned, both in reality and in the financial reports.

Although now based in California, Transit still came under the Missouri regulatory authority, who performed an audit every 3 years. In September 1984, work began on the audit for the years 1980–1983. The brave souls from the Missouri Division of Insurance waded through the reports and, finding them to be materially incorrect, ordered the company to stop writing business in April 1985. It was too late: the company was liquidated in December.

The whole debacle was an exercise in poor risk management: from the lack of underwriting rules, measuring success only in terms of policies sold, and ultimately, not understanding the risks that were being held. But there was also a failure on the part of the regulators: triennial reporting allows for big changes in the intervening period. Thankfully, much has changed since 1985—not least in part due to Transit's failure. The risks that Transit had written included large amounts of **long-tailed** liability risks, plus almost every landfill site, lorry, tobacco company, and breast implant in the USA (to give some sense of the variety of risks taken!). This left the company with a huge amount of environmental clean-up, asbestos, and medical negligence claims, to name just a few of the big claim costs.

But all those risks would have been manageable or at least, partially managed, if the capital modeling had included some chance of reinsurer default. As it was, many of the reinsurers themselves became insolvent, or became difficult to get hold of. It took until 2001 for the administrators to receive $930m owed from reinsurers, although some was inevitably written off. Total claims paid to date by Transit for the five-year period of chaos were up to $640m (in 2001 terms), with policyholders receiving around 70% of what was owed to them, with the rest written off.

There was undoubtedly some illegal activity by agents, but also by the owners of Beneficial Standard, who in 1984 separated their good assets from Transit, with contracts that meant that the other assets and the owners' personal wealth could not be used to pay Transit's liabilities. That year, the two owners of Beneficial Standard cashed in their stock options for $647,000 and $1.3m; a huge fortune, especially in 1984 terms. These two, and those like them, are referred to as 'pirates and dolts' in the 1990 US House of Representatives' paper, *Failed Promises: Insurance Company Insolvencies*, which also called Transit the 'Titanic of Insurance Insolvencies'.

It takes more than a bit of archaic name-calling to fix a broken system, and regulation around capital requirements is now much more comprehensive. I highly recommend reading the *Failed Promises* paper, or at least the scathing introductory statement. For now, let's continue to Chapter 13.

13

Reinsurance

Chapter Objectives:

- Be familiar with reinsurance products, and why each might be bought
- Understand the London Market Excess of Loss (LMX) spiral of the 1980s, why it mattered and what stops it from happening again
- Understand what Reinsurance To Close is, and why it is used
- Be familiar with different financial reinsurance products, and the benefits of each product
- Be familiar with how risk can be shared outside of insurance markets, and how risk can be transferred to government and financial sectors

I experience each moment like baklava: rich in this layer, and this layer, and this layer.

—Ram Dass, American yogi and psychologist

As we all know, sharing is caring. From the moment we can physically hold an item, we are told, reminded, and sometimes scolded that we need to share with others. Even now, you might be reminded to 'like and share' content on social media, or feel judgment from a 'sharing bag' of M&Ms that you eat by yourself. As a younger sibling, I have always been aware of the benefits of sharing: my sister always had more interesting toys and clothes than I did, so in order to benefit from her excellent taste, I would often invoke the rule of

© The Author(s), under exclusive license to Springer Nature Switzerland AG 2022
C. Townsend, *A Risky Business*,
https://doi.org/10.1007/978-3-031-11673-5_13

sharing (with varying degrees of success). Sharing has wider benefits too. It is like a social glue: we share cake on birthdays, we share ideas in the office, and we share our favorite TV shows as a bonding exercise.

As we've already seen, insurance started as a way to share risk: ship owners would huddle together in Lloyd's coffee shop and, if there was a shipwreck, all would share in the financial loss. Nowadays, policyholders often wish to transfer the majority of the risk to the insurer rather than share it as equals, although most insurance policies require that policyholders retain a small bit of the risk through the use of policy **excesses** or deductibles. Insurers take on risk, but even they have a limit as to how much they want to take on. They may have inadvertently written too many policies, be very exposed to potential large losses, or perhaps they were unable to find enough similar risks to create a pool in the way they hoped. Is there someone who would insure the insurer?

A **reinsurance** company does exactly that. Some reinsurers are also insurers themselves: they provide policies to the public, and also provide insurance to other insurance companies. They operate in a similar way to insurers, are bound by the same capital requirements, and their actuaries will be concerned with the same activities: pricing, **reserving** and capital modeling. Similar to multinational insurers, reinsurers are big companies needing lots of capital so they can back very big risks. The main difference is the way that the products work, so let's take a look at what the products are.

13.1 A Problem Shared: An Introduction to Reinsurance

How do we share? It might sound like a trite question that you'd ask a naughty toddler, but actually reinsurance products share risk in a surprising variety of ways. The most common split of reinsurance products is between proportional and non-proportional products.

13.1.1 Proportional Reinsurance

Proportional reinsurance shares risk in a way that is, well, proportional. The insurer and reinsurer agree a percentage split, and both premiums and claim costs are split in that way. Proportional insurance is particularly good for retaining exposure to a wide range of risks, thus diversifying your portfolio and reducing variation in your losses in a way that is similar to **pooling**, but takes a reduced level of exposure across a wide range of risks.

The simplest proportional reinsurance product is called a **quota share**. A quota share policy covers many risks, and the risks are all shared in the same percentage. The percentage that the reinsurer will be responsible for is helpfully included in any references to the policy. As an aside, reinsurance policies that cover multiple risks are usually referred to not as policies but as treaties, and we'll use that term frequently in this chapter.

Example 13.1: An insurer buys a 30% quota share treaty. The insurer writes $50m of business covered by the treaty, and experiences claim costs of $32m. What is the monetary loss and **loss ratio** for the insurer? What is the monetary loss and loss ratio for the reinsurer?

The insurer will retain 70% of the premiums and 70% of the claim costs. In monetary terms, that is 70% × $50m = $35m of premium and 70% × $32m = $22.4**m** in claim costs. As a loss ratio, that is $22.4m ÷ $32m = 64%**.**

Now for the reinsurer: they will receive 30% of the premium ($15m) and pay 30% of the claims, totaling **$9.6m**. Their loss ratio is $9.6m ÷ $15m = 64%.

Quota shares are administratively simple, which is a big advantage. However, simply sharing profit or loss means that the insurer could still experience a really big financial loss. On the plus side, it directly reduces capital requirements (in particular the **solvency ratio**), because it reduces the net premiums, as some premium is passed straight to the reinsurer. The solvency ratio is calculated as the company's capital[1] divided by its premiums net of reinsurance.[2] So, it is relatively easy to be in excess of a prescribed solvency ratio by using a quota share treaty to reduce net premiums. This immediately relieves the business of holding a higher absolute amount of capital.

But what happens if the insurer wants to vary the percentage of the risk that is shared? Perhaps some of the risks are riskier or add to an **accumulation** of risk and so they want to pass more of those onto a reinsurer, but wish to keep a higher percentage of some of the risks that give them more diversification or that they have more experience with. They could draw up several quota share agreements with a reinsurer, although this would create a lot of duplicated administration. Instead, it could make one agreement but specify the sharing percentages for each risk (or set of risks). This is another form of proportional reinsurance, called surplus reinsurance. The reinsurer will usually put some limits on the kinds of percentages it would accept and the kinds of risks that can be covered by the treaty, but it is more a flexible

[1] Technically the 'free capital'—the capital not earmarked for paying claims or other costs.

[2] Premiums net of reinsurance means the premiums that the insurer doesn't give to the reinsurer.

contract than a quota share, saving the insurer from creating multiple quota share contracts.

Although surplus insurance is proportional as both the premiums and claims are split in a given percentage, it is usually described by two parameters: the amount of the risk (in monetary terms) that the insurer keeps, and the multiple of that amount that the reinsurer is willing to take on. This multiple is referred to as a line. The amount of risk is usually given as a sum insured or an **expected maximum loss.**

Example 13.2: An insurer wants to reinsure their property portfolio. They traditionally write residential property, but want to branch out into small commercial properties, and have started to write this line of business. However, they don't wish to be exposed to an expected maximum loss of more than $100,000 for any one portfolio.

They agree a surplus treaty with a reinsurer. For property portfolios covered by the treaty, the reinsurer is willing to accept up to 8 lines with a retention of $25,000.

The insurer could then decide to share the risks in the following way:

- Residential property portfolio—insurer wishes to retain an expected maximum loss of $100,000. The total EML for this portfolio is $125,000, so it must cede one line of $25,000. Therefore, the insurer retains 80% of the risk.
- Warehouses and manufacturing premises portfolio—insurer wishes to retain an EML of $50,000 but the portfolio EML is $200,000. It therefore cedes 6 lines of $25,000 so the insurer retains 25% of the risk.
- Business premises excluding warehouses and manufacturing portfolio—insurer to retain an EML of $100,000 but the portfolio EML is $200,000. It cedes 4 lines of $25,000. The insurer retains 50% of the risk.

As this example shows, proportional reinsurance can be a useful tool to gain experience in new products. When a new product is launched by a company, it is unlikely they have sufficient data to price the product accurately. Sharing the risk means that not only are they transferring some of that risk, but the reinsurer (who will experience the same loss ratio as the insurer) also has an interest in making sure the product is profitable. It therefore might be willing to provide expertise and insight, and share some historic claims data from a similar product to inform the price. So, reinsurance may not just transfer the risk of underpricing but reduce the risk of mis-pricing occurring altogether.

13.1.2 Excess of Loss Reinsurance (XL)

Excess of loss reinsurance shares the risk in a non-proportional way. Instead of sharing premiums and losses using percentages, with XL the reinsurer

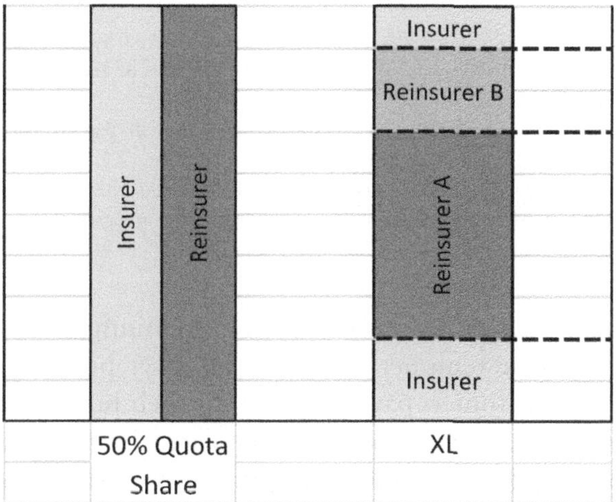

Fig. 13.1 Risk sharing in quota share vs XL reinsurance. The different colors represent different parties paying the claim cost

agrees to cover the amount of a loss that is above a given level: it covers the loss in excess of a certain threshold. This is usually not unlimited; instead, the reinsurer will usually cover the cost of a claim between two given points. If a claim exceeds both points, the remainder of the cost will be covered by the insurer or, more commonly, the insurer will buy another layer of XL from another provider until they are satisfied with the amount of risk they are holding. When stacked like this, insurers can all but guarantee a maximum loss,[3] which removes a lot of uncertainty and variability from their financial results. These layers are defined by the width of the layer in excess of the lower point, e.g., $10 m in excess of $2 m means that the reinsurer will pay for the portion of claims that fall between $2 m and $12 m (Fig. 13.1).

Example 13.3: Insurer A wants to cap the losses from a certain risk to $25 million. It has XL reinsurance from reinsurer B: $10m in excess of $25m. This however only covers the portion of the claim that is more than $25m but less than $35 m. If a claim was over $35m, this reinsurance wouldn't be sufficient as the insurer would be liable for the claim cost above $35m, as well as the amount under $25m.

So, the insurer buys another layer from reinsurer C, this time $30 m in excess of $35 m. Now the insurer will only pay $25m for any claims up to $65m. If there was a small possibility of a claim over $65m, it may wish to buy another XL layer.

[3] It is not certain of course, because reinsurers can default on their treaties, although the reputational damage of this means that it would only happen if absolutely necessary.

Example 13.4: An insurer experiences a claim of $10m, covered by their reinsurance treaty.

Under a 50% quota share treaty, the insurer will pay 50% of the claim, so the reinsurer will pay $5m and the insurer will pay $5m.

Under a $5m in excess of $2m XL treaty, the insurer will pay the first $2m and the reinsurer will pay $5m. However, the insurer could buy another XL treaty, this time for $2m in excess of $7m, which would pay the next $2m. This means the insurer would pay $3m, reinsurer A would pay $5m and reinsurer B would pay $2m, as per Fig. 13.1.

Excess of loss doesn't take a share of the premium written. Instead, XL reinsurers receive their remuneration through a set premium upfront, in a similar way to how customers pay an insurer for their home or auto insurance. There's more on how that premium is calculated by reinsurers in the next section, but for now, let's explore some common variations on XL. There are three main variations, where the variation comes only from the type of risk being covered.

- Risk XL covers an individual large loss occurring on a portfolio. If a single claim from a portfolio exceeds the reinsurance threshold, the reinsurer will cover the part of the claim that falls in their layer. For example, an insurer might have a risk XL treaty covering their property portfolio for $8m in excess of $2m. If a large fire blazed through a property, the owner might submit a claim for $3m. This is over the $2m threshold, so the reinsurer would pay the amount that falls in the layer (here, $3m - $2m = $1m) and the insurer pays the rest. Risk XL usually has a limit on how many times it can be used in a year for a specific portfolio, and will only pay up to a certain number of claims in a year before it requires an additional premium from the insurer.

- Aggregate XL, unsurprisingly, covers aggregated losses for a given portfolio. This is particularly good for lines of business where although a single claim is very unlikely to hit a reinsurance threshold by itself, the risk of a series of mid-size risks is too high for the insurer to keep. The claims could be aggregated by a claim cause, such as asbestos-related claims, or it could be by a line of business or product. For example, an insurer might have an aggregate XL treaty for fire claims on its property portfolio for $8m in excess of $2m. It experiences three fire claims: one for $3m, one for $1.8m, and another for $1.5m. The insurer reports that it has an aggregated fire loss of $6.3m. The reinsurer pays $4.3m and the insurer pays $2m. Under a risk XL treaty with the same layer, the insurer would have had to pay \$5.3m ($1m + 1.8m + 1.5m$) themselves.

- Catastrophe XL is a form of aggregate XL, where the aggregation is based on the claim cause being due to a single catastrophic event. Most commonly, this is a natural catastrophe such as a hurricane or flood. The difficulty with catastrophe XL is that the reinsurer is only paying claims over the reinsurance threshold in respect to one particular event, for example a storm. But when did that storm start and finish? It isn't uncommon for the same storm to die down a bit and then pick up, but it also isn't uncommon for a second, separate storm event to follow quickly after the first. These two events would be treated differently under a catastrophe XL treaty: the treaty only covers claims relating to a single event, so the first scenario should be covered but the second storm would count as a new event and a new reinsurance claim. In reality, these two scenarios may be very similar. To avoid arguing over such points, catastrophe XL treaties usually have an hours' clause. This clause states that the reinsurer will cover all catastrophe claims occurring within a certain period—usually 72 hours. Under this clause, a storm that drops and then picks up again within 72 hours will be covered as one claim, as will a storm that is followed within 72 hours by another storm. The insurer can choose when the 72 hours begins, so they will wait for the claims to be reported and then pick the 72-hour period to maximize the amount the reinsurer pays. Any claims related to the event that don't fall within the 72-hour period are not covered at all by the reinsurer.

We've already seen that **quota share** reinsurance is priced by sharing the premium from the business covered, in exactly the same split as the claims. If you are looking to impress your non-actuarial friends, tell them that you can calculate quota share reinsurance premiums in your head[4]!

XL reinsurance is priced using original loss curves, as described in Sect. 8.5. However, in addition to this, reinsurers look to gain extra information from their customers such as their approach to pricing, underwriting, and claims management. As a result, reinsurance pricing is much more of an art than a science. Insurers and reinsurers will cultivate relationships in order to enhance the proposition: insurers will be looking for reinsurers to lower their premiums and provide exactly the cover they want, and reinsurers want to encourage insurers who have robust underwriting and pricing and have strong claims management to manage costs effectively.

[4] The author does not guarantee that this will actually impress your friends...

Claims Management.
A claims department who processes and manages claims quickly and efficiently have a huge positive impact on lots of other business functions—and the bottom line!

- With fewer delays to repairing the damage, costs associated with interruption cover (such as business interruption claims or hire cars) are minimized.
- Fraudulent claims are spotted and not paid.
- Fewer open claims means that there is less uncertainty in the reserves.
- Reducing operational risk will be reflected in lower capital requirements.
- Reinsurance will likely be cheaper if claims costs are persistently lower.

13.2 The LMX Spiral

When we think of invention and innovation, financial products aren't usually the first thing to spring to mind. In the early 1980s, most people might have been more interested in the first mobile phone (1983, a large Motorola that weighed almost 800g), the new Apple Mac (1984), or CD-ROMs (1985), but in Lloyd's of London, another new product had great promise. Although reinsurance was well-established, this was usually for specific, large risks. But a new invention, a reinsurance policy that covered a portfolio of risks, was nonetheless exciting (and in hindsight, has had a lot more longevity than CDs). These policies were similar to treaty reinsurance, but there was no requirement for the risks grouped together to be similar in any way. They could mop up any risk that the insurer no longer wanted to hold, and so were known as generals. Instead of confining reinsurance to large, specific risks or risks that were suitably homogeneous, now there was almost no end to how much sharing of risk could be achieved!

Like in the early days of any new invention, companies rushed to jump on the bandwagon. Existing reinsurers (and insurers) started writing these generals, and some reinsurers were explicitly set up to write them. Within the framework of Lloyd's, it was relatively easy to set up new **syndicates** backed by third-party capital. These new syndicates had no real business of their own and relied on reinsuring business already within Lloyd's, writing **excess of loss** policies for an insurer's entire portfolio, taking an increasingly thin slice of business that already existed.

Pricing generals was not given too much thought: techniques such as Original Loss Curves sufficed. Of course, with more insurers and the same amount of insurance business, competition was fierce. But in this soft market, if you were an insurer or reinsurer holding some risk, it was easy and cheap to transfer all but a slither of risk away to a willing reinsurer. This meant that reinsurance layers became increasingly narrow. Commission (and spirits) ran high in the newly built Lloyd's building.

This XL reinsurance in the London Market was not a particularly efficient form of risk transfer, but it wasn't a spiral. The spiral element occurred when the same risk was reinsured a second time by the same reinsurer. This was largely accidental: no one wants to be confronted with something they thought they disposed of ages ago! But, in all the passing and parceling risk, some reinsurers accidentally managed to reinsure themselves[5]!

Example 13.5: A Reinsurance Non-Spiral.

Insurer A writes a policy. They take out reinsurance for any claim above £25 million from reinsurer B. A is happy that they have effectively capped their exposure to loss. Reinsurer B takes out some reinsurance themselves: For any claim to them above £10m, they have reinsurance from reinsurer C that pays the excess.

If A receives a claim of £50m, the split of the cost will be as follows: A will pay the first £25m, B pays the next £10m, and C pays the remaining £15m.

Example continued: A Reinsurance Spiral Begins...

Insurer C feels a little uncomfortable at the prospect of a large loss, so it buys a general from reinsurer D: D will pay the excess of any loss over $5m to their portfolio. Reinsurer D decides to reinsure the risk in their portfolio too: insurer A (who has decided that reinsurance might actually be quite lucrative...) will take on any excess of D's portfolio losses over £12m.

Insurer A is informed of a claim costing £100 million. Fortunately, it has reinsurance from B, so A assumes they will only pay the first £25 million and B will pay the rest.

A must pay the first £25 million, then B pays their layer's cost of £10m. C pays the next £5m, and D pays their £12m. Now, D passes the cost to their reinsurer, who must pay the remaining £48 million. Unfortunately, that is (re)insurer A, who is stung with a total claim cost of £73 million. So much for reinsurance!

The London Market Excess of Loss (LMX) spiral was not a neat helix; mapping all the relationships between insurers and reinsurers would look more like a plate of spaghetti than any geometric shape! To imagine quite how complicated this could become, imagine watching a game of 'pass the parcel' in reverse. It would start with the child acquiring the prize, then wrapping it up and passing it to the next child along. This child would add another layer

[5] Some underwriters did manage to write spiral reinsurance profitably, although they are by far the exception, rather than the rule. They managed by following a strict underwriting plan of what they would and wouldn't write, careful portfolio monitoring and more accurate pricing.

of gift-wrap and perhaps a candy or two, then pass it to the next child. This would be repeated until the music stops, when a child has no idea what is in the parcel they now hold. While reinsurance was largely confined to individual risks, it was much easier to track the initial insurer. After packaging and repackaging risks, the transfer was almost totally opaque.

The other problem with the opacity of this process was that it was impossible to track not only whose risk, but what risk was being transferred. As we saw in Chapter 8, understanding the underlying risk, and the associated claim **frequency**, **severity** and distribution of claim costs, is fundamental to accurate pricing. Without knowing what risk was being transferred, it was not at all clear to the **underwriters** whether a very large claim was very likely or not. As such, the highest layers were unwittingly underpriced, but it also meant that reinsurers at the highest layers weren't expecting to get hit very often (if at all). Therefore, reinsurers with the highest layers often didn't have much reinsurance of their own. This led to two capital problems: firstly, reinsurers that didn't understand the risks they were holding didn't properly model extreme events and the capital required in such situations. And, when the risk itself was understood, the number of reinsurers the risk had passed through and the associated layers for each of them was poorly understood. Therefore, these spiral reinsurers were hugely under-capitalized. Secondly, even if they had been able to model accurately, the ultra-low premiums charged meant that they didn't have much capital to hold in reserve anyway. These upper layers were usually unlimited liability, so they were on the hook for any claim cost passed to them.

It is hard to imagine that underwriters thought this was a good idea, but most genuinely did. It was conventional wisdom that reinsurance was about sharing risk, and writing generals and layers of reinsurance was seen as a quick way to gain exposure to a variety of different risks. However, in this model, it wasn't so much sharing risk, but passing risk to the least informed.

I'm sure by now you can see where this was leading, even if the insurance industry at the time did not. It wasn't even that insurance spirals were an entirely new concept: after Hurricane Betsy in 1965 (a deadly hurricane that breached the flood levees in New Orleans[6]), those with inside information retained the good reinsurance risks and passed the bad ones to the less informed, causing a series of property insurers to collapse. Unfortunately, in 1988 there wasn't just one bad event, it was the beginning of a series of catastrophic losses.

[6] A very similar circumstance and, tragically, a similar outcome to Hurricane Katrina exactly forty years later.

It started on July 6th 1988, when a mechanical failure created a series of explosions on the Piper Alpha oil rig, about 180 km off the coast of Scotland. 163 crew members died; almost 75% of those on board. As well as the loss of life, the damage to the oil rig, and the business interruption costs, insurers also had to cover the immediate clean-up costs. The summer also saw a slightly above average US hurricane season and the year ended with another major oil spill, this time off the coast of Canada.

To get an idea of how much each risk was passed around, we can calculate the magnification of a claim. Insurers are obliged to report their gross losses, that is, their losses without any deductions for reinsurers' contributions, so by comparing the gross claim cost to the loss reported by the policyholder (called the **original loss**), we can measure how much the risk is passed around the market.

Did you know…?
The original claim that Occidental Petroleum made for Piper Alpha was $1.4 billion. The gross claim cost to all parties of the same claim was estimated to be $15 billion—a magnification of a whopping 1071%!

Example 13.6: The magnification of the £50m original loss in Example 13.5 is as follows:

A receives a claim of £50m, so their gross claim cost is £50m. A will pay the first £25m and passes the remaining £25m to B, so B has a gross claim cost of £25m. B pays the next £10m and passes the rest to C, meaning C's gross claim cost is £15m.

From a reporting perspective, the gross claim cost of this claim is now

£50m + £25m + £15 m= £90 million, i.e., the cost at and above the threshold at which they assume cover.

The magnification of the claim is £90m ÷ £50m = **180%**

1989 wasn't much better: another poor hurricane season, with Hurricane Hugo receiving the dubious honor of being the costliest ever Atlantic hurricane (this title was later usurped by Hurricane Andrew), and another major oil spill. The Exxon Valdez spill, caused by Exxon not providing a sufficient and capable crew, is still known as one of the biggest manmade environmental disasters in US history. 1990 and 1991 also experienced their share of hurricanes and smaller oil spills, and in 1992 Hurricane Andrew blew into the South Atlantic and blew all previous hurricane records out the water: causing $27.3 billion of damage, it is still in the Top 10 most costly Atlantic hurricanes ever.

The unfortunate mix of generals, a lax regulatory environment for petroleum companies, five years of heavy hurricane activity, and the oversupply of reinsurance meant that insurers leant heavily on their reinsurance policies: claims would be passed around Lloyd's syndicates until someone ran out of reinsurance cover. Suddenly, the game of pass the parcel was running the other way: insurers were paying for the claims in their layer, then quickly passing the rest onto their reinsurer. When the music stopped and there was no one else to pass to, the reinsurer holding the unlimited claim cost would pay what they could, but if they ran out of money, they folded. And so, the risk was transferred back to the previous holder: they thought they were off the hook, but once their reinsurer went bankrupt, they were legally liable for that claim again. They paid what they could but if they ran out of money, they too would have to close. And it continued, chaotically, as many syndicates were unsure of the relationship between the bankrupt and themselves: would the claim suddenly land in their lap?

So, from the top layer down, a series of reinsurers folded. The 1988–1992 losses were so crippling that no one wanted to provide capital to syndicates anymore, as any profit would just be swallowed up by those liabilities. It is hard to overstate the panic in Lloyd's at the time: would this be the end of the historic Lloyd's of London?

After a couple of years trying to disentangle all the relationships (this still hasn't completely been achieved, even 30 years later!), Lloyd's set up a group that would assume all the Lloyd's claims incurred in or before 1992. It was called Equitas and, although it didn't write any new business, it started in 1996 with £15 billion of liabilities.[7] Equitas wiped the slate clean for syndicates and they could start to build their businesses again, notably with some additional requirements around understanding and disclosing risks written, and also the introduction of capital requirements, including the **Lloyd's Chain of Security.** The LMX spiral affected at least 400 participants and 3,000 non-London market insurers. It was the closest that Lloyd's has ever come to declaring bankruptcy and its legacy is the foundation of insurance regulation.

The LMX spiral is a story with a rather tragic ending. Less than a decade after Equitas was set up, it was happening all over again. Companies were bundling risk in a new product that was not fully understood by all participants, spreading risk throughout the market, while underpricing and not reporting their risks adequately. All it would take was a spark. This time, it was not a large claim but a fall in house prices. These new instruments were

[7] At the end of 2020, Equitas' annual report stated that it still had £4.6 billion of outstanding liabilities.

mortgage-backed securities, and it was bankers that were raking in the profits while house prices continued to rise. As soon as they fell, these new products were passed between the players, causing a string of bank collapses in 2008. There is one main difference between the two crises: insurers did pay for the mess made by the LMX spiral. By setting up Equitas, the industry is paying the claims that it promised it would, even 30 years later. Lehman Brothers, Bear Sterns, and a host of other banks that failed have left their obligations to be picked up by those they owed money to, or by the taxpayer.

From the LMX spiral, insurers learnt the importance of understanding risk. However, a less noble lesson exists too: that sharing risk is incredibly profitable if you can find someone less knowledgeable than you to share with.

13.3 Reinsurance to Close

The previous sections have covered reinsurance as a means of sharing risk. Ad extremum, a 100% quota share could be agreed: the reinsurer would receive 100% of the premiums and pay 100% of the claims. This is called fronting, where the reinsurer is really acting as an insurer, and the insurer is just a front. This might be advantageous for tax or regulatory reasons, but isn't particularly common.

Transferring a risk in its entirety using reinsurance is very common in Lloyd's of London, but for accounting purposes. Recall from Chapter 2 that Lloyd's syndicates raise money to back their insurance ventures by raising money from Names. Names buy a percentage of a syndicate's activity for just one calendar year: this is the **Annual Venture** for that syndicate. However, the profit or loss can't be calculated on December 31st of that year because claims relating to that year will take some time to be reported, and then to be paid and settled. The syndicate creates a pot of money, called the Premium Trust Fund (PTF), using the Names' capital with which it will pay claims arising from that Annual Venture.

You will by now be aware that claims are not well-behaved. They may be reported late, subject to legal battles, or waiting on medical treatment or recovery to find out the final claims cost. How long must Names wait to get their money? A **latent claim** could crop up decades later—must they wait decades to receive the return on their investment? Such a wait would be administratively difficult, let alone unappealing to investors. Syndicates need a way to transfer all the remaining risk from an Annual Venture from the Names to some other party, in order to effectively close the Annual Venture and release the profit belonging to the Names. What it needs is **Reinsurance To Close (RITC)**.

Let's assume we are interested in a syndicate's 2022 Annual Venture, which starts on the 1st of January 2022. At the end of 2022, the syndicate's actuary will estimate the liabilities of this Annual Venture. If the estimate of the liabilities is bigger than the amount in the Premium Trust Fund, Names would be required to put in an extra bit of money, in proportion to their share of the Annual Venture. If the PTF is bigger than the liabilities, all is well. The same process occurs the following year, at the end of 2023—two years after the Annual Venture opened.

At the end of 2024, three years after the Annual Venture opened, the actuary estimates the liabilities. Having done so, they look to find someone to assume all of this liability and the associated risk that the liability might increase. This increase could be because new claims could emerge, or because existing claims increase in cost. Of course, if the claim costs were certain, then the syndicate actuary could set aside that amount without issue. However, because of the extra risk, the cost of this product, called Reinsurance To Close, will be higher than the **best estimate** of the liabilities.

The cost of the RITC will be taken out of the Premium Trust Fund. If there is still a little leftover in the PTF, this is profit and can be shared among the Names. This is a very compelling reason to participate in insurance: in 2019, Lloyd's made a profit of £2.5bn.[8] If the RITC is bigger than the PTF, the Names have to stump up more money to cover it, and will have made a loss on the Annual Venture.

But who is the mysterious buyer of the 2022 Annual Venture's outstanding risks? The syndicate's 2023 Annual Venture!

So, at the end of 2024, the 2023 Annual Venture will receive a big boost to their PTF as the RITC payment rolls into their account. However, they now have to pay the outstanding claims for both the 2023 and the 2022 Annual Venture.

But wait—the 2021 Annual Venture might not yet be fully settled yet either. It bought RITC from the 2022 Annual Venture in 2023. So, by the time the 2022 Annual Venture was closed, it was in fact liable for the 2022 and 2021 Annual Ventures. And 2021 was liable for 2020, who was also liable for 2019….[9] The history of the syndicate suddenly becomes very relevant, as

[8] In 2020 it declared a loss of £0.9bn, due to a mixture of COVID-19 and a bad hurricane season. It reported a £1.4bn profit for the first half of 2021—the full year results aren't available at the time of writing.

[9] This rolls all the way back to 1993, to the creation of Equitas which assumed the liabilities for the 1992 and prior years in order to end the LMX spiral.

every Annual Venture will eventually resume the liabilities for all previous Annual Ventures.

The benefit of RITC is that Names can realize their profit in a (somewhat) timely manner, which is crucial to the continued availability of capital for Lloyd's syndicates. Names are also fairly consistent in which syndicates they invest in. The process of the capital auctions means that it is expensive and time-consuming to jump in and out of a syndicate every other year. As a result, the Names in the 2022 Annual Venture will likely be a similar group to those in the 2021 Annual Venture, and so resuming a previous year's liabilities largely keeps the risk with those who originally underwrote it initially. If the claims from the 2022 year suddenly skyrocketed after 2024, the Annual Venture now holding that liability could make a big loss. This means that Names will scrutinize the history of a syndicate very closely before investing their money. This scrutiny acts as a powerful incentive for syndicates at Lloyd's to consider the legacy of every action and helps avoid myopic decision-making.

13.4 Sharing Risk Outside of Insurance

But why only share risk between insurers and reinsurers? Is there any benefit, or indeed any appetite, to share risks more widely?

We have so far considered risk transfer as a way of reducing risk, in order to smooth results and lower capital requirements. In this section, we will explore two further sectors which risk can be transferred to: the financial sector, and the public sector.

13.4.1 Risk Transfer to the Financial Sector

During the insurance cycle, there are periods when demand for insurance outstrips supply: there is too much risk and not enough capacity to take it on within the insurance industry. After the LMX spiral, insurance capacity dropped as insurers tried to manage their historic losses and understand their risks again, so insurers were forced to look to other sectors to take on risk. The increasingly complex financial sector was willing and able to take on large risks, particularly catastrophe reinsurance.

It may not be obvious that risk can be moved outside of insurance, but by reframing catastrophe XL reinsurance as money contingent on a binary event, we can draw parallels between reinsurance and financial options. To do so, we need to redefine the claim, not in terms of its cost, but in terms of the underlying event. Instead of making the reinsurance pay-out contingent on a certain claim cost, we map the underlying event to an objective scale—perhaps a category of hurricane, a certain depth of rainfall, or a certain average temperature.

> **Example 13.7:** An XL treaty, \$5m in excess of \$10m, pays out \$5m if a claim is over \$15m, and the claim amount less \$10m if the original loss is between \$10m and \$15m.
>
> A bet on the weather pays out \$5m if a Category 5 hurricane hits the US this year, and \$2m if a Category 4 hits.

This is a very crude way to approximate reinsurance. Paying out big chunks of money contingent on events is not particularly consistent with what banks normally do—this is firmly within the realms of insurance! While the catastrophe risk might be very attractive to the financial sector, as it is largely uncorrelated with systematic market risk and therefore provides some diversification for them, the pay-out structure is less so. Instead, financial instruments, particularly options and securities can be altered to be contingent on the occurrence of catastrophe events.

One of the most common techniques is to use securities as reinsurance. The insurer will issue a bond to an investor who pays the insurer a lump sum for it. The insurer will pay back this sum of money plus interest, unless a specified event, such as a Category 5 hurricane, occurs. If a hurricane does hit, the insurer can use the lump sum to pay claims. It's a pretty risky bond for an investor, but if the event doesn't occur, they will be compensated by receiving the lump sum in full plus interest.

If securities are like a catastrophe bond, another instrument—post-loss funding—is like a catastrophe option. Post-loss funding is when an insurer and financial institution agree the terms of a loan or company equity that can be used if a catastrophe occurs. If it does, the insurer has the option to raise money on the pre-agreed terms, which it can use to pay catastrophe claims. This is beneficial because after an event, the insurer's share price might drop, or financial institutions might feel able to charge higher interest rates on a loan. The cost of post-loss funding is quite low upfront: payments on the loan or equity are paid after an event, so it is initially cheaper than regular reinsurance. Conventional options based on weather or catastrophe metrics are increasingly popular too, although less so with insurers. However, regular businesses that have some dependency on the weather, temperature, snowfall, or windspeed can now buy a normal option that is based on these metrics.

These instruments aren't transferring the catastrophe risk itself, but the financial impact of it. By using these instruments, insurers aren't reducing claim costs but they are improving their financial position and alleviating strain on capital should an event occur, which will help to smooth their results and manage their solvency and liquidity. As such, instruments such as these are often called financial reinsurance (shortened to FinRe[10]). It doesn't necessarily reduce the risk itself, but it does ease the financial consequences.

13.4.2 Transferring Risk to the Public Sector

Throughout this book, I argue that insurance is a social good and is necessary for our society to function. Insurance needs to be available and affordable so that risks can be managed and mitigated. But this social good is largely administered by private companies. What happens when it is not available?

13.4.2.1 Case Study: Flood Re

The UK is a series of small, rather rainy islands in northern Europe, lashed by the Atlantic Ocean to the west and the icy North Sea. As Brits will tell you (and they *will* tell you), it rains a lot. So, in terms of catastrophes in the UK, floods have always been one of the top **perils**. But that wasn't a problem necessarily, until the late 1990s when data and computing power allowed for more complex insurance pricing structures. As insurers gained more granular information about flood risk, an increasing number of UK homeowners were unable to get flood insurance. Insurers deemed their property too high a risk, and wouldn't offer them cover. This wasn't a small group, but one sixth of British homes.

In 2000, flooding caused £1 billion of damage to UK homes. Many of the flooded homes were uninsured and left uninhabitable. Councils faced a surge in homelessness. Some homeowners wanted to move immediately, but found their homes now deemed uninsurable and hence almost worthless: no one wanted to buy a house if their life savings could be wiped out overnight, without insurance to protect against such a loss. This was terrifying to the government: would this lead to a breakdown of the housing market? And how much would an emergency response cost?

---

[10] Say 'Fin-Ree'. It also comes under a broad umbrella of Alternative Risk Transfers, which is abbreviated as ART (say 'art'). There's a lot of acronyms in reinsurance—even reinsurance itself is commonly abbreviated to RI!

Unfortunately, the 2000 storms weren't a one-off. In 2007, UK floods caused £4bn of damage, only £3bn of which was insured. In 2015, they caused another £1.3bn. Flooding is set to increase in both frequency and severity as climate change creates wetter winters and gives rise to more tropical storms,[11] and that these supposedly '1 in 10 year' floods have happened three times in 15 years shows that the effects of climate change are already being felt in the UK.

But it's not just coastal properties or riverside homes that are at risk. Even more homes are at risk of surface flooding, caused by sudden heavy rainfall. This is more likely in urban areas, especially London, which is dependent on a Victorian plumbing system that was not designed to cope with the number of London properties it now supports. Understandably, insurers were reluctant to insure those houses: insurers are always looking for good risks, the ones that will generate more premium than claim costs over time. Some houses were so likely to flood and indeed, had flooded multiple times in recent years, that insurers refused to generate quotes. As homeowners grew increasingly anxious, the government was forced to intervene. It had to find a way to make these bad risks palatable for insurers.

In 2016, the UK government and insurance industry set up a not-for-profit reinsurance company called Flood Re—short for Flood Reinsurance. Under this system, insurers write flood insurance policies for homeowners as before but they can choose to reinsure any of these policies with Flood Re, who will provide XL reinsurance. If a reinsured policy has a flood claim, the insurer will pay the first £250 and the rest will be paid by Flood Re. The premium charged per policy is determined based on the council tax band (a proxy for the value of the property).

The benefit of Flood Re is that insurers are able to make a profit on very poor risks, and therefore are happy to provide insurance for poor risks. The expected flood risk cost for the reinsured policy will not be based on the severity of the flooding as calculated in Chapter 8, but based on the costs that the insurer will incur:

$$\text{Risk Cost of Reinsured Policy} = \pounds 250 \times \text{Annual probability of flooding} + \text{Flood Re Premium}$$

As very poor risks are now profitable to insurers, companies want to compete on price. Therefore, there is no temptation to inflate premiums for very risky customers, as insurers now actively want that business. But if these

[11] When ex-tropical storms hit the west coast of the UK, they cause heavy rainfall and high winds—the flooding in 2000 was partly due to the bad hurricane season that year.

policies are now profitable for insurers, where is the money coming from to pay the claims? Flood Re expects that its claim costs will be higher than the premiums it collects, as if this wasn't the case, the insurer would not choose to reinsure those risks. To make up for this shortfall, the government imposes a levy on all UK home insurance policies which goes straight to Flood Re, totaling about £180 million a year.

Flood Re has been successful in ensuring that the worst flood risks have been able to get affordable cover: about a quarter of a million homes, representing the worst 2% of flood risks, are covered under the system. It is also cost-neutral for the government, as it is funded by all insured homeowners. However, Flood Re has two main limitations: first, it is only a temporary solution as the scheme will close in 2039. The idea is that by then, there will be more flood resilience (be it flood defenses or individual property preparations) but there isn't currently a clear path to de-risking all homes. Secondly, it only covers houses built before 2009. This was to discourage building on flood plains, but about 10,000 homes a year continue to be built at a high risk of flooding. This means that although Flood Re reinsures about 250,000 homes, since 2009 about half as many again have been built that will not be covered by the scheme.

One criticism of Flood Re is that flood-prone properties are usually cheaper than other properties, and so under this system homebuyers are being protected from the consequences of their decision to buy a cheaper but riskier house, largely funded by other homeowners. However, as the UK's climate changes and becomes wetter, more properties will become flood-prone, which arguably homeowners could not have anticipated.

At the moment, there is not a clear plan for flood-prone properties post 2039, and it is unlikely that flood risk will decrease. In 2020, the government announced it would spend just under £890 million per year on flood prevention, including maintenance of current defenses. Spending is currently focused predominantly on building and maintaining concrete flood barriers in a bid to protect against river and coastal flooding. Insurers already take into account flood defenses such as these (as well as the risk that these might fail), and yet Flood Re has still been necessary, so it doesn't follow that the government can build its way to widening insurance cover. The flood defenses also don't protect against surface water flooding, nor against floods that breach the barriers, so without another government-led insurance or reinsurance scheme there is likely to be a significant number of houses in 2040 with no flood cover.

Flood Re currently shares the risk of flood-prone houses among all home insurers. But it doesn't address the underlying problem, which is that people continue to live in flood-prone areas. It's not clear what the government will do to address this, but it will need to be a significant step change from the current policy of approving housing development on floodplains. As climate change exacerbates this problem, insurers will be increasingly less tolerant of a risk that government home-building policy has encouraged.

13.4.2.2 Other Public Sector Flood Insurance Models

However, the government directly sharing flood risk is not necessarily the answer. The National Flood Insurance Program in the USA is a government-run scheme that, unsurprisingly, provides flood insurance. It can do this directly, but in the majority of cases it acts as a reinsurer to private insurance companies, who are paid a fee for writing the policy, and a fee for handling any claims. The NFIP is then responsible for the losses.[12] But this scheme, set up in 1968, has also failed to encourage homeowners to move. It thought that by making people aware of the risk of flooding, people would move elsewhere. But as the federal government assumes the risk, homeowners of flood-prone properties don't lose financially by staying. With a frequent history of flooding, the properties are often nearly worthless, and so even if they want to move, many homeowners can't afford to. Unlike Flood Re, this scheme has no specified end date, although it does receive periodic authorization (that could theoretically be denied). When claims exceed premiums, the NFIP borrows money from the US Treasury to cover the shortfall. To date, the scheme owes the Treasury (and by extension, US taxpayers) over $20 billion.

Other natural catastrophe reinsurance schemes exist around the world. The French *Caisse Centrale de Réassurance* is a state-backed catastrophe reinsurance scheme. Like Flood Re, insurers can opt to reinsure risks with the scheme, or not. However, this French scheme is more like a quota share, with most insurers opting to share the risk 50:50 with the scheme, which in recent years has hovered around a 100% **Combined Operating Ratio**. Creating a market for affordable but sustainable catastrophe reinsurance is only going to become more pressing in the coming years, and the creativity and expertise of actuaries will hopefully guide governments to a solution. But the *Caisse Centrale de Réassurance* suggests that state-backed reinsurance could be a long-term

[12] This fee structure has the effect that when there are lots of flood claims, private insurers tend to make higher profits, as they receive a fee for handling each claim. For example, after Hurricane Sandy, private insurers made a profit of $400 million Business Of Disaster: Insurance Firms Profited $400 Million After Sandy: NPR.

solution to affordable catastrophe insurance—a soundbite only slightly less catchy than 'sharing is caring'.

14

Catastrophes

Chapter Objectives:

- Understand the claim features of catastrophe events
- Understand the challenges of pricing for catastrophes and the basics of catastrophe modeling
- Recognize opportunities and challenges for different kinds of insurance in different geographical areas
- Consider the ways in which manmade catastrophes can be insured against, and the challenges to insurers of providing cover for manmade catastrophes such as cyber-crime and terrorism.

Speak Through the Earthquake, Wind and Fire, O Still Small Voice of Calm.
—'Dear Lord and Father of Mankind', Hymn by John Greenleaf Whittier

This chapter is about catastrophes. We could equally use the word disaster, but catastrophe is more frequently used in insurance. We will consider these terms in greater detail shortly, but we all will almost certainly agree on the characteristics of such events: that they are big or very big, they cause widespread harm, and are difficult to control and mediate.

It might seem surprising that some of these events and situations (whether pandemics or the environmental instability of the earth) can be insured at all. As this chapter reveals, insurers apply many of the same principles to the

© The Author(s), under exclusive license to Springer Nature Switzerland AG 2022
C. Townsend, *A Risky Business*,
https://doi.org/10.1007/978-3-031-11673-5_14

insurance of catastrophes as they do elsewhere. Certainly, the insured cannot always get what they want—as we saw in the previous chapter in regards to flooding, insurance cannot be provided for any eventuality. There is a bottom line: for insurance to be sustainable and resilient, disasters cannot be disastrous for the industry itself. Yet, the fact that insurance is applied to large, complex events with high societal and commercial impact represents a further demonstration of the maturity and value of the insurance world, and reinforces its claim to be just as essential to modern life as a good legal system.

Because of their scale, uncertainties connected with catastrophes have the potential to generate very substantial risks. Not only do we not know enough about the weather to predict it long term, we can't predict with any certainty, even in the medium term, when the next devastating event will occur and whether it will be in my street or yours. There are uncertainties in scientific understanding which change the models used to predict both the probability and severity of events. These uncertainties change, but they are always with us. The insurance industry, like governments, must take heed of scientific developments and of the maturity of scientific research in the relevant disciplines.

We can no longer talk about these events without mentioning climate change, as it provides particular and extensive challenges to insurers. Climate change has both localized and regional impacts and it triggers events through a complicated series of relationships that control our environment. Because this causal chain is so complicated, there are disagreements within the scientific world about its nature, likely impact and causes. It is not for us to reproduce these arguments here, but a consensus has emerged which has been coordinated by highly respected and authoritative international scientific organizations. The consensus of scientific opinion is that the earth is warming and that such increases in temperature are measurable. Further (and quite different) modeling indicates that human activities have and continue to make a major—or *the* major—contribution to the global warming. Climate change is believed to cause changes that can be modeled to various degrees of accuracy: changes in biodiversity, crop production, and a rise in extreme weather episodes. For insurers, it is these symptoms of climate change that are most of interest. These are believed to be long-term effects, and even if the causes of climate change were magically removed today, it would take many decades before the consequences were significantly attenuated. We needn't go beyond this level of detail to provide the backdrop for the role of the insurance industry, and so these features will be our starting point.

Even faced with such potentially damaging long-term effects, actions must be balanced with other consequences. For example, despite commitments to reduce climate change, the countries of the world must continue to feed their populations, maintain their health services and keep their population warm in winter. These aren't necessarily competing priorities, as ignoring climate action is likely to mean that current food crops are untenable, but rather we should acknowledge that there are many considerations in taking any large-scale action. Even in a pandemic such as COVID-19, there are risks to closing down the economy that are just as real as failing to control infection. The message is that the decision-making landscape is very complicated and variable between different countries. Uncertainty in climate change outcomes shouldn't be an excuse for policymakers' inaction, but future governmental policies and industry regulation is itself an uncertainty that adds to the challenging environment for the insurance industry.

14.1 Modeling Natural Catastrophes

For an insurer, a catastrophe[1] could range from a debilitating cyber-attack, a natural disaster, a terrorism incident, or a bout of disease. With such variety, the Oxford English Dictionary comes to our rescue by identifying what these events all have in common: they are all '*a sudden event that causes many people to suffer*'. The United Nations adds that disasters are events '*which exceed the ability of the affected community or society to cope with using its own resources*'.[2] The key point here is that they are all single events—one hurricane, one cyber-attack, one disease—that cause multiple claims, so give rise to an exceptionally large total claim cost. This definition is based on the event's effect, rather than the cause. So, what does a catastrophe claim look like, and what does this mean for insurers?

As previously alluded to, a catastrophe is not usually one large claim but a cluster of smaller claims arising from the same event. This leads to operational challenges and costs for insurers in responding to sudden events: insurers need more claims handlers to deal with the surge of claims and the ability to rapidly deploy loss assessors to a focused area.

As Superman noted, '*aeroplanes are the safest way to travel*'. But when one crashes, the human and material cost—and so the size of any insurance

[1] The industry-standard abbreviation of 'catastrophes' is 'cats', and this chapter will make use of this. However, if you search online to find out more about this topic, 'cat modelling' may return very different (although amusing) results.

[2] Risks and Disasters | UN-SPIDER Knowledge Portal. https://www.un-spider.org/risks-and-disasters

claim—are huge. It is the same with catastrophes (indeed, a large plane crash might even be considered a catastrophe): they are characterized by low event frequency but very high severity.

Act Natural

Catastrophes such as floods and storms were historically called 'Acts of God', to denote that they were unexpected and not caused or controlled by human hands.

Nowadays, extreme weather events are more frequent and more severe due to anthropogenic climate change. Calling them an 'Act of God' is rather disingenuous (perhaps slanderous?) in this case.

To avoid this, catastrophes are often split into natural catastrophes (shortened to nat cat) and non-natural catastrophes. Here, 'natural' refers to the context rather than the cause: a flood is 'natural' because it occurs in nature, not because it would have occurred identically in an uninhabited planet. Non-natural catastrophes are the ones directly caused by people, whether accidentally or deliberately, such as a radiation leak or cyber-attack.

The term 'Act of God' is still sometimes used, mostly for earthquakes and other non-weather-related events.

Catastrophes are primarily a concern in property insurance, although there is often an impact on financial loss **perils** such as business interruption as well. Liability lines are largely unaffected, as parties are often temporarily freed of their contractual obligations during an Act of God. As an example, delivery dates for goods and services can be suspended without penalty during a hurricane. For non-natural catastrophes, there is often a *force majeure* element in a contract. Literally meaning 'superior force' (although more commonly translated as 'overwhelming force'), it frees parties from liability during catastrophes such as war, riots, crime, strikes, or epidemics.

Catastrophes are not common—by definition they are exceptional events. This means that unlike non-catastrophe claims, when a catastrophe does occur, the claims will be reported at roughly the same time. This leads to difficulties for **reserving** actuaries, as the historic development pattern will not be applicable to a development period with a cluster of catastrophe claims in. As we noted in chapter 11, cat claims will often have to be removed from a claims triangle in order to give sensible estimates. However, cat claims still need a reserve! There are three commonly used methods for this:

1. Use the loss adjustor's estimate as the ultimate claim cost. Of course, this relies on the loss adjuster being able to assess the damage.

2. Another method is to increase all catastrophe **FNOL**[3] estimates by a constant, for example, by 20%. Then the reserve for these claims is the current estimate plus an extra 20%. This is particularly suitable for early stages in the catastrophe. It isn't scientific, but it is relatively prudent and so can be a good stop-gap until the claims can be assessed.
3. If there is sufficient data, a catastrophe claims triangle can be created. The development period for this might be measured in weeks or even days, and the cohort would be an individual catastrophe. This can be really useful for business planning as well as reserving: if you can estimate how many days after an earthquake your policyholders will report their loss, extra claims handlers can be provided.

Catastrophes being present in the data poses a problem for actuaries, as seen above. But as they are uncommon events, it's likely that the dataset won't contain any events at all. This poses another challenge: without data, traditional statistical techniques can't be used for pricing or capital modeling. Instead, a specialist cat model needs to be created, based not on historic data but on hypothetical scenarios and the judgment of expert earth scientists. These are produced by cat modeling companies and sold to insurers. Insurers very infrequently create their own cat models due to the cost and expertise (of people who don't traditionally work in insurance companies, such as geologists or seismologists) required. Cat modeling is very expensive, but it is essential to understanding the company's risks—and if that's not incentive enough, it is also mandatory for regulatory capital requirements! Buying proprietary models from an external provider is expensive, although often cheaper than building a cat model from scratch, but the cat modeling companies provide frequent updates and can tailor to individual insurer's exposures. The role of monitoring scientific papers then modeling and turning them into quantitative predictions of possible liability is an interesting and increasingly important role—for both actuaries and earth scientists alike!

Cat modeling, which began in earnest in the 1990s, is now very sophisticated, largely due to advances in computing power, better scientific understanding, and—unfortunately—a number of these events actually occurring in the past three decades. The structure of a cat model is generally based on five parts: event, hazard, inventory, vulnerability, and financial analysis. Combining these parts, as in the diagram below, produces an estimate of the loss due to that event (Fig. 14.1).

[3] First Notification Of Loss estimate i.e., the first estimate made by a claims handler.

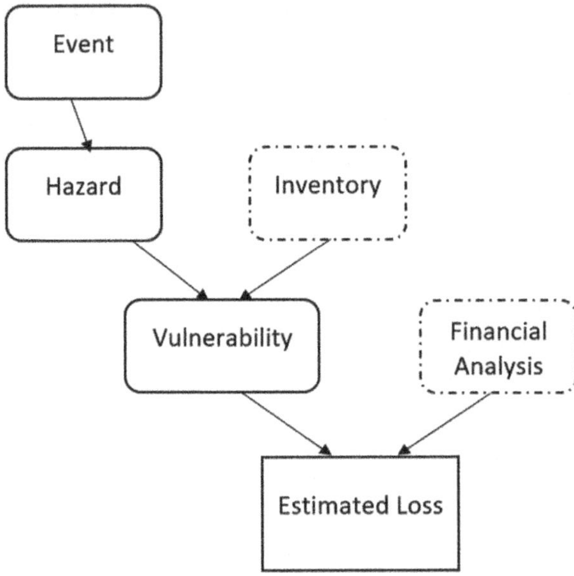

Fig. 14.1 Diagram of a Catastrophe Model. The dotted outline denotes the modules that are specific to the insurer, and the estimated loss is the final outcome

In the diagram above, there are five parts that are required for the calculation of the estimated loss. Only two of these (inventory and financial analysis, shown in the diagram in a dotted line) are inputs from the insurer: the rest of the module will be the same for all users. Let's look at each part in more detail.

Event: As you might expect, the event module defines the catastrophe being modeled. The input is a database of events that need to be modeled. Some of these might be historic, such as Hurricane Katrina, but some might be possible but hypothetical events. Each event needs to be assigned a probability of occurrence and a location.

Hazard: The hazard module contains data on the consequences of the event. It quantifies the problem. For example, what are the consequences of a hurricane? Perhaps you'd expect winds of 80 mph, 15 cm of rainfall in a 24-hour period, and a storm surge of 0.5 meters.

Inventory: This is the first dataset that the insurer, rather than the cat modeling company, needs to provide. The inventory contains data on where the insurer is exposed to risk, and some risk details. Location data is crucial, but the type of building, construction type, age of building, building value, and the number of storeys can be included too.

Vulnerability: By combining the hazard and inventory datasets, the vulnerability module calculates the degree of loss for each policy for each event. For example, a large flood could cause a total loss for a bungalow on a flood plain, but the same event would not cause any damage to a penthouse flat. The vulnerability module calculates an estimate for the full loss. However, this is unlikely to be the cost to the insurer as policies tend to have **excesses** or other limits.

Financial Analysis: For each policy, there is likely to be a limit to the cover, perhaps some exclusions or excesses to pay. These terms and conditions mean that the cost of the event is not the cost to the insurer. As we are interested in the latter this policy information should be combined with the vulnerability information to give an estimated loss to the insurer.

Example 14.1: An insurer wishes to model losses from a Category 3 storm. They cover three buildings, the details of which are:

Policy A: A warehouse in Georgetown, valued at $3m. It is insured for $2.5m, and there is a policy excess of $50,000.

Policy B: A tower block Georgetown, valued at $7m. It is insured for the full value, with a policy excess of $100,000.

Policy C: The village hall in St. Helen's, valued at $1m, insured for $900,000, with no excess. Theft claims are excluded.

What might each of the cat model modules contain and what is the loss to the insurer?

Event dataset: Category 3 storm; location: Georgetown; storm; probability of occurrence = 0.07.

Hazard dataset: Category 3 storm: wind speeds of 65mph, 5mm of rainfall daily; no storm surge.

Inventory dataset: Policy A, warehouse, Georgetown, $3m; Policy B, tower block, Georgetown, $7m; Policy C, village hall, St. Helen's, $1m.

Vulnerability:

Policy A: Storm damage estimated cost = $1m, Flood damage estimated cost = $2m (these numbers would be given by the cat model)

Policy B: Storm damage estimated cost = $4m, Flood damage estimated cost = $0.5m.

Policy C: Storm damage estimated cost = $0, Flood damage estimated cost = $0 (notice that this policy is in a different location to where this storm will hit, so will be unaffected).

Financial Analysis dataset: Policy A, Sum insured = $2.5 m, Excess = $50,000, no exclusions; Policy B, Sum insured = $7m, Excess = $100,000, no exclusions; Policy C, Sum insured = $900,000, Excess = $0, theft excluded.

Estimated loss to the insurer (combining the financial analysis with the vulnerability dataset):

Policy A: Minimum of modeled loss and sum insured = $2.5m, less excess of $50,000, gives a loss of $2.45m

Policy B: Minimum of modeled loss and sum insured = \$4.5m, less excess of \$100,000, gives a loss of \$4.4m.

Policy C: Minimum of modeled loss and sum insured = \$0.

So, the estimated loss to the insurer of this Category 3 storm in Georgetown is \$2.45m + \$4.4m = **\$6.85m.**

The event dataset will usually contain many scenarios. Some will be historic, but some will be hypothetical. As with all statistical models, it can only model the stated events. But this is much more of a problem in catastrophe modeling as you have far fewer events—even one extra event could materially change (even double) your frequency. Also, each event has such a high severity that missing just one event could understate severity by millions. Events may not have been seen in the past but may also be missed from this hypothetical list and so the model cannot reflect the range of possible outcomes associated with the disaster in question. There's no statistical way of checking if possible outcomes have been missed, so this is a considerable potential source of bias in our models. The probability of each event is also determined by expert judgment, so can be affected by recency or availability bias.[4]

As cat models feed into the regulatory capital requirements for insurers, it's important that the event scenarios included in the cat model are relatively consistent throughout the market, at least for large events. Lloyd's of London annually produces a set of events that must be included by all Lloyd's **syndicates** in their reporting. These event scenarios are called **Lloyd's Realistic Disaster Scenarios.**[5]

Lloyd's Realistic Disaster Scenarios are updated annually, and with good reason. Firstly, the estimated cost to the market of each event needs to be updated yearly to reflect inflation and urban development. If more buildings are built, the higher the potential insured cost. As part of the location dataset in the event dataset, Lloyd's includes a map of the affected area, listing the estimated value of all property, and individually listing all points of interest, such as airports, as well. The event set also changes to incorporate new events.

[4] Recency bias is the name given to the effect whereby people favor recent events over older ones. In this case, it could lead to experts overstating the probability of an event similar to one which happened recently, while (more worryingly) understating the probability for an event that hasn't occurred for a while. Availability bias is where events that people more easily remember are given higher probabilities of occurring, which can be heavily influenced by the media coverage.

[5] Non-Lloyd's insurers, particularly those who write US and European business, will also keep an eye on what the Lloyd's RDS are and perhaps use some or all in their own modeling.

For example, after the flooding in Thailand in 2011, Lloyd's added a South-East Asian flood event to its scenario set. Before this, insurers hadn't really considered the potential for a flood in this region as it hadn't previously been a big insurance market. The flooding caused about $30 billion of damage, of which only $12 billion was insured. However, this loss (including the $2.2 billion paid by Lloyd's of London) made insurers take note: relying on old lists of events is not sufficient.

Cat modeling uses a scenario-based approach, but actual events still provide valuable learning opportunities. The 2011 Thai flooding was unexpected as it was totally missed from the event set, but it was Hurricane Katrina that forced insurers to improve the underlying model sophistication.

14.2 Lessons from Hurricane Katrina

Since it was founded in 1713, New Orleans has become known for its vibrant music scene, delicious beignets, and European architecture. On the 29th August 2005, 'The Big Easy' became known for one more thing: being the victim of the biggest windstorm loss ever as part of the longest and most severe hurricane season in history.

Being so close to the Mississippi river, New Orleans was used to flooding. The town was surrounded by levees—much needed, as the town is 1.8 m below sea level. It was no stranger to hurricanes either: the New Orleans Hurricane of 1947 caused floodwater 1 m high in some parishes, and Hurricane Camille in 1969 hit Louisiana as a Category 5 storm, although its path swerved at the last minute and it narrowly missed the city.

As a Category 5 windstorm, Katrina was a severe event meteorologically speaking, with the highest recorded wind speed of 175mph. But it was the loss that shocked the world: both the $125 billion (in 2005 terms) of financial loss, and the 1,816 deaths recorded (Fig. 14.2).

The insurance industry was simply not prepared for a natural disaster of this extent. Why didn't the catastrophe modeling of Atlantic hurricanes predict such a large loss? Many lessons have been learnt from Katrina, particularly lessons about preparation, government response, and levee maintenance, but I will focus only on the lessons that the insurance industry learnt.

14.2.1 Lesson 1: Run for Cover

Although the loss to insurers from Katrina was over $60 billion—a catastrophic loss by any measure—, it could have been double that: the damage

Fig. 14.2 New Orleans, 29th August 2005 (Photo taken by and reproduced with the permission of AP Photo/U.S. Coast Guard, Petty Officer 2nd Class Kyle Niemi)

was valued at $125 billion. Insurers could have been hit a lot harder if there had been sufficient coverage in place. A raft of exclusions, policy terms, government-run flood insurance, and insufficient cover meant that a swift recovery enabled by insurers (like in the 1906 San Francisco earthquake) wasn't possible. So why didn't people have sufficient cover?

- Flood insurance is specifically excluded from most home insurance policies in the USA, and so insurers on the whole didn't have to pay for damage related to flood. But should a storm-caused flood be covered by storm cover? This question went to court, but the answer was that it depends on the exact policy wording, and exactly how the water entered each building. Of course, if the homeowner has evacuated (as they are told to do), providing evidence of how water got in is practically challenging.
- Flood insurance in the USA is provided by the National Flood Insurance Program, a government-run scheme that provides flood insurance. It can either be bought directly or through private insurers, but the maximum

cover is $250,000 for buildings and $100,000 for contents. This creates a coverage gap for homes that are worth more than $250,000 to rebuild.

- Cover was squeezed at the bottom end too: policyholders who hadn't carefully read their terms and conditions were surprised by high deductibles (similar to policy **excesses**—the insurer pays the cost of the claim less the deductible amount), meaning they had to pay a larger amount out of pocket than then were expecting.
- A lot of people just didn't, and still don't, have insurance. In Louisiana, the Insurance Information Institute estimates that only 46% of people have flood insurance. In Mississippi, it is less than 20% of people (Mississippi filed almost 30% of insurance claims for Hurricane Katrina). Businesses had more cover: $40 billion of the $60 billion insured loss was paid to businesses, not homeowners.

There has been a lot of criticism of the insurance industry's response to Katrina, particularly in regards to the poor communication of policy terms and exclusions. In the years that followed, there were a slew of lawsuits (mostly consolidated cases that represented tens of thousands of policyholders against dozens of insurers) regarding exclusions. However, criticism should also be aimed at the poor communication of local governments, meteorological agencies, and insurance bodies for failing to communicate the level of risk, and if that is too high for an individual, the importance of risk transfer.

14.2.2 Lesson 2: Count the Seconds

One problem for insurers was that their catastrophe models hadn't included everything that could have gone wrong. In their windstorm models, the hazards included strong winds and heavy rain, but hadn't accounted for any of the subsequent hazards.

Hurricanes are systems of low atmospheric pressure. Like people, water spreads out into the space it can easily take over, and when the air is low pressure, the water spreads upwards into that space increasing the volume of water. This is known as storm surge and, during Katrina, the height of the water increased by up to 3.5 m. Clearly, 3.5 m of water can cause a lot of damage! Storm surge wasn't routinely modeled in cat models pre-2005, although (unsurprisingly) it now is an option for hurricane models. This is an example of the insight that insurers and cat modelers take from previous disasters.

Claim Costs Go Nuclear

Nuclear plant damage after an earthquake, although exceedingly rare, provided one of the biggest scale of disasters for the insurance industry when the Fukushima nuclear reactor facility in Japan was damaged in 2011. The earthquake was measured at magnitude 9.0, and resulted in a 650 km of sea floor being shifted by 10–20 m. This caused a tsunami that crippled the reactors' coolant equipment, leading to three nuclear reactor core meltdowns and the evacuation of over 150,000 people.

The decommissioning of the Fukushima reactors was estimated to cost $15 billion, with many other large costs associated with compensation and rebuilding. The ramifications of the disaster are much greater than the physical or human disruption or harm, since it has brought into focus how the nuclear industrial processes (particularly its energy production, reactor decommissioning and nuclear waste storage) can be insured when the scale of potential damage is so great. In fact, many countries (including the US and Japan) the government effectively subsidises potential liabilities in order to make the nuclear industry financially viable, as it is unlikely that any private company would agree a contract to provide the full costs of any claim.

In the US, this subsidy operates by specifying a maximum liability for private companies of only $350 million. In the event of a disaster, any further costs would have to be paid by the government. When the cost of nuclear power is reported, an allowance for this huge additional potential liability (albeit for governments) is not always included in cost per kw calculations.

Storm surge is an example of a secondary peril. It is not the main hazard of a windstorm, but in conjunction with heavy rain and high wind speed, it can cause a lot of additional damage. But not all secondary perils are meteorological.

The problem with cat models pre-2005 is that they assumed claim costs per policy as independent. For example, if your house costs $100,000 to rebuild and your neighbor's costs $120,000, the cat model suggests the total claim cost is $220,000. But if both houses needed urgent repair, it is easy to imagine that you and your neighbor could get into a bidding war for the builder's time. The builder is so busy, she quotes double the usual price for her time, so your rebuild cost is now $160,000.

Imagine this on a city-wide scale. All the builders are busy, all the skips hired. The cost of materials jumps in price too, as the roads are so damaged it is difficult to get new materials into the town. These secondary costs due to increased demand are called demand surge. It's hard to estimate exactly how much demand surge will add to the cost of claims, but estimates for Katrina range between 10 and 40%. This wasn't modeled by insurers but is now a common optional module in the cat model. Other secondary perils include losses due to looting, fire following an earthquake, a tsunami following an earthquake, and increased loss assessor expenses (as getting insurance officials into a disaster zone might not be logistically easy!).

One secondary peril that is very difficult to spot is insurance fraud. After a catastrophe, claims handlers are deluged with claims, so it can be easy for them to miss a fraud indicator. For example, a policyholder unaffected by a local natural event who suffers a loss due to an uninsured peril, but fraudulently claims the loss as being due to the natural catastrophe. Fraud can even be committed by insurers. For example, after Katrina, the State Farm Fire and Casualty Company ordered claims handlers to misclassify hurricane claims as flood claims. As there is a state-funded National Flood Insurance Program in America, this claim categorization shifted the liability from State Farm to the government. For the Katrina incident, the company was forced to pay $750,000 in damages, but it was only the first of many investigations. All in all, the company was found to have evaded $522 million of claims!

14.2.3 Lesson 3: Take a Break (into Consideration)

The flooding in New Orleans was exacerbated by the town's levees (flood barriers) breaking in several places. This shouldn't have been a surprise during the Category 5 storm, as the levees were only built to withstand the pressure of a Category 3 storm. Local budget cuts also meant that $71 million that was planned to be spent on levee maintenance was cut. All in all, the town's levees were broken in 50 places by Katrina, allowing flood water to gush into New Orleans.

Further investigation showed that a lot of the damaged levees were filled with earth or sand, which (although it was a common filler in the 1960s) led to these levees being almost totally obliterated. Other failures had occurred where two levees joined together.

Lessons should be learnt about resilient construction, engineering, and maintenance of flood defences. But for insurers, the fact that levees could break had not been modeled, so these levee failures from Katrina presented an unexpected additional hazard and cost. Nowadays, insurers have a lot more data about where exactly flood defences are. Sophisticated insurers will have models that show the expected flood depth at a very granular geographic resolution. To allow for the possibility of a breach, locations will often be given two flood scores: one including the effects of a flood defence, and one undefended score, alongside some measure of the defence's adequacy.

14.2.4 Lesson 4: Social Good, but Profit Better

The tragedy of Hurricane Katrina is that the losses didn't need to be this high. There was a lack of appropriate building codes for New Orleans, so despite being below sea level, few areas in the city had made adequate preparations for hurricanes and flooding. The federal government, the state of Louisiana, the local governments, and the city of New Orleans itself all failed to create and enforce building codes to reduce or prevent flood and storm damage.

After a catastrophe of this size, insurers are forced to reconsider the profitability of writing business in the affected area. Will similar events occur? Will they become more common? Without strong action to decrease potential storm or flood losses, insurers might feel the risk is too high, and stop providing cover to the area. Insurance is a social good—as demonstrated by the $60 billion paid out to enable policyholders to rebuild—but if it isn't profitable, it won't continue.

Here is the implicit contract between insurers and society: society needs to take action to manage its risks, and insurance will pick up the portion that cannot be controlled further. Economists might see the parallels here between systematic and unsystematic risk: investors who hold unsystematic risk (or unmitigated risks that could have been mitigated) might lose their money, but that is not the fault of the market. A society that doesn't manage its risks (which can range from locking your front door, building flood defences, carefully disposing of toxic waste, or enforcing safety rules) should bear the cost of that themselves. Without this mitigation, claims will be more frequent and potentially higher cost, reducing the profitability of insurance. Eventually, this may cause insurers to stop writing those risks. This not only affects the policyholder, but anyone else who wishes to buy that type of insurance, and the victims of the policyholder's actions (for example, employees who experienced harm working for the policyholder, or a third party whose property was damaged) who may not receive adequate compensation if the policyholder is not insured. Insurance works best as a social good—as a safety net—when society is good at reducing the risks wherever possible.

After Katrina, Louisiana brought in 'Rule 23'. This opaquely named piece of regulation '*suspends the right of any insurer to cancel or non-renew any personal residential, commercial residential or commercial property insurance policy covering a dwelling, residential property or commercial property located in Louisiana that sustained damage as a result of Hurricane Katrina or its aftermath, or Hurricane Rita or its aftermath.*' This artificially kept the supply of insurance high in New Orleans for a fixed time period after 2005, allowing a state-run non-profit insurer, the Louisiana Citizens Property Insurance

Company, to set up. This insurer covers those who have been rejected from private insurers, but is more expensive than a private policy. A failure to effectively reduce risk means that the state needs to find a different solution, usually in the form of a state-run insurer or reinsurer. And this is just the financial impact: there is also a human benefit to minimizing disruption and event deaths. Katrina gave us a glimpse of a society without insurance. Simple risk management such as enforcing better building regulation is the very least that can be done to prevent that dystopia.

14.2.5 Lesson 5: It Never Rains but It Pours

Hurricane Katrina was devastating to New Orleans and Louisiana, but it was only one of four Category 5 hurricanes that season to hit North America, with Emily (hitting Mexico), Rita (again hitting Louisiana, hindering the rebuild and worsening the effects of Katrina), and Wilma. Wilma is the most intense storm ever to hit the Atlantic basin, and her most severe effects were felt in Mexico, Cuba, and Florida.

When events are modeled, they are given probabilities of occurrence. But these events aren't totally independent: one bad storm in a season is a strong indicator that another bad storm might occur in the same season. Models that assume that events are independent[6] are not suitable for catastrophes!

It's common for actuaries involved in catastrophe modeling to be aware of El Ni ño years and have different probabilities assigned to events in those years. Alternatively, the cat model could use a Poisson distribution to model the number of events, which assumes that each event is independent. However, if certain events occur (for example, a Louisiana Category 5 hurricane), it is marked as an event cluster and another Poisson counting process begins to determine the number of similar events in that event cluster.

Example 14.1: Hurricane Katrina could have been modeled as one of a cluster of four Category 5 Atlantic hurricanes on 2005, rather than an independent event itself.

That so much was missing from cat models pre-Katrina shows that statistical models are always subject to bias. Scenarios we remember, or perceive

[6] For example, any model using a Poisson process to model events.

as being likely are included, but events beyond our imagination, even if they are similarly likely, are unwittingly excluded. Statistical models will get more complex, and hopefully lessons for the future will be learned from every natural disaster. Actuaries have a key role to play in identifying bias and creating solutions that are viable for insurers but still provide a crucial service to society.

Did you know…?
Historian Gustav-Adolf Schoener, of Hanover University, claims that the flood of 1524 may have been the first mass media event in European history.
http://esoteric.msu.edu/VolumeIX/EsotericaIX.pdf
(p167).

14.3 Promising the World - the Challenge and Benefits of the Globalization of Insurance

After 2005, catastrophe models were improved with the lessons learnt. But if there were shortfalls in modeling US windstorms (the USA is the biggest insurance market in the world), the problems for modeling newer markets were just beginning.

As we have seen, insurance started in Europe, and the insurance markets of North America and Australasia are particularly strong now. But in the past 20 years, more people in less developed countries have demanded insurance too. As the middle class or middle-income group grows, people have more property that they wish to protect and enough money to afford an insurance premium. Global insurers are now truly global, and write policies all over the world. But it's not as easy as just taking a pen and starting to underwrite your usual policy elsewhere: new markets need new products. Homes might be built of different materials (consider the fire risk of a thatched house versus a slate roof), be subject to different weather, and each market is subject to different regulatory standards.

Expanding to new markets was a big opportunity for insurers, but the flooding in Thailand in 2011 tested their strategy. Many insurers, including those within Lloyd's of London, had got a small amount of business in the country but insurers hadn't fully considered their risk modeling techniques, especially their catastrophe scenarios. Despite the monsoon season being a well-known risk to the Thai people, European-based insurers hadn't included severe monsoons in their cat modeling. In 2011, particularly heavy monsoon rains led to 5.5% of Thailand being flooded, with water 3 meters deep in several locations.

The damage cost $30 billion (in 2011 terms), of which only $12 billion was insured. Like Hurricane Katrina, the majority of these payments—roughly $11 billion—were made to businesses, not individuals. Compared to Katrina, the size of the loss is small, but it did force one important realization: insurers were missing a lot of probable catastrophe scenarios. Lloyd's of London, who paid $2.2 billion for the Thai 2011 floods, immediately added Thai flooding to their **Lloyd's Realistic Disaster Scenarios**, forcing all syndicates from 2012 onwards to consider the cost in case a similar event happened again.

However, a much bigger and potentially more concerning question emerged: other than experiencing huge losses, how can we identify what we are not modeling? The actuarial community and the Association of British Insurers puzzled over this, and came up with four areas where non-modeled perils often lurk:

- Geographic regions not modeled (these usually coincide with regions where historically there has not been much insurance demand)
- Secondary perils—Hurricane Katrina brought a few to international attention, but the ABI listed some others, as well as mapping secondary perils to previous events.
- Lines of business not covered by catastrophe models—catastrophes tend to relate to buildings and contents insurance. But motor vehicles, marine vessels, wind farms, or oil rigs could also be susceptible to bad weather for example.
- Extra cover or policy details—catastrophe models usually cover the big components of loss, like rebuilding the property or replacing contents. But there may be other benefits to the policy that could be claimed on in the event of a catastrophe as well, for example, an insurer providing alternative accommodation or premises or paying business interruption claims.

The damage in Thailand in 2011 cost $30 billion, ($12 billion paid by insurers, $13 billion paid by the Thai government, the remainder paid privately) but the floods had ramifications for the whole world. Thailand manufactures a lot of goods for the rest of the world, and companies on the other side of the planet suffered economic losses due to this supply chain failure. These costs have been estimated at $15 billion and are insurable, called interests abroad coverage. Hurricane Sandy that hit the East coast of North America in 2012 also incurred similar losses: although New York is not a manufacturing hub, it is a global hub for financial trade. Financial service companies on the other side of the ocean claimed for business interruption.

Assessing risk for the entire world is therefore a lot more complicated than 'new market' type problems of pricing and demand, as the world is increasingly interconnected. Modeling supply chain risk is a hot topic in actuarial science at the moment, as more insurers try to get to grips with the risks that their policyholders in one region might be affected by a risk in a totally different area. It is a small world, after all.

It is easy to be wise after the fact, and the flooding in 2011 ignited demand for more **general insurance**. **Gross Written Premiums** in Thailand rose 29% in 2012 compared to the previous year. Life insurance in Thailand has also dramatically increased in popularity: the market has increased tenfold since 1999 according to the Thai General Insurance Association. This can partly be attributed to the awareness of insurance and of the risks from climate change, but also the increasing number of households who can afford insurance.

Access to insurance has vastly increased for most of the world's population. Traditional insurance products work best for the middle classes, who have sufficient property and cashflow to pay a premium to protect their assets. But what about poorer individuals: must they bear all risk by themselves?

14.4 A Micro Introduction to Microinsurance

For those who are very poor or very rural—or both—traditional insurance products are largely irrelevant, unaffordable, and unobtainable. But, to give people in these circumstances some financial protection from risks, some innovative companies are offering microinsurance. As the name implies, microinsurance is a miniature version of regular insurance in a lot of ways: smaller sums insured, smaller premiums. For example, for an annual premium of $10 a year, a product may pay $450 for the death of the policyholder, or $250 for fire or flood damage. The aim for microinsurance is therefore not necessarily to indemnify, but to reduce the immediate financial shock of disaster.

But small doesn't mean easy. Microinsurance isn't just a scaled-down version of traditional insurance; it requires some flexibility from the insurer. Actuaries in microinsurance need to be continually communicating with the communities they are trying to serve, and design products that address the specific needs.

Did you know…?
Seven out of ten people in Africa depend on small-scale farms. That's a big market for microinsurance!

Exercise 14.2: What parts of traditional policies might need to be changed for microinsurance purposes?

Some of the challenges in providing microinsurance are:

- How to actually sell the policy to the customer. Customers may have limited internet access, although they may have a mobile phone.
- There is unlikely to be much data about the insured risk. Micro life insurers could use the country's census data (or data from global organizations) related to mortality, but general insurers might use a farmer's own estimates of how often floods or droughts occur.
- Traditional checks on claims might not be available. For example, whereas in the USA a life insurer needs to see a death certificate before paying a life insurance claim, in Bangladesh, perhaps confirmation from a respected member of the community is sufficient. This will also speed up the time it takes to pay claims, which makes insurance more appealing.
- Exclusions will erode trust and make products more complicated (not ideal when customers may have a lack of knowledge about insurance products), but the micro-insurer still needs a way to avoid **anti-selection**.
- Markets may suddenly become untenable if political instability means that brokers or actuaries can't travel safely, or if violence means that life insurance cannot be sustainably provided.

Microinsurance can be profitable, but it also provides a social good. By receiving even a small claim payout after a disaster, policyholders can often continue farming (investing in more seeds or animals), preventing them from falling into poverty. In 1998, a government-funded microinsurance program in Mexico was created to improve food security and encourage economic development after extreme weather events. Like the original insurance at Lloyd's, groups of farmers were encouraged to form small pools of similar risks, with the government subsidizing premiums. Whether it is government- or privately run, microinsurance is one tool to encourage sustainable development and risk mitigation techniques in developing countries. This isn't charity, this is effective risk transfer.

Parametric Insurance

Parametric insurance is one way to run microinsurance. Instead of defining cover using the loss to the insurer, it defines it using the **loss event** itself. For example, instead of being covered by a flood policy (which would pay if any water that came from outside caused havoc in your home, be it from a river, coastal flooding, rainfall or storm surge), parametric insurance defines the parameters of the loss event ahead of time, so you might be insured if there was more than 5 mm of rainfall in one day at your nearest weather station. It also defines what the payout will be ahead of time, rather than waiting to assess the damage. This means that claims can be paid much faster, and the claims assessment process can be written as an algorithm.

The payout can also be a function of the objective measurement. In our rainfall example, this might be a payout of $1000 for 5 mm of rainfall, but $2000 for 7 mm and $5000 for 10 mm. Due to its potential for automation, parametric insurance can be simple and fast. However, for complex risks there are many parameters to define, so it is unlikely to become mainstream for developed insurance markets.

14.5 Cyber Insurance

If you thought predicting the weather was hard, then predicting people's behavior is even harder. Cyber-attacks and cyber-terrorism are examples of non-natural catastrophes—events with a direct human protagonist. Other examples include terrorism, riots, and even large oil spills. Like other catastrophes, they are low frequency events, but can have very high severity.

Cyber risk is a particular concern for insurers as it is a comparatively new risk, and so many industries and individuals are reliant on computers and therefore have the potential to be affected. This section focuses predominantly on cyber insurance for this reason, but it is left as an exercise for the reader to consider the similarities between terrorism or oil spills and nat cats.

But how similar are cyber events to nat cats?

- Like hurricanes, cyber-attacks usually happen in clusters. We've already discussed the 2005 hurricane season, where not one but four Category 5 hurricanes hit. A similar situation is seen in cyber: one weakness is exploited, and then more weaknesses are identified once the attacker is in the system. Once a cyber-attack occurs, a subsequent attempt is likely to occur soon. As with hurricanes, this means that statistical distributions that assume independent events (such as the Poisson distribution) are not suitable choices, but such a distribution could be used in conjunction with one that models the size of the cluster given an event is happening.

Did you know…?
It is estimated that by 2025 there will be 200 zettabytes of data, half of which will be stored in the cloud. That's over 186,264 billion gigabytes!
http://cybersecurityventures.com.

- Like the Thai floods and Hurricane Sandy, large cyber-attacks can affect multiple policyholders. However, as computer systems are virtually connected, rather than co-located, it is hard to judge **accumulations** of risk. **Underwriters** may ask questions to judge this kind of risk by mapping which companies or customers a business is regularly in contact with, and asking questions related to the system itself (which cloud provider is being used, for example). Quantifying supply chain risk is still in its infancy, but judging these accumulations is a concern for both nat and non-nat events.
- Unlike nat cats, which are sometimes described as 'Acts of God', there is usually some human error in cyber events: a weak password, someone accidentally clicking a dodgy link, responding to a spam email etc. In traditional insurance, claims that arise due to the policyholder's fault are often excluded or severely limited in order to reduce **moral hazard**. It's much harder to determine fault in cyber events. Perhaps an employee had a weak password, but would an extra character or two really have prevented the claim entirely? Because of this, cyber-attacks are often treated in a similar way to natural disasters: they are expected to occur with varying severity and cannot be entirely prevented, but the damage can be reduced with good planning and security strategies. Of course, some cyber-attacks are performed or encouraged by governments, and perhaps should not be modeled as isolated and unpredictable events.
- Like nat cats, cyber-attacks can cause secondary perils. We think of these events as causing purely financial damage but depending on the attack location, there can be real-world implications as well. There can often be a component of property damage caused by the attack, especially if a railway or water station is hacked. On February 5th 2021, a hacker gained access to a water treatment station in Florida and changed the level of sodium hydroxide in the water from 100 parts per million to 11,100 parts per

million.[7] Fortunately, this attack was quickly spotted, but it could have caused substantial damage to property, animals, and humans. In October 2020, ransomware (a cyber-attack where the user is asked to pay money in order to release their files) targeted a hospital in Duesseldorf, Germany. This attack caused patient records to be unavailable, so patients were being sent to an unaffected hospital 20 miles away. One woman, taken to hospital with a life-threatening condition, died as a result of this delay to her treatment.

- Both weather systems and cyber-attacks are continually evolving. Natural catastrophes are being altered in frequency and severity by climate change, and cyber-attacks are continually evolving to exploit weakness in new systems and technologies. As such, the past is not necessarily a good guide to the future, so pricing for these risks is a real challenge for actuaries.

There are clearly a lot of similarities between traditional catastrophes and cyber risks, but one similarity is that governments are increasingly needing to play a bigger role in risk mitigation. As we saw after Hurricane Katrina, insurers wanted to exit the market but were prevented from doing so (at least for a short time after the event) by regulators and governments. For natural catastrophes, governments are being forced to take on their share of the risk, either by operating schemes such as Flood Re, the Mexican microinsurance program, or through emergency aid after an event. This seems reasonable, particularly in light of the huge subsidies that states give fossil fuel producers.[8] So far, governments' intervention in cyber security has largely been to encourage training and employing a suitably skilled cyber security workforce,[9] but that may change in the coming years. Intervention has already occurred in terrorism insurance, with some governments setting up government-backed terrorist reinsurers. For example, the UK has Pool Re and the USA has the Terrorism Risk Insurance Program.

Previous nat cats have shown that insurers need to be transparent about the cover that they are providing. In recent years, clarity over whether cyber risks are covered, and to what extent, has been demanded by both regulators and customers in order to avoid a post-event realization of underinsurance.

[7] Sodium hydroxide is also known as caustic soda, and is extremely combustible. It is commonly used as commercial oven or drain cleaner. Needless to say, ingestion is not recommended.

[8] Governments spend $423 billion per year subsidizing fossil fuels (estimate as at October 2021, United Nations Development Programme).

[9] For example, the UK government has launched an apprenticeship scheme in cyber security programming, but not made any policies relating to liability for state-backed cyber damage. https://www.abi.org.uk/products-and-issues/choosing-the-right-insurance/business-insurance/cyber-risk-insurance/

For cyber risks, companies have perhaps relied too heavily on insurers paying claims after the event, rather than putting preventative risk measures in place. However, in December 2021, Lloyd's of London implemented new exclusions on its cyber policies. Lloyd's, acting as a marketplace, has effectively created four standard exclusions relating to cyber-attacks that syndicates can use if they wish. This firstly standardizes the cover and exclusions available, but has the important secondary effect of getting insurers to think about the risks involved in cyber policies. One excludes damage that occurs because of '*war, or a cyber operation that is carried out in the course of war*'. It also excludes retaliations from countries, and where it is not possible to find concrete evidence of an attacker, it also excludes cover based on '*an inference that is objectively reasonable*'. These wordings, although optional, are likely to be adopted by most Lloyd's syndicates. They are controversial because they seem to exclude state-sponsored hacking outside of a full-on war scenario. With these exclusions in place, the costs of damage arising from state-funded hacking will no longer be borne by insurers.

Whether it's hacking or hurricanes, actuaries are key in ensuring fairness between those who cause the risk and those who bear the consequences.

15

Climate Change and Biodiversity Loss

Chapter Objectives:

- Consider how climate change affects insurance companies, and potential ways of mitigating the risk
- Understand the risks of biodiversity loss and the role of actuaries in this emerging area

In our hands now lies not only our own future, but that of all living creatures with whom we share the Earth.

—David Attenborough

15.1 Climate Change

The main assumption that lies behind almost all actuarial work is that the past is a good guide to the future. Imagine the challenge to actuaries then if that assumption is almost certainly false! Natural catastrophes are increasing in both frequency and severity due to climate change, and this is not only a problem for general insurers but also for life and health insurers: the WHO estimates the direct cost of climate change to health by 2030 will be up to an

© The Author(s), under exclusive license to Springer Nature Switzerland AG 2022
C. Townsend, *A Risky Business*,
https://doi.org/10.1007/978-3-031-11673-5_15

additional $4 billion USD a year, and cause an additional 250,000 deaths a year.[1]

The good news is that there are many statistical models of the effects of climate change. Many excellent professors of climate science and meteorology are already using modeling tools similar to those of actuaries. Most climate models are based around an assumption of Equilibrium Climate Sensitivity—basically, how much does the Earth's temperature increase if the carbon dioxide concentration doubles. The models are very sensitive to this parameter, so as estimates are refined there is considerable parameter risk[2] here. Another assumption is how climate change impacts temperature. It may be that mean temperature is shifted, but the rest of the distribution remains the same. If this were true, for insurers, as long as countries can adapt to the new, warmer temperature, the variability of temperatures and therefore the risk of extreme events would remain the same as it is now. On the other hand, it could be that the mean temperature remains the same, but the variability increases. Then, both hot and cold extreme events would be more frequent, even though the average temperature would be the same. This situation would be very difficult for countries and insurers to transition to, as systems would need to be robust for both floods and droughts, snow and heatwaves. Alternatively, the relationship between climate change and temperature may affect the distribution in another way—maybe skewing the distribution, so extreme cold or extreme heat events are more common, but not both. Most likely, the real effect is a mixture of these possibilities: the mean temperature will increase, but the variability will also increase, perhaps also to being slightly skewed. Each of these will have different effects for insurers, and gives a glimpse of how difficult parameterizing climate change models can be (Fig. 15.1).

Catastrophe modelers or pricing actuaries may want to incorporate some elements or assumptions from these models into their own work. For example, actuaries might use recommendations from the Intergovernmental Panel on Climate Change as a starting point for scenarios for regulatory change. Here, actuaries are not the expert on the exact mechanics, but they are experts in modeling. Amidst the plethora of climate models, an actuary's judgment and discernment can help choose an appropriate one for the specific business case.

An actuary's role in the climate emergency is not just as a passive data-gatherer, who models and reacts to the unfolding disaster. There are lots of

---

[1] From the World Health Organization, Climate change and health (who.int), estimates as at 31 October 2021.

[2] Parameter risk—the risk that the model's parameters are incorrect.

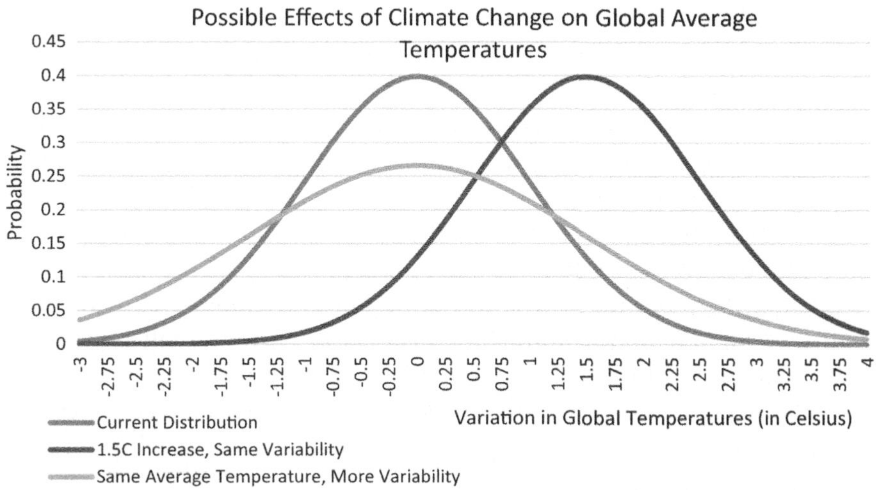

Fig. 15.1 Possible effects of climate change on the distribution of effects on average temperature

roles where an actuary must act as a leader on this topic. Here I have listed just a few specific activities where an actuary adds value. There are two vital skills required here: the ability to quantify risk, and the ability to communicate it. Fortunately, actuaries can do both, and so an increasing number and variety of roles are now open to those with an actuarial background.

- Asset valuation: actuaries have traditionally worked in investment, trying to find assets that match the liabilities they have just quantified. Actuaries can work powerfully in this area by valuing assets ethically: climate change flips the traditional 'risk vs. reward' scenario, where bearing the highest risk gives the highest reward. Instead, under current valuation systems, the investor holds only a small amount of climate change risk, but others (generally those in the poorest regions) bear the majority. By including these additional stakeholders, the assessment of risk vs. reward changes.
- Holding capital: capital requirements suggest that some assets are low-risk. Climate change may change which those are: shares in an oil company used to be almost a guaranteed profit, but increased regulation may mean that the company faces restrictions on activity in future, impairing profitability. Oil reserves may become stranded assets, where they still exist but can't be sold to realize their value due to regulation. Regulation change is a risk in itself that actuaries will need to manage.
- Advising **underwriters**: as we navigate into a low-carbon economy, insurers who currently insure polluting industries will need to create a

strategy to move from their current business mix. This is not just the right thing to do, but also critical to a business' profitability: if all your policies are oil and gas, then you may not have any customers in 20 years. Businesses are used to converting actuaries' assessments of risk into underwriting limits; factoring in the increased risk of extreme climate change if these industries continue as they are is just a tweak to the calculation. 35 insurers have already declared that they plan to stop writing insurance for coal production, but only 3 have done the same for oil and gas.[3]

- Designing new products: it is not all doom and gloom—climate change will lead to new opportunities for insurers as batteries will be vital in a green energy revolution. Energy that is effectively free may in turn spark new industries and processes that need insurance, such as electrolysis. Actuaries will need to have a say in designing, pricing, and **reserving** these new products which enable others to act sustainably.
- Managing transitions in pensions and insurance providers: moving to a new investment or business model has some risks. There is a risk of the new system failing or being unprofitable, there is a risk of getting the pricing wrong, or that other companies will continue to prop up the old system and the company will be an outlier. However, non-insurance businesses will also face **transition risks**, and may wish to insure them, creating another opportunity.
- Government actuaries or those with **reinsurance** knowledge can help set up funded, well-structured natural catastrophe risk transfer programs. Some examples include the National Flood Insurance Scheme in the USA, and the Flood Re scheme (as discussed in Chapter 13) in the UK. These schemes should not only look to provide insurance, but also to encourage and support industries and individuals to move into less flood- or hurricane-prone areas.

Even without one of the above roles, actuaries can still use their valuable skills to lead the way on climate change by joining an environmentally-focused professional working group. The Institute and Faculty of Actuaries have set up several variations of these groups to create an industry response to the challenges, including transition risk, measuring biodiversity loss, and asset valuation. Roles that encompass any of the activities above are starting to be called 'Climate Actuary' positions, and as more companies set up teams

[3] As at COP26, November 2021. If you are interested in finding out more about what insurers are (and aren't!) doing in this space, the COP26 talk 'The Insurance Industry: Climate Leaders or Climate Laggards?' is interesting, accessible and free.

focused on sustainability, the demand for actuarial skillsets to be applied to the problem of climate change risk is only going to increase.

But a warmer world is not the only environmental risk that must be quantified and managed.

15.2 From A(ctuary) to Z(oonotic)

Actuaries use models to explain and predict all the time. Finding an explanation does not have to lead to attributing blame, but it does give valuable insight about the likelihood of future events. It may even help us to prevent them entirely.

This is why Coronavirus-19 was so surprising: it was predictable, and yet, very few countries were prepared. Many viruses such as SARS (severe acute respiratory syndrome), MERS (Middle East respiratory syndrome), HIV, and Ebola have sprung up in the past two decades, becoming national crises and threatening global health. In this sense, the only surprise is that it took this long for a disease to take hold globally.

But what is causing this apparent increase in viral diseases? One thing that the above list of diseases has in common is that they are all zoonotic diseases. The World Health Organization (WHO) defines a zoonotic disease as any infection naturally transmissible from vertebrate animals to humans. These examples are just the big ones in recent memory. Although these types of infection have always been a possibility (the Black Death[4] was a zoonotic disease) the WHO estimates that 60% of emerging infectious diseases in humans are zoonotic, and so-called cross-over events, where an infected animal transmits the virus to a human, have tripled in the past decade.[5] This increase is due to anthropogenic habitat loss. Animals are forced to live closer to humans, and so the chance of an encounter increases. Coupled with this, as fish stocks decline globally, some people become more dependent on wild land animals for food. Eating an infected animal, or being bitten by an infected animal, are the two most common causes of a cross-over event.

Coronavirus-19 is not currently classified as a zoonotic disease as a host animal has not been found. Pangolins are being investigated as a potential host based on circumstantial evidence. There have also been cases of mink

[4] The Black Death or the Great Pestilence are the common names for the bubonic plague pandemic that spread across Eurasia and North Africa between 1346 and 1353, killing between 75 and 200 million people. This is the same plague suffered in London in the 1600s, leading to the creation of the London Bills of Mortality.

[5] WHO EMRO | Zoonotic disease: emerging public health threats in the Region | RC61 | À propos de l'OMS. http://www.emro.who.int/fr/about-who/rc61/zoonotic-diseases.html

infecting humans with Covid,[6] so the virus does have this cross-over ability, and so at time of writing it has been classified as an *emerging infectious disease of probable animal origin.* Regardless of whether it is zoonotic, the pandemic has highlighted the effect that a virus can have globally. This of course affected pensions companies, life assurance providers, insurers, and investment firms as well. As the recovery from this pandemic continues, companies will want to include the risk of a similar event occurring again.

As is the case with climate change, actuaries are not passive data-gathers here. As well as making changes in their own lives and at their own places of work, actuaries can encourage the protection and promotion of biodiversity through their work. Thinking on biodiversity is currently well behind the progress on climate change and decarbonization, but no less important: it was ranked one of the top 5 risks globally in 2021.[7] Companies and governments are currently focusing on climate, but losing biodiversity often starts a feedback loop: losing vegetation cover on moorland can lead to the loss of peatland, which are huge carbon sinks. Their loss releases carbon dioxide, which warms the planet and increases the rate at which peatland is lost. You might be aware that rainforests are the lungs of the planet, but every habitat performs a vital role in keeping the Earth at a stable temperature, as well as providing a home for animals and plants. Whether controlling infectious disease or reducing climate change is the goal, retaining and promoting biodiversity is a core part of that.

Encouraging biodiversity and slowing climate change can induce the same actions. For example, halting deforestation not only allows trees to continue absorbing carbon dioxide, but also maintains a safe habitat for biodiversity. However, the two goals are distinct. For example, in order to absorb as much carbon dioxide as possible, quick-growing woody trees such as maple, oak, or ash would be the most effective. However, planting a monoculture of any of these would not foster biodiversity as effectively as a mixture of species.[8] In fact, it may even reduce biodiversity. By allowing pests a field of their favorite meal, and with few predators living in the monoculture to reduce their numbers, unwanted insect populations can flourish. If this is an invasive

[6] Mink don't live in China, where the original outbreak began, so are unlikely to have been the original host species.

[7] The AXA Future Risks Report 2021 | AXA UK. https://www.axa.co.uk/newsroom/media-releases/2021/axa-future-risks-report-2021-climate-change-biggest-concern/?gclid=8606d861da0b1bdb2bf3ba4c1f9d019d&gclsrc=3p.ds&msclkid=8606d861da0b1bdb2bf3ba4c1f9d019d

[8] Mixed-species versus monocultures in plantation forestry: Development, benefits, ecosystem services and perspectives for the future - ScienceDirect, https://www.sciencedirect.com/science/article/pii/S235198941830088X.

species that competes with native species, planting a monoculture forest or crop could be actively harmful to native plants and wildlife.

The role of actuaries in promoting biodiversity is therefore similar to the role of actuaries in mitigating climate change: advocacy, product design, and ethical investment. However, there are some additional challenges to biodiversity risk. Here are just three hot topics in actuarial work on biodiversity. I stress that thought in this area is emerging, and this is only my personal selection of the many interesting discussions in this space.

15.2.1 Underwriting Exclusions

It is simple, but not over-simplified, to say that industries pollute because they get away with doing so. If they had to withstand the costs of doing so, they would not. So, insurers can encourage their customers to be more environmentally friendly by refusing to pick up the tab if they routinely pollute or have poor waste control practices. Specifically, they can exclude some risks, such as environmental liability risk or biodiversity risk, from policies. This only works if the wording is specific, and it can be hard to define exactly what was caused by biodiversity loss. It can also be difficult to collect data on which companies are likely to be polluters, as historic spills and lawsuits may be covered by non-disclosure agreements. There are two very good examples of where exclusions really should apply:

- Teck Resources, the world's second largest exporter of steel-making coal, has been fined multiple times in the past decade by Canada's Ministry for the Environment. The largest fine was for CA$1.4 million, for dumping waste water contaminated with selenium into the Elk Valley River. Selenium is a naturally occurring chemical that is released during coal extraction, and causes deformities and fertility loss in aquatic species. Similar effects are found in animals that eat contaminated fish, including black bears. Insurers trying to promote biodiversity should not be enabling customers to evade the consequences of their own actions, especially if there have been multiple fines levied for the same activity.
- Exclusions could also apply for business interruption due to environmental protests. In Peru, protestors have blocked access roads to mining company MMG's Las Bambas mine. Locals who rely on the biodiversity in the area were unhappy with the arrangement: a multinational mining company degrading the quality of the local environment while paying a low environment levy to the central government in return. The local area, who had to put up with the pollution, noise, and dust, received no benefit. If this

interruption and legal risk sits with the mining company, there is more incentive for them to change to a more inclusive and less polluting way of working.

15.2.2 Creating a Biodiversity Loss Metric

Part of the reason that biodiversity loss has not been part of strategic conversations is that its value is hard to measure. Therefore, it doesn't easily fit into traditional investment decision frameworks: it's easy to compare the value of two projects if you have a monetary **Net Present Value**, but what if the decision is between making \$2 million and destroying 10 acres of forest, and \$1 million and 1 acre of forest?

Firstly, actuaries (as risk-quantifiers) can help devise ways of measuring biodiversity loss, akin to the ways we currently have of measuring carbon emissions. There needs to be at least two metrics here: one relating to the goal (for carbon dioxide, this is average global temperature), and one relating to the progress (carbon emissions). Biodiversity doesn't have these yet, and so although we are peripherally aware that some animals are endangered,[9] it is hard to create a specific, actionable strategy.

Metrics would also allow regulation for financial institutions. In 2021, the Bank of England required banks and insurers to stress-test their assets with respect to climate change: effectively, to see how each institution would fare in a set of situations related to climate change. This is called the Net-Zero stress-test, and the scenarios only relate to carbon emissions and global temperature (and their effect on natural catastrophes). Because of the lack of a metric, there isn't a similar exercise for biodiversity loss.

Discussions are ongoing about the creation of a biodiversity loss metric. We are used to hearing about the deforestation in terms of acres (or football fields!), but that doesn't account for the amount of biodiversity in that space. An acre of rainforest supports more species than an acre of ex-industrial brownfield: how should that be accounted for? Perhaps some function of the number of species supported would be advantageous (although this advantages the areas that have been more thoroughly explored, and scientists discover new species regularly). Even if only one important species lives in a space, if it is uniquely found in this location, what then is the value? Finally, are all individuals of a species equal? What is the value of losing a breeding

[9] In fact, 1 in 4 species (including plants) are at risk of extinction.

pair of Red Kites,[10] as there are only 4,600 pairs in Britain? What would the value of that loss be if there were fewer than 20 pairs, as there were in 1960? Unlike carbon, where all carbon dioxide is equal, biodiversity metrics are complex because we are sometimes literally comparing apples with oranges.

Climate change already has some international agreement. As well as a defined metric, the Paris Agreement has set clear targets for countries to align to. This has helped countries and companies create a roadmap to Net-Zero. Biodiversity doesn't yet have such global accord. This is a problem because biodiversity loss has a global impact. It not only increases the likelihood of zoonotic diseases causing pandemics, but it also creates climate feedback loops. Climate feedback loops are like vicious or virtuous cycles. Biodiversity reduces the effect of high carbon emissions by giving the planet ways to absorb and store carbon to stabilize its temperature. When we lose biodiversity, not only does our planet lose these beneficial mechanisms but they actually start to work against us, increasing global warming. Without measuring and reducing biodiversity loss, there is a risk that efforts to reach climate change targets will be adversely affected too.

15.2.3 Calculating the Value of Biodiversity to Insurers

Increasing biodiversity is not purely altruistic. There is a real and increasingly large cost of not doing so. Firstly, insurers and life assurance companies can count the direct cost of zoonotic diseases (£2.5 billion for UK insurers in 2020 due to Covid, more than 5 million global deaths). Then, insurers should take into account the costs for their customers as potential claim costs. For example, North American bats are estimated to be worth $22.9 billion[11] for their crop protection and pest control services. If bats are lost, this cost could be passed to insurance companies writing crop insurance.

Monoculture also brings a multitude of risks due to the lack of biodiversity. As well as potentially increasing pest numbers, monoculture is also more prone to loss due to bad weather. It is all-or-nothing for the farmer: they are more likely to suffer a total loss as their entire crop is susceptible, whereas in polyculture farming the exposed crop would only be a proportion of their income. Crop insurers will be most interested in these effects, but there are wider risks here. If poor weather is experienced in an area of monoculture farming, especially when it is an essential crop for locals, food shortages may

[10] A graceful reddish-brown bird of prey with a distinctive forked tail. It has been the focus of a slow but successful conservation program in Wales, and now they are so numerous that the RSPB doesn't count them individually (4,600 breeding pairs is only an estimate).

[11] Department of Ecology and Evolutionary Biology, University of Tennessee, 2011.

ensue. Food shortages can destabilize local economies. Although this may seem unlikely, twin crises of more extreme weather caused by climate change coupled with an increase in monoculture farming leads to a very real risk of food insecurity—and with it a host of political, economic, and health risks.

More immediately, life and health insurers should value biodiversity for its beneficial effects on physical and mental health. Research shows that spending even 20 minutes in nature, especially by water, reduces stress, blood pressure, and heart rate.[12] It benefits mental health too: access to green space increases mental well-being and reduces mental health diagnoses, particularly depression.[13] More directly, exposure to wildlife has been found to significantly increase life expectancy. Life and health companies sometimes provide access to preventative healthcare such as annual physical exams; providing access to nature could also reduce claims costs while potentially costing very little to provide.

Measuring value in this way is not ideal: it is entirely anthropocentric, based around what nature can do for us. It does not value nature in its own right, and assumes nature has no inherent value other than what we can gain or extract from it. It also fails to include the value of human life. We need to eat a variety of plants to stay healthy, we need a system to provide clean water, and we need something to stabilize the level of oxygen in the air we breathe. However, financial valuations can work in the short term to increase awareness of the role that ecosystem services play in stabilizing and enriching our lives. It should also be understood that the monetary value is only ever a partial valuation. We haven't yet learnt everything from every plant and creature, and we can't value what we don't know exists. For example, the blood of horseshoe crabs is antibacterial and has been used as a test for contamination of vaccines, including Covid-19 vaccines. If scientists weren't aware of this incredible property, these ancient arthropods would likely be undervalued and under-appreciated. As species are being driven to extinction, there is a loss not only of ecosystem services, but natural properties of wildlife that disappear unnoticed.

Deriving a monetary value of natural capital is just one way in which actuaries are relevant to this developing field. Advising governments on biodiversity regulation and quantifying and managing the transition risk of moving

[12] Japanese Forestry and Forest Products Research Institute.

[13] University of Exeter.

to a new, eco-friendly economy are also crucial to success in the medium term.

We are inextricably linked to biodiversity. Our food depends on it, our jobs depend on it, our health depends on it. In short, our society depends on it, and so its loss is one of the greatest, but largely yet unquantified, risks facing the world.

But that's enough of the urgent and immediate global problems of today—in the next chapter, we'll look at problems for the future!

16

Latent Claims: A Problem for Another Day

Chapter Objectives

- Understand the definition of latent claims and the stages and challenges of modeling latent claims
- Identify the ways in which latent claims can be mitigated against and priced for
- Consider possible future latent claims
- Understand how insurance companies interact with the legal system
- Understand the effect of environmental claims on insurers

Cause you don't know what you don't know

—Taylor Swift, Mean

16.1 An Introduction to… (Wait for It) …. Latent Claims

Some people are always late. Whether they spend ages getting ready, can't read a watch, or are eternally optimistic about traffic conditions, I'm not sure. I'm not one of them. If you know in advance that someone is a late person, you can at least mitigate the situation: bring a book, grab a coffee, or give them an earlier time to aim for so they actually attend at the time you need. The

© The Author(s), under exclusive license to Springer Nature Switzerland AG 2022
C. Townsend, *A Risky Business*,
https://doi.org/10.1007/978-3-031-11673-5_16

most frustrating thing about late people is that you are never sure if they are just running late, or if they're not coming at all.

For all of the modeling, estimating, predicting, and quantifying that is part and parcel of actuarial work, actuaries don't know the future. In Chapter 3 when we considered 'What could possibly go wrong?', we were mainly thinking of things that are likely to go wrong. Sometimes, something way outside of our expectations occurs, many years after the policy began. These **latent claims** are part of the job, but they aren't expected. In fact, one definition of latent claims is that they are claims arising from a risk not specifically considered when the policy was written.[1]

Latent claims most commonly occur in liability products. There are three main categories of latent claims: ones that relate to bodily injury, usually due to a workplace environment; pharmaceutical claims (when a drug has unintended consequences), and environmental damage. But why does it take so long for these claims to emerge?

Previous chapters have discussed reasons for reporting delays, including the possibility that the policyholder is not themselves aware of the claim event for a period (for example, if they were burgled while on holiday). Latent claims are an extreme version of that scenario: either the damage was not clear until a later stage, or the damage was not attributed to the actual cause until many years later. A clear cause is required for latent claims as usually the person who suffers the damage is not the policyholder: the policyholder will be a company responsible for causing the circumstance where the victim suffered loss. So, the policyholder might not be aware that there was a claim event until the damage is evident to the victim and the cause understood enough that the case is reported to the policyholder, who is potentially liable.

Example 16.1 Pregnant women in the late 1950s were offered thalidomide to reduce morning sickness. Testing showed that there was no lethal dose for animals, so it was deemed safe for humans. The drug was licensed in 49 countries (notably not the USA—see below). Some of the women who had taken thalidomide had children with disabilities: some had shortened limbs, some had organ damage and some suffered eyesight and hearing difficulties. However, it took five years for the link to be made between taking thalidomide and fetal birth defects, for a number of reasons. Firstly, it caused a wide range of disabilities, some of which were similar to symptoms of genetic conditions. Secondly, these were seen across many countries. In pre-internet times, it wasn't easy to collaborate long-distance in a timely manner. Finally, even at an individual level it takes 6–9 months from taking thalidomide to having a baby.

[1] The alternative definition is that a latent claim is any claim with a long period between the claim event and claim settlement. As we will see, the surprise element in the first definition alludes to some specific challenges for actuaries, so this chapter will assume that definition. It is also more in line with the commonly used definition of the word *latent*, meaning hidden or concealed.

This caused a lag between seeing the effects, hypothesizing and then proving that thalidomide was the cause.

About 10,000 babies suffered birth defects because of thalidomide, half of whom died as infants. The drug manufacturers were brought to trial in 1968, ten years after it had been made available for sale.

Did you know?

In 1960, there was a new recruit at the US Food and Drug Administration. A pharmacologist called Frances Oldham-Kelsey had just been hired. Much to the surprise of her manager, on the first day in the office, it transpired Kelsey was a woman!

When Kelsey was given the application for thalidomide to review its safety, it was already being used in 40 other countries, so Kelsey's manager thought that this application would be an easy one to start her off with. The application relied heavily on doctors' testimonials rather than clinical trials, so Kelsey denied the application. Instead of supplying evidence of further clinical studies, the drug company demanded Kelsey approve their application six times.

Kelsey had experience questioning drugs' effects on women, especially pregnant women. In 1938, she had tested the safety of quinine, and despite many trials showing it was safe for adult males (the default test subjects), Kelsey found it was not broken down effectively by pregnant rats, nor by rat embryos.

On behalf of the FDA, Kelsey rejected the application to take thalidomide to market in the USA, saving thousands of babies from avoidable birth defects. In 1962, Kelsey was awarded the President's Award for Distinguished Federal Civilian Service by President Kennedy.

Let's look at the claims journey of a latent claim event in more detail. Like all risks, it starts when the policy begins. During the period of cover, the policyholder inadvertently exposes their employee or member of the public to a harmful substance or situation, but the damage is not immediately obvious. At this early stage—where all latent claims and their damage is unknown—, no latent claims are expected (or the policyholder would immediately stop exposing the public to hazardous substances, and stop exposing themselves to a lawsuit).

Just after the policy cover ends, the insurer and policyholder (and probably the victim too) still has no idea that damage has occurred. It may take years for enough people to come forward with similar damages, and scientists to investigate what all these injuries have in common. This is called the potential stage—the damages are only potentially related to the policyholder. It is at this evidence-gathering stage that the uncertainty around claim costs starts to increase, although the reported claims are likely to be small. For example, perhaps one or two customers have sued a company for damages after using their product, but the link is unproven. So, the insurer may be facing two claims to be paid (with a suggested claim settlement), but this might be revised down to zero if the product was not found to be at fault or, if the case is successful, other plaintiffs may come forward. At this stage, the insurer can only keep an eye on the news. Notice that it is not the insurer who would be going to court here, it is the policyholder. The insurer's role is to wait until the policyholder is deemed legally liable for the damage, and indemnify them. Due to the time required for comprehensive scientific studies and legal challenges, this stage can be many years—even a decade or more (Fig. 16.1).

The potential stage answers the question 'Is there a cause to claim?'. If the damages are found to be due to some other (non-insured) factor, then the latent claim fizzles out to nothing, and insurers breathe a sigh of relief. Sometimes, the link is unproven due to lack of evidence and so the claim remains in the potential stage until science or the law can make a ruling either way. The assertion that A is caused by B must be properly supported by scientific evidence, and that takes time. The delay in confirming a link, or otherwise, is not only due to the desire of companies to minimize payments (although they clearly have a vested interest in casting doubt on such links);

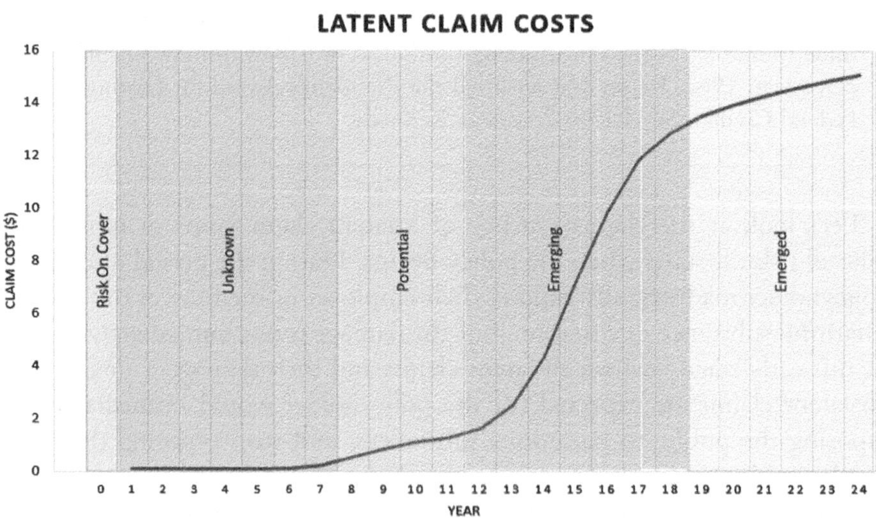

Fig. 16.1 The stages and costs of a latent claim

it may also represent a prudent scientific approach. But if there is cause to claim, that still leaves many more questions than answers; the most obvious being 'how much will this cost?'.

As we noted in Chapter 8, instead of estimating a total claim cost, we can instead estimate the claim **frequency** and **severity**, and then multiply them together. The issue for latent claims is that some people who suffered loss may not have realized that the loss was due to exposure during some previous experience. Let's be realistic—if you suddenly developed a rare form of cancer, your first thought might not be to list all substances with which you've had contact in your lifetime! Instead, most victims will only realize the link once there has been a well-publicized successful claim, and then put forward their own case.

History Repeats Itself
To gain a better understanding of how long and convoluted the process of evidence-gathering can take, I highly recommend reading *The Radium Girls* by Kate Moore. This is an account of the consequences of absorbing radioactive radium in paint. The victims were young female painters of watch dials, who licked the tip of their brushes to narrow the brushstroke, but the connection between this activity and the illness and death of the painters was contested for many years.

The pattern of company denials, obstruction of the publication of evidence, industry lobbying and a refusal to voluntarily compensate victims (especially minority groups) detailed in *The Radium Girls* might feel familiar: readers can draw their own parallels to modern industries.

This is the first stage where any meaningful reserve can be calculated. There are two ways of looking at it: top-down, or bottom-up. The top-down approach is to consider the market-wide loss. For example, by estimating the number of users of a certain product or number of employees of a certain profession in the past 25 years, and then applying the insurer's market share to that number in order to estimate how many injured parties the insurer might need to pay. Then, using the initial cases to give a reasonable range for average severity, an actuary could multiply the expected severity by the estimated number of victims and calculate a loss cost. Alternatively, if there is a market view of a total insured loss due to the latent claim (sometimes industry bodies or actuarial professions might do this for a particularly large or topical issue), an insurer could assume they take their market share of that cost.

Example 16.2 If smartphones were found to give rise to a latent claim and an insurer had a product liability market share of 25%, a top-down reserve could be made as follows:

Industry figures suggest there are 5.3 billion smartphone users in the world. We assume that the insurer therefore covers 25% × 5.3 billion = 1.325 billion smartphone users. If each claim cost $10,000, the insurer's top-down latent claim estimate would be **$13,250 billion**.

This top-down approach ignores any differences between insurers: any two insurers with the same historic market share would estimate the same cost under this method. However, like other industries, some insurers are specialists in certain areas: one insurer might target younger customers or self-employed people, or certain lines of business. If the data was available for them to calculate their market share of the affected demographic, the top-down method could give an accurate estimate, however this is unlikely to be easily accessible.[2] Instead, an insurer may need to review every policy to decide if each one might be affected by the new latent claim type—this is the bottom-up method. Of course, this is very time-consuming, particularly if policies were written with different wordings or terms and conditions. For each policy that may be affected, the sum insured or the limit of the liability could be assumed as a worst-case scenario. This would give a maximum claim cost for the insurer. A proportion of that worst-case scenario would be held, depending on the outcomes of various studies and cases.

In the emerging stage, claimants will come forward and sue the policyholder for damages due to historic exposure to a certain product or environment. The damages caused would likely have been unheard of at the time of writing the policy. Drama and uncertainty ensue—we've all seen legal battles unfold on courtroom TV shows. But after the historic cases have been reported and start to be paid, the latent claim loses much of its interest: insurers wise up and either exclude the latent claim from cover, or they price it into their policies. It is no longer a surprise, and instead is just like a regular claim cause. The latent claim is said to be emerged: the extent of the historic damage is known, and future costs will develop in a predictable way. Notice that the total claim costs due to this substance might still increase, especially if insurers continue to write policies covering it (but price those policies in a way that includes the cost of the latent claim), although as the historic cases are settled, the costs related to those policies will no longer increase.

[2] I am reminded of the story of one insurer who tried to calculate their market share of abattoirs. They counted up the number of business insurance policies on cover for a given year, and realized that they appeared to be insuring 120% of the abattoirs in the country! Upon investigation, it seems that some underwriters had selected 'abattoir'—the industry at the top of the alphabetically ordered list—as the trade if they couldn't be bothered to scroll down the list and select the real answer.

Not all possible latent claims will move through all the stages in this process: some will remain unknown forever as we misdiagnose or misattribute their damage to other causes; and some will be rejected in the potential stage, either due to lack of evidence or because the damage is proven to be due to something else. Now we will look at the most famous example of a latent claim (now fully emerged in the Western world, although emerging in developed nations): the one-time 'wonder-fabric' called asbestos.

16.2 Asbestos: The Wonder Fabric with a Dark Secret

It's a shirt. It's a sock. It's a glove. It's a hat. But it has other uses. Yes, far beyond that.

You can use it for carpets. For pillows! For sheets! Or curtains! Or covers for bicycle seats!

—Dr. Seuss, The Lorax

Since the Stone Age, humans have seen the amazing properties and potential of this substance. 100% natural and fire resistant, this fibrous material can be spun into a fabric (and made into napkins, sheets, clothing, sofa covers, and curtains), or into blocks. It has been used to make cooking utensils, oven mitts, fertilizer, insulation and much, much more. In fact, this material—actually a family of chemical compounds—seemed to be the perfect all-purpose substance.

Asbestos has been found in Europe for thousands of years, with Italy and Russia having the biggest deposits in Europe. But this was soon swamped by the discovery of asbestos in Canada in the 1870s. The growth of the railways during this period greatly increased both the ease of mining, but also demand for asbestos: its fire-retardant properties were ideal for use on steam-trains. But wider demand was growing too: a deluge of new inventions using asbestos was created by enterprising Victorians. The Canadian mines produced around 50 tons a year in 1878, but once the mines were mechanized in 1895, production soared to over 10,000 tons annually. Whole towns sprung up based on this 'white gold'. But the gold rush came with a cost.

In 1899, senior physician Dr. Hubert Montague Murray from London's Charing Cross Hospital saw a 33-year-old man who seemed to be dying of bronchitis. This was not unusual in Victorian London, but the patient mentioned that nine of his friends had also died in a similar way. These ten men had all been asbestos spinners—men who spun the fibers into a usable fabric. When the patient died only a year later, Dr. Murray found that his lungs were heavily scarred. This was not bronchitis, but seemingly something related to their occupation, possibly asbestos. It took a further six years for a confirmed death due to asbestos to be reported. Nevertheless, the Principal Inspector of Factories in Britain, Dame Adelaide Anderson,[3] included it on a list of harmful substances in 1902.

However, doctors didn't fully understand the mechanism by which asbestos harms people until 1924, when the first case was reported in a medical journal. Nellie Kershaw, a 33-year-old who had worked as an asbestos spinner in Rochester factory since the age of 12, died with severe lung scarring. Asbestos fibers were found in her lungs, and so she was deemed to have died from 'asbestos poisoning'.[4] In 1932, the first piece of UK regulation around asbestos came into force.

Although regulation around asbestos continued to be tightened throughout the twentieth century, Nellie's employers continued to make asbestos goods until the 1990s. They, and other manufacturers tried to suppress and deny evidence of asbestosis, even though Nellie's own employer (the same one who repeatedly denied her compensation in 1932) noted in 1937 '*All asbestos fibre dust is a danger to lungs*'. Even today, the factory site where Nellie worked is still contaminated with fibers.

There are three main medical symptoms of exposure to asbestos. Asbestosis, the disease Nellie Kershaw died from, seems to be dependent on the amount of exposure to asbestos: the more prolonged the exposure, the more severe the lung scarring. A less severe response to exposure is the formation of a pleural plaque. Like the plaque that forms on teeth, these lung plaques don't cause any symptoms or suffering themselves and so don't receive any insurance compensation. The deadliest form of asbestosis poisoning is caused mesothelioma. This is a form of lung cancer that isn't related to the amount of asbestos inhaled—in theory, exposure to a single fiber can cause it. Most victims of mesothelioma don't survive more than two years, even with modern medicine. However, the cancer doesn't seem to form quickly: mesothelioma can develop up to 50 years after exposure to asbestos.

---

[3] Dame Anderson was quite an extraordinary activist. Born in Australia and raised in London, she studied at Girton college Cambridge before joining the civil service. She worked tirelessly against child labor and to promote workers' rights. Later in life, she worked in both China and Egypt to campaign against child labor.

[4] Like most early asbestosis victims, Nellie's company refuted her illness and refused to pay her any missed wages or compensation. She was buried in a pauper's grave. However, the legacy of her short life was her 4-year-old daughter and eventually, industry regulation of asbestos.

> **Did you know?**
> The first known use of asbestos was to make pots and cooking utensils in Finland over 4500 years ago!

Although Nellie didn't receive any compensation, many after her did. By the mid-1960s asbestos companies had realized that the time had come when they could no longer wriggle out of their liabilities. Asbestos production continued to increase into the 1970s in the western world, and with it did the number of asbestos deaths. By the 1980s, demand was dropping due to concerns over safety. But the death toll kept rising. In the North East of England, where Nellie worked, asbestos deaths peaked at about 360 deaths a year in the mid-1990s. It now hovers around 200 deaths a year in that region alone,[5] despite production being halted for almost thirty years. Compensation is now paid to victims (although not paid to those only with pleural plaques) who suffer ill-health due to asbestos, and most of this money has been claimed through insurers under employer's liability or product liability schemes. Workers' compensation insurance in the USA tends to have low limits to cover, and so claiming against a product liability allows for a higher level of compensation. Around $65 billion has been paid out by insurers for asbestos claims so far, and estimates suggest that the total cost could be $100 billion globally.

Asbestos exposure is not solely a historic problem either. Although more than 50 countries have closed their asbestos factories, aging ex-employees, children of employees and local residents continue to claim. The cost of these secondary exposures is harder to estimate: while the number of employees at a certain time may be known, the number of children they had is unlikely to be recorded by employers. Asbestos exposure nowadays can also be due to the removal or disruption of existing material. After 9/11, Manhattan was covered in a gray dust, much of which was asbestos fiber from the World Trade Center. Several first responders have already died of lung cancer that is thought to have arisen due to World Trade Center asbestos exposure.[6] With over 110,000 people exposed to potentially lethal fibers due to that event alone and a lag time of 30–50 years for problems to develop, many more asbestos claims could reveal themselves at a huge cost to insurers, and a significant cost to life.

---

[5] Asbestosis, mesothelioma, asbestos related lung cancer and non-malignant pleural disease in Great Britain 2021 (hse.gov.uk).

[6] 9/11's delayed legacy: cancer for many of the rescue workers | September 11 2001 | *The Guardian.*

You will not be surprised to learn that insurers routinely exclude asbestos from cover in most regions now. But insurers experienced an administrative challenge during these claims too: it is near impossible to know which insurer should pay for claims based on the exposure period for asbestos. In the case of Nellie Kershaw who was employed for almost 15 years in one factory, the company may have had several insurance companies in that time. Which should be liable? Rather than squabbling among themselves, a new way of writing liability insurance emerged. Instead of providing cover based on the time when the exposure occurred, there was a shift to providing cover for claims that occurred in the policy period. This is called a **claims-made basis,** where cover is provided for any claims that are reported in the policy period. Professional **indemnity** policies are usually written on this basis now. Insurance industry agreements set out the procedure for the situations where a claimant can't find the details of the relevant insurer to claim on policies with a **losses-occurring basis** (where the policy pays claims if the harmful exposure or accident occurred while the cover was in place, no matter when the claim is reported).

Although a claims-made basis reduces the requirement for information about risk exposure, for the policyholder it means that insurance may be required for a period after manufacturing has ceased, to cover risks not yet reported. If a company neglects to do this, they would have to pay all claims reported after this point in full. In practice, the company is usually bought by another company which would buy insurance, or it tries to release itself from its potential obligations by creating contracts, as Occidental Petroleum did in the case of Love Canal (see Love Canal, Hate Waste in Sect. 16.4.)

Example 16.3 A factory is open from 1950–1979. It used insurer A in the 1950s, insurer B in the 1960s, and insurer C in the 1970s. Policies were written on a claims-made basis. Three claims were reported: one in 1968, one in 1974, and one in 1985. Insurer A is liable for no claims (although would likely be had the policies been written on a losses-occurring basis, i.e., based on exposure). Insurer B is liable for the 1968 claim, as the damage was made known in the 1960s. Insurer C is liable for the 1974 claim. The 1985 claim would either be payable by the factory itself (if it still had some assets), or perhaps a government fund would give a limited payment to the victim.

Due to historic and continued potential exposure, asbestos claims continue to be a concern for insurers in the developed world. These countries now have tight Health and Safety regulation and victims have recourse to compensation, even if their former employer goes bust. With an overwhelming scientific consensus that asbestos fibers are severely harmful, the days of mining, spinning, and surrounding ourselves with asbestos might seem like ancient history.

So, it might surprise you that over 2 million tons of asbestos are still produced every year, mainly from Russia, China, and Brazil (a far cry from the 10,000 tons from Canada in 1895!). These hazardous materials are then exported, mainly to developing Asian nations to be manufactured into products. The top consumers of asbestos[7] represent more than 50% of the global population, and so asbestosis and mesothelioma deaths have likely not peaked yet. Because of the well-known effects of asbestos, global insurers (and many local insurers too) won't cover asbestos claims when providing liability insurance to employers or companies. But people in countries without regulation are likely to be exposed to its deadly effects. Without the capital backing of insurers, victims are dependent on companies to hold enough money, and to exist long enough, to be sued for damages. This is unlikely to be sufficient, leaving thousands of potential victims without compensation.

Did you know?
Asbestos was used in the filters of cigarettes in the 1950s.

With an active lobby of a few asbestos producers and the backing of large states, there is unlikely to be a global ban on asbestos production or imports any time soon. The challenge for actuaries is to find a solution that transfers the risk of using this hazardous material to the companies and governments who encourage its use, without transferring the risk that these companies might evade their responsibilities to their victims. Remember, the actuaries' code means that actuaries have a duty to the public first and foremost so, where they can, they should create innovative risk transfer products to bring about social justice.

The story of asbestos does not end there though. This 'wonder fabric' has been found to be remarkably effective as a carbon store, so it may be used in future after all…

[7] These are (in order of consumption, highest first): Russia, China, India, Brazil, Kazakhstan, Indonesia, Sri Lanka, Thailand, Ukraine, and Mexico.

16.3 The Blame Game: Insurance and the Law

Deciding who pays is always a tricky subject. Whether you are splitting a $50 dinner bill ('well Emily had three cocktails but Isobel had dessert...') or a multi-million-dollar compensation claim, who should pay what can require some difficult negotiations and painful decisions. You might be relieved to know that actuaries aren't responsible for this; you might be less relieved to hear that the legal system will likely determine it. The potential stage of latent claims is led by scientists to determine a clear cause and effect, but the emerging stage is for lawyers to argue and the judicial system to decide who bears the cost.

Latent claims usually emerge through a small number of individual cases before snowballing into a large cost. A victim sues the policyholder for the damages they have suffered. The policyholder's first thought is likely to be to 'lawyer up' and defend the case. They should, however, also inform their insurer that there is a potential claim, indeed, this is likely to be a requirement of their policy.[8]

Exercise 16.1 Why might the policyholder not want to tell their insurance company about a potential claim?

At this stage, the lawyer acts as the claims estimator for the insurer: they will provide the insurer with their estimate of the cost of settlement or court award and, if legal fees are covered by the policy, the estimated cost of the legal fees to defend this case. In some jurisdictions and in certain circumstances, the loser of the court process pays the legal fees for both parties. In those jurisdictions, if the policyholder wins, they won't have to pay anything. But if they lose, they not only have to pay the court award, but also two sets of legal fees. Legal cover is usually an add-on to another type of insurance, often marketed as a perk or bonus for the customer. However, there is often a benefit for insurers to providing legal cover. As the cost of legal fees might be recouped from the insurer, the policyholder might opt for a better lawyer, and pay for more time working on the case. This might reduce the court settlement, or help the policyholder (and the insurer) avoid the claim cost altogether.

[8] In the most extreme circumstance, not informing your insurer of a potential claim against you could invalidate the claim.

The next difficult decision, particularly in the case of clear damages such as mesothelioma claims, is who should pay. There may be several companies that the victim could try to sue for compensation. The victim will likely choose the cases that they are most likely to win, and the companies where they will receive the highest sum. This means that companies who are more able to pay such compensation, or have paid out in the past, will tend to be hit by more, similar claims. In the case of Nellie Kershaw, the directors of her company denied her compensation because they realized it would start an unfortunate (for them!) precedent.

Courts inadvertently compound this issue by trying to ensure that once liability is confirmed, the victim does actually go onto receive compensation. A common way to avoid paying compensation is for the company to declare bankruptcy, or try to argue down the award because it would create 'financial hardship' for the business. Although the court theoretically bases the compensation on the degree of responsibility, in reality, who has the deepest pockets is a consideration. It not only impacts who pays the award, but who lawyers will choose to take to court in the first place. In the more common scenario where the case is settled out of court, lawyers will again determine the paid amounts.

Because of the power of lawyers in determining these claims, liability insurance is particularly sensitive to the litigiousness of society. In regions where suing is fairly commonplace, even encouraged, that is priced into the insurance premiums. The USA is very litigious, although all regions of the world have become more litigious over the past 50 years. One way of quantifying this trend is to count the number of lawyers: in 1961 the USA had 200,000 lawyers, compared to over a million in 2016.[9]

There are lots of potential and emerging latent claims at present, with scientists and lawyers working on cases at various stages. For the rest of this section, we consider a few potential contenders. Remember that not all of these will become an asbestos-sized problem, they are only ghosts of claims yet to come. For some, the scientific evidence may show that the exposure is not related to the damage at all. Some claims may be defeated in court (at least for now…). As is the way with latent claims, there are probably some not yet envisaged that will emerge and shock the industry. I encourage you to make your own predictions.

[9] Data from the US Census.

- **Contact sport head injuries (emerging)**

  In 2013, the National Football League, the top American Football league in the USA, paid $765 million to settle a lawsuit with over 4500 ex-players. The ex-players reported symptoms of Alzheimer's, depression and dementia that they attribute to repeated blows to the head during their time playing professional sport. They accused the NFL of profiting from their injuries (skull-cracking blows are often played and replayed in slow-motion), downplaying the risks of head injuries, and rushing them back to the field after a hit. As part of the settlement the NFL denied any wrongdoing, but said that paying the settlement was '*the right thing for the game*'.

  In the UK, in 2020 a group of rugby players started legal action against World Rugby after a number of them were diagnosed with early-onset dementia which has been linked to repeated head injuries and concussion. Unlike American Football, rugby is usually played with no protective head-gear. The lawsuit is ongoing and has grown to represent 150 ex-players. It is currently expected that, like the NFL case, a settlement will be made out of court.

  Although the NFL and the National Hockey League (the top North American ice hockey league) have already made payments, more players from these and other sports are expected to come forward over the next 20 years.
- **Talcum powder (emerging) and silica (potential)**

  In 2006, a woman with mesothelioma received $3.35 million in compensation. But she wasn't a factory worker; her exposure to asbestos had occurred from using talcum powder containing asbestos. Subsequent cases were filed, and the lesson was learnt that one shouldn't put a substance known to be carcinogenic in a product made for babies. But that wasn't the end of the problems for talcum powder.

  In 2018, big pharmaceutical company Johnson & Johnson was sued by a group of 22 women, all of whom had experienced cancer after long-term use of talcum powder. They won the case, splitting $4.7 billion between them. Johnson & Johnson appealed to reduce the amount and number of plaintiffs. It went all the way to the Supreme Court in June 2021, and the verdict (albeit at a lower sum of $2 billion) still stands. Subsequent scientific studies found that even without asbestos, there is a consistent association between use of talcum powder and developing ovarian cancer.

Since the initial court case, over 35,000 cases have been filed about talcum powder in the USA alone.

While talcum powder is an emerging risk, the use of silica in cosmetics is also a potential latent claim risk. Silicosis, traditionally known as stonemason's disease, is caused by breathing in silica, and causes breathing problems and chronic obstructive pulmonary disease (COPD). Health and safety regulation around silica now applies to construction jobs. However, cosmetics containing silica are still sold. When applied around the mouth and nose, they can easily be inhaled or ingested. Currently, there is no evidence that at this level of exposure silica is harmful. However, many cosmetic users have begun to question this, and lawsuits and scientific studies may follow.

- **Cell phones (potential)**

  In April 2021, a 49-year-old pastor, Frank Walker, died from an aggressive type of brain tumor, called a glioblastoma. He'd used cell phones for 25 years. Could the two things be connected? His wife and sons filed a lawsuit against the telecoms company (although other companies and even regulatory bodies are also co-defendants). There are currently over 70 lawsuits filed in the US regarding cell phones and brain tumors, but this case, called Walker v. Motorola, could be set to get into the courts first due to the efficiency of Louisiana's state justice system.

  The reason this is a potential latent claim rather than an emerging claim is that there isn't currently a scientific consensus. Many studies have shown correlation between cell phone radiation and tumors, including the USA's Food and Drug Administration's (FDA) own commissioned research, a study by the National Toxicology Program and even Motorola's own commissioned study.[10] However, the FDA concluded that their study, which was based on tumors formed in rodents, was not applicable to humans.[11]

  Perhaps science will find another cause for Frank Walker's glioblastoma, and this latent claim won't make it past the potential stage. On the other hand, if cell phone radiation is found to be carcinogenic, this could move very quickly into an expensive and emerging latent claim. Not only that, but a widescale rejection of wireless networks or mobile technology could

[10] The filing for the Walker vs Motorola case paperwork gives a comprehensive list of studies that show correlation between radiation and brain tumors (it doesn't, however, give a list of those that don't). walker-lawsuit-vs-ctia-etc.-2-21-cv-00923.pdf (wordpress.com).

[11] Statement from Jeffrey Shuren, M.D., J.D., Director of the FDA's Center for Devices and Radiological Health on the National Toxicology Program's report on radiofrequency energy exposure | FDA. https://www.fda.gov/news-events/press-announcements/statement-jeffrey-shuren-md-jd-director-fdas-center-devices-and-radiological-health-national

Fig. 16.2 A Nazi-era anti-smoking ad titled 'The chain-smoker' reading: 'He does not devour it, it devours him' (from the anti-tobacco publication Reine Luft, 1941;23:90)

> follow—creating a high level of **transition risk** (the risk created while transitioning from one system to another). This latent claim could change everything—but not quite yet.

These claims could fizzle out to nothing, or they could be incredibly expensive. In 1998, America's four largest tobacco companies paid $206 billion so that in 46 US states they would no longer be liable for any claims from smokers, or from the 46 states on behalf of smokers. The connection between smoking and disease follows the same pattern as outlined in this chapter: denial (and even concealment[12]) of the dangers of smoking by the tobacco companies, followed by a slow but ultimately successful set of claims in the 1960s and 1970s by groups representing victims. The dangers of smoking were known much earlier, but it was left to governments to raise awareness. Interestingly, Nazi Germany had an active anti-tobacco movement, even though other developed countries didn't at the time (Fig. 16.2).

Trying to predict the next big latent claim runs the risk of falling into time-consuming conspiracy-theory rabbit holes, but for companies who write a lot of liability business, it is well worth the wait.

[12] Inventing Conflicts of Interest: A History of Tobacco Industry Tactics - PMC (nih.gov).

16.4 Environmental Claims

We've seen that exposure to harmful substances can be expensive for insurers. But cleaning them up can also be costly! The removal and disposal of asbestos in buildings is a long process, but because it can be done in a controlled way (with many layers of safety precautions for those doing the removal), the likelihood of harmful exposure is relatively low. Although costly for the building owner, these controlled clean-up operations aren't particularly complicated for insurers.[13]

Unexpected clean-up operations are altogether more costly, although the area in which spillage occurs (heavily populated areas vs rural areas, the amount of wildlife in the area, etc.) and the substance that has spilled will also affect the cost. Substances which trigger immediate clean-up claims include oil, radioactive material, sewage, and milk.

Some of these spills are obvious: spilt milk will certainly attract the attention of neighborhood cats. An oil spill will often be noticed by the oil tanker itself, by satellites/drones, or the local population. The clean-up might be expensive,[14] but it is not a latent claim.

> **Did you know?**
> In some countries, businesses can be fined for pouring milk down the drain.

Latent environmental claims however—like all latent claims—are subject to long reporting delays. These delays occur because the pollution is difficult to see or detect, and the harm is not obvious, or takes a long time to come to light. Like other latent claims, there might not be a single polluter: as there were many asbestos factories, there might be a multitude of companies whose waste could have seeped out. Usually, the largest company will be sued (another instance of deep-pockets syndrome), and then that company will sue smaller ones to get them to share in the liability. On some sites, it is not

[13] Although insurers may put in extra checks to ensure that risk mitigation and health and safety procedures are being strictly adhered to.

[14] The Exxon-Valdez oil spill cost $3.5 billion in 1989, and was one of the most expensive oil spills in US history. In 2010 it was surpassed by the Deepwater Horizon oil spill, the current largest ever spill in history, which cost $62 billion.

accidental spills or leaks, but poor historic waste disposal that is only now creating claims.

Love Canal, Hate Waste

With lots of beautiful homes near Niagara Falls and two newly built schools, Love Canal seemed like the perfect town to live in. The name came from an unfinished canal, started in the 1890s. It had been left, owned by some company but never developed on, until the school board bought it in 1953. They built two schools, and sold the rest of the land to developers, who sold the land to homeowners.

Residents had noted the strange black fluid seeping out of the Love Canal site, but didn't think much of it. But then, resident after resident started to receive diagnoses of cancer: 9 women on one street alone needed mastectomies. Still, the town was ranked the 4th best for social well-being.

In 1977, heavy snow hit the town. When it started to melt, water began to flood basements. The state's departments for health and environment thought they should test the water, for any bacteria and the like. What they found was shocking.

The water had toxic levels of many different chemicals, including benzene (an extremely potent carcinogen), PCBs (a family of carcinogens), chloroform, toluene (causes nausea and memory loss) and a host of other toxins. Before the sale of the land in 1953, the Hooker Chemical Company had been dumping waste into the empty canal cavity. The barrels had started to rot and chemicals were leeching out, in some cases vaporizing, causing cancers, birth defects, seizures, and rashes. A third of residents had chromosomal abnormalities.

Residents were finally evacuated at a cost of $15 million, but this chemical waste wasn't a surprise to everyone. The education board, who'd bought the land for only $1 in 1953, had signed to release the Hooker Chemical Company from any liability due to exposure to toxic chemicals. Hooker, which by this time had been bought by Occidental Petroleum, asserted that it wasn't certain that it was their chemical waste, and even if it were, the school board had released them from the liability.

The federal government paid to clean up the land to the cost of $400 million and took over twenty years (it was declared clean enough to live on in 2004). In 1995, Occidental Petroleum were sued by the federal environment agency and paid $129 million in restitution, including $375,000 for the damage caused to local wildlife. Occidental's liability insurance may have covered all of these costs, subject to cover limits and exclusions. If the name Occidental Petroleum rings a bell, the Piper Alpha oil rig, which exploded in 1988 and was mentioned in Chapter 13, was also owned by them.

More recently, Occidental Petroleum was sued by a tribe of indigenous Amazonians in 2007, as many had suffered horrific deaths after drinking water contaminated with cadmium, lead and arsenic by Occidental's oil drilling in the area. In 2015, a settlement was reached (the amount is confidential). Given the scale of Occidental's drilling, many more claims may yet to come to light. Occidental have been named as at least partly responsible for six US sites that are still to be cleaned.

For insurers, excluding environmental liability claims may be the most risk-averse option, but many continue to write them as oil, gas, and chemical manufacturers are global companies who can pay large premiums. However, governments need to find a better way to manage this risk as well. It is a basic rule of economics that externalities (such as air or water pollution) should be paid for, but this is largely still theoretical. 'Polluter pays' taxes have been long argued for by some parties, but industry lobbying means that these haven't been put into practice everywhere.[15] In December 2021, US Congress approved a deal that included 'polluter pays' taxes for US chemical waste sites, 37 of which are still awaiting clean-up. In August 2021, Democrats proposed a carbon dioxide equivalent; where companies who historically produced more than a certain level of carbon dioxide should be taxed. The fund would then be available to pay compensation to those most affected by climate change, therefore moving the risk from victims, taxpayers, and the government back to the polluters themselves. This is not a new idea; a similar tax was in place until 1995, but it seems that there is renewed motivation to deal with historic environmental waste.[16]

Although the effects of chemical waste materials have been known for some time, new concerns are now being raised about liability claims relating to other waste materials. In particular, the prevalence and damage caused by both nanoparticles and plastic pollution has led to many insurers refusing to insure nanotechnology companies. Insurers' refusal to insure companies

[15] At the time of writing, a 'Polluter pays' tax is being considered in the UK for the construction industry as a reaction to the 2017 Grenfell Tower disaster, where the use of combustible cladding killed over 70 people. If passed, this would stand in stark contrast to the response after the 1966 Aberfan disaster, where the government insisted that the monies donated from across the world and the UK would be used to make other coal tips safe, rather than take on the financial burden itself or impose a burden on the company or industry responsible.

[16] Environmental agency reports have found that climate change will exacerbate toxicity of these sites, and so departments are keen to clean them up before they worsen.

suppresses innovation in these sectors, as companies are forced to self-insure,[17] or may even only pursue low-risk inventions. The presence of affordable, effective risk transfer encourages innovation in society, and is vital for inventions to take the leap from lab to real life.

Plastic pollution has received lots of media attention in the past few years. Although the focus is usually on marine life and the oceans, the air pollution caused by burning plastic is a more immediate threat to human health. It would not be surprising if insurers (and potentially even life insurers) take more action on writing risks relating to plastic in the next decade. For the same reason that lack of insurance suppresses innovation, it can also suppress an industry or process. By limiting underwriting appetite for the production and disposal of plastic, the companies would have to retain the risks of the process. They would pay for any costs related to operational failures or even just the day-to-day activities[18] of their own company, and so reducing pollution might therefore become more worthwhile financially. This needn't be done by harsh underwriting exclusions either: there could be discounts for companies who use lower polluting processes, or who have particularly stringent waste capture or disposal processes. One particular plastic of interest is polytetrafluoroethylene (PTFE), which is what the non-stick coating in pans is made from. It is particularly unreactive, so it hangs around in the environment for a long time and so the world is slowly accumulating PTFE in the environment. However, when it does break down, the fumes can cause flu-like symptoms and can be lethal, both to animals and humans. People who live near PTFE factories have a higher-than-average incidence of lots of different types of cancer and thyroid disease, so insurers are starting to consider PTFE a potential latent claim and include it in extreme claim scenario modeling.

Latent claims have been a bit of a 'gotcha!' moment for insurers. The current product design of liability cover has allowed some companies to knowingly cause harm, but leave governments and insurers financially responsible. As more and more liability claims relate to environmental concerns (rather than the specific chemical or radioactivity claims historically seen), actuaries working both in the public and private sector have a real opportunity to re-design and reform products to encourage lower waste processes, thoughtful risk mitigation, and social justice.

---

[17] This usually means the policyholder setting some money aside as an emergency fund instead of buying insurance.

[18] Environmental liability policies usually don't cover pollution caused by day-to-day processes, but these could be covered by other business insurance, potentially public liability.

17

Modeling Coronavirus-19

Chapter Objectives:

- Understand the role of actuaries in population modeling and as government advisors
- Predict disease spread with stochastic models
- Understand the issues and data limitations in a crisis, and how to mitigate some of these
- Understand the other effects of the pandemic and how actuaries modeled this uncertainty
- Consider how the world has changed due to the pandemic and the effect on actuarial work

Prediction and explanation are exactly symmetrical. Explanations are, in effect, predictions about what has happened; predictions are explanations about what's going to happen.

—John Rogers Searle, American philosopher

© The Author(s), under exclusive license to Springer Nature Switzerland AG 2022
C. Townsend, *A Risky Business*,
https://doi.org/10.1007/978-3-031-11673-5_17

17.1 Modeling in a Crisis

It was 4 pm on a Monday and I'd just received a call from my son's nursery to say he was ill. Just a cough and a temperature, but could I possibly pick him up immediately just in case? I closed my laptop and shoved it in my bag, downed the rest of my coffee (leaving the empty mug on my desk) and rushed off, leaving my jacket hanging on the back of my chair. A few days later, just as my son was getting better, the whole country went into lockdown. When I next returned to my desk, some twenty-two months later, my desk had changed. The office had changed. The world had changed. And my jacket had vanished.

Actuarial science rests on the assumption that the past is a good guide to the future. Without this assumption, modeling using historic data would be a pointless exercise. Similarly, actuarial science also assumes that there is some relevant objective data. Without this, there is no actuarial *science* to be done, only speculation. Modeling in the circumstances of March 2020 was by no means standard actuarial practice because of the lack of both historical data and current objective information, and it was by no means perfect.

We want to cling onto certainties in a crisis. What will happen? How will this unfold? When can I see my loved ones? Who, if anyone, is safe? But with no short answers and no certainties, the only response was a flurry of statistical models. These early models were fairly general, based on very early scientific data, and had a high level of uncertainty. Actuaries did not create all of the models, but they were asked to interpret and advise.

The one thing to understand with these epidemiological models is that they were never going to match how Covid-19 would unfold in reality. This was partly due to the bold assumptions[1] in the models, but partly because the models were meant to inform public health policy. They highlighted the dangers of not taking any action, and so action was taken, meaning the models' predictions did not come to pass. This is not a model failure; this is using the models to make informed decisions—exactly what they were designed to do. Two years after these early predictions, it is clear that some of the predictions from a 'take no action' scenario were wrong, but understanding *why* is part of the learning process that always takes place in actuarial work. If models were perfect first time round, actuaries would be out of a job very quickly!

---

[1] Assumptions in some initial models around severity and transmissibility have turned out to be incorrect, although with little data on a new disease, in the words of Cardinal John Henry Newman, '*Nothing would be done at all if one waited until one could do it so well that no one could find fault with it.*'

While ostensibly it seems that actuaries were never trained to model the spread of Covid-19, in fact, they were. In their arsenal were years of building statistical models, albeit for other purposes, and that experience still held true. Broadly, the steps in building statistical models are:

1. Specify the problem
2. Acquire data
3. Make assumptions
4. Develop a solution
5. Monitor the experience

How does this work in practice?

1. Specify the problem

With so many questions and unknowns during a pandemic, it is essential to decide precisely what your model will predict. Being too ambitious means defining more parameters that need estimating, which can introduce the risk that the parameters are wrong—called parameter risk. But being too cautious runs the risk of having too simple a model, and not predicting or explaining the reality well. This is the most important step because it will inform all others. A model is a simplification of a system or process in order to explain and predict. Therefore, some simplifications are necessary, but it is vital to determine the model's purpose upfront so you don't simplify the crucial part. Some models focused on the numbers of people who would catch Covid, and we will look at those models in more detail in the next section. Some only wanted to consider how many people would get seriously ill and require intensive care or hospitalization. Some wanted to estimate the effect of lockdown on the economy, or the likely outcomes under different public health guidance. Some modeled what would have to be done to reach a target outcome, such as limiting the number of deaths. When it comes to modeling, you can have 99 problems, but you have to pick one (at a time).

Case study

A large UK hospital wanted to model Covid-related attendance during the first lockdown in order to optimize staffing and capacity. The model needed to not only predict how many people will require intensive/critical care, but also when patients will arrive at the hospital. The key output would be the number of people in intensive care at every time period.

2. Acquire data

Recent historical data is usually the most relevant to the task at hand. Failing that, industry benchmarks or expert judgment might be required. Here, modelers may be able to leverage industry knowledge or contacts to gain access to information, but that data might not have been scrutinized thoroughly. If the data is too flawed or irrelevant, there might need to be a data collection exercise.

Case study continued

The hospital had some predictions from previous modeling exercises. One modeled an Ebola event and one modeled a particularly virulent and lethal form of flu. The flu scenario was deemed inadequate and irrelevant on the advice of respiratory consultants in the hospital. The Ebola scenario was also deemed inadequate as it only considered five cases of Ebola in the hospital, but rough government prediction for Covid-19 was that there would be local cases in their hundreds. There was no relevant historical data, as recent coronaviruses SARS and MERS didn't affect the UK severely. So where else could data come from?

The hospital analyzed some data from a respiratory consultant based in a hospital in China, which had already seen Covid-19 spread for a few weeks. The data was shocking. It showed a very high proportion of cases that required intensive care. First, it was crucial to estimate how relevant the data would be for the hospital's population. The area was less densely populated than the Chinese hospital's location which may reduce Covid cases, but early in the pandemic it was not clear how compliant the UK population would be with lockdown, whereas it was known that China's population had a very high compliance rate.

3. Make assumptions

Assumptions largely stem from flaws in the data, and the places that you are willing to make simplifications in your model. Any kind of adaptation of the obtained data will require some assumptions. Assumptions for any model should be stated clearly and tested as much as possible. One way to test assumptions is to do a sensitivity test; once the model has made predictions, rerun the model with one assumption varied by, say, $\pm 10\%$, and see how much that changes the result. This tells you how sensitive your model is to that assumption. If it is very sensitive (i.e., a small change in your assumption causes a big change in your results), the uncertainty in the assumption should be minimized, and the final model predictions should be communicated very carefully and clearly, perhaps with confidence intervals.

Case study continued

Early research suggested that older people were more susceptible to being severely ill with Covid, so the hospital wanted to include that information in their model. They compared the demographic of the local population from the Chinese hospital to their own demographic and made some assumptions about how the Chinese data would scale on a slightly older population. However, the data showed a very high proportion of severe cases—was it possible that the less severe cases weren't tested for Covid-19? Based on some early research, PCR testing[2] seemed to only pick up 80% of positive cases, with those with less severe symptoms less likely to be confirmed positive. Positive radiology of the lungs was a more accurate way of diagnosing Covid, but much slower and more staff-intensive. An assumption was made that because of these flaws in the diagnosis, the percentage of severe cases from the Chinese data was likely to be an overstatement of the severity.

4. Develop a solution

In non-pandemic circumstances, and with plenty of time, this step can be multitudinous and very grand indeed. In times of crisis, both time and data are short, so the solutions had to be quick and actuaries had to work with the data available. Of course, it must also answer the specified problem. Actuaries have lots of technical experience and modeling approaches in their arsenal, so the first port of call is usually to run through those as potential solutions, thinking of adaptions if necessary. Solutions that we have met in this book include cashflow models, **reserving** methods, **generalized linear models**, original loss curves, and scenario-based models, although there are many more that are used in practice, including stochastic models, which we will meet applied to this topic in the next section.

Case study continued

As the specified problem relates to both number of admissions and the timing of admissions, we need a model that estimates the number of people in the hospital at every period. We also know from epidemiologists that hospitalization generally occurs a few days after infection. So, if patients are grouped by when they became infected (roughly), we can use reserving methods, specifically the Chain Ladder method, to predict for each time period how many people will be in hospital. This method also allows for the number of people

[2] Polymerase Chain Reaction testing: a form of testing that detects the genetic material of the relevant organism, here, of the Covid-19 virus.

in intensive care to decrease as well as increase,[3] so reflects discharges and deaths that reduce the number of patients in that cohort.

Some assumptions are made about the pathway of the disease: that it takes some days to reach peak severity, before the patient either dying or recovering. Now, using the data from China, the hospital starts to create claims triangles, grouping Covid patients by their admission to hospital. The hospital knows that patients are usually first admitted to a general ward before being moved to a critical care ward. If their aim is to predict critical care requirements, they can use the general hospital admissions as a cohort. If general hospital admissions are the goal, more assumptions about the wider population's infection will be necessary. Using development factors, actuaries estimate the number of patients in intensive care each day per cohort, then they can sum across all cohorts to give a total bed requirement.

5. Monitor the experience

Although it meets our stated outcome, the model prediction is rarely the end of the process. Instead, it is merely a 'starter for ten': the model must be repeatedly tested and refined, incorporating new research, new data, and new external environments. In some actuarial models, external environments may include the economic or market environment. In epidemiological models, external environment includes the number of daily cases in the wider population. Government action (or inaction) on public health guidelines is a key input here which has the ability to create a step change in the predictions. Monitoring the model against actual experience requires new data to be collected, which can be expensive and time-consuming in itself. However, regular monitoring with a view to adapting and refining the model is an important step that should not be skipped: it can highlight deficiencies in the data and the model which may lead to poor predictions.

Case study continued

After creating some projections, the hospital realized it needed ongoing data on how many patients were being admitted with Covid, as well as data on when Covid patients were entering and leaving intensive care. To gain data on the number of patients with Covid, senior respiratory consultants were tasked with triage (a job often left to the most junior). By putting experienced staff on triage, they were more able to accurately assess whether a

[3] Recall that the Bornhuetter-Ferguson method does not allow downwards development, so would not be suitable for this purpose.

patient did have Covid or not, giving more certainty on Covid rates in general hospital admissions. They also could ask questions directly to patients before the patient became critically ill and unable to talk, and so could determine when the patient started to feel unwell, and therefore how quickly the illness was progressing. This meant the new data could be more accurately grouped into disease cohorts, giving more confidence in development factors.

The insights gained from statistical modeling in this hospital during the first wave of Covid gave them superior projections of requirements for the rest of the pandemic compared to the pro-rated government projections which overestimated the number of beds and staff required. This meant that this hospital was able to allocate resource more effectively and, in particular, it performed more (non-Covid related) operations during the pandemic than it would have otherwise.

Did you know?

Nottingham University Hospital improved on the government's own projections by leveraging their established data system, in which the physiological indicators of severity of illness were collected (heart rate, oxygen levels, blood pressure, etc.). They combined this with international data on Covid-19 from Australia and China.

The outcome was a surprisingly accurate prediction of the number of existing patients who were likely to enter critical care (intensive care). This proved very useful in predicting staff demands and so in maintaining medical treatments and operations on patients from pre-Covid waiting lists.

This endeavor was largely due to a committed hospital team under the direction of Professors Joe West and Richard Hubbard, and their teams.

Modeling in a crisis is not the textbook ideal-case for actuarial work: being short on both data and time is not how actuaries can do their most technical work. But it is very important work, and the iterative process I describe above helps overcome these deficiencies. The pandemic made the public more aware of risk and more interested in the quantification of risk, especially transmission risk. This interest thrust actuarial science into the spotlight in a way that had not been the case previously. I was equally surprised and proud on the first day that I heard strangers at the supermarket discussing the relative merits and assumptions of two statistical models. The models guided politicians, hospital leaders, councilors, and the general public, informing them about likely outcomes of the future. Let me be clear: actuaries do not and cannot predict the future. Even ignoring the simplifications made in order to create a model, there is a random component to the actual outcome that cannot be modeled exactly. But they do make predictions about the future. More importantly, as we have seen during the pandemic, predictions can alter policy or behavior and, in that sense, actuarial predictions sometimes change the future.

17.2 Modeling Coronavirus-19 Spread

The previous section gave the steps that an actuary would take during modeling. This section considers a family of models, called stochastic models, which can be used to model Covid-19, and aim to predict how many people in a population will catch the disease by a certain point.

Stochastic models are models where there is a random element in at least one model parameter. What this means in practice is that the results of a stochastic model contain a bit of randomness. To make the model outcomes reproducible, stochastic models are run hundreds or thousands of times so that a distribution of outcomes is given, similar to simulations.

A really simple stochastic model of the population first involves defining different 'states' of being. For an epidemiological population model, the states might be the following: susceptible to the disease, currently infectious, immune from the disease, and deceased. The key thing is that the states are clearly defined and mutually exclusive (you can't be in two states at once) and there must be a way for members of the population to move between the states. We will quantify that movement in a moment, but first let's draw a diagram of the possible ways people can move between these states. Arrows means that a person can move from one state to another, in the direction the arrow is pointing to. You might be interested in movement back as well, but we'll ignore that here and assume that people can only move in the direction of the arrows, and only where there is an arrow drawn. To make the calculations even easier, let's assume time is discrete: people have the opportunity to move only once per time period (we'll use days for our Covid model). But they don't have to move from a state on any given day, so we also need to give them a way to remain in their current state. It makes the calculations easier if we describe this as a movement from the state to itself, and drawing a looping arrow that moves from one state to itself. Now we have engineered a situation where every individual must move at the end of every time period, even if that move is to stay where they are.

In this 4-state model, the whole population starts off being susceptible to the disease, and no one is infectious, immune, or deceased.[4] People may then catch the disease and move from the susceptible state to the infectious state. From the infectious state, they either move to the deceased or immune states, and stay in that state for the rest of the modeling exercise.[5] In stochastic models, it is the rate of movement between states per period that is parameterized. Let's define our parameters in the following way:

- The rate of moving from Susceptible to Infectious (the probability of catching Covid) is c.
- The rate of moving from Infectious to Deceased (the mortality rate) is m.
- The rate of moving from Infectious to Immune (the recovery rate) is r.

These three parameters are all we need for our model. We can now calculate the probabilities of not moving state on a given day by taking the complement of the rates moving out of a state. Remember, probabilities must sum to 1.

- The probability of not catching Covid in one day is $1 - c$.
- The chance of remaining infectious in a day is $1 - m - r$.
- There is no chance of moving out of the deceased state (i.e., you move out with a probability of 0), so the chance of staying dead in a period is $1–0 = 1$.
- Equally, (for now) we assume no chance of moving from the recovered state, so the probability of staying recovered is 1.

We can add these parameters and their associated looping arrows onto our diagram to give Fig. 17.1:

It is important to realize that we might easily have chosen a different model, different states, or more states. But this 4-state model will do nicely to illustrate the principles of stochastic modeling, and we will consider some variations in the next section.

You can imagine this system as if the arrows are pipes moving small balls between four different buckets. The balls can only move from a state if there is a pipe out, and the parameters show how big the pipes are: the bigger the rate of movement, the quicker the balls will flow out and get to their end states. As you can see from the diagram, in our simple model all balls will

[4] This is the first simplification; that Covid somehow enters our population.

[5] This assumes that people can only move in the direction of the arrows, so no one can move from Recovered to Susceptible (although we will relax this assumption in the next section).

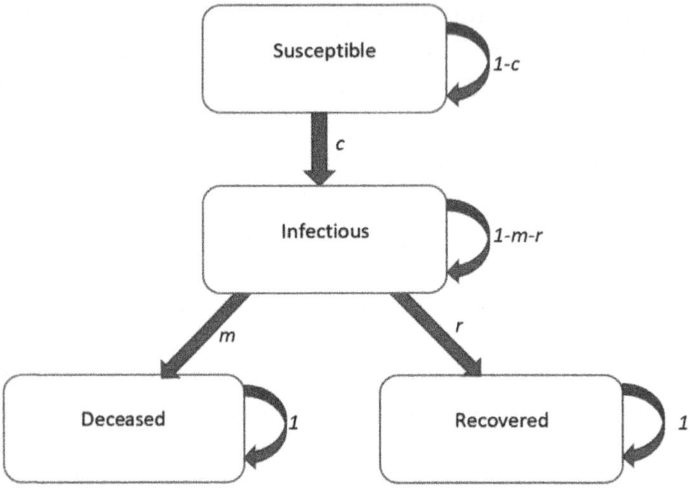

Fig. 17.1 The basic 4-state stochastic Covid model

eventually end up in the deceased or immune states as there is no way out of these states. Equally, given enough time, no ball will stay in the susceptible or infectious states. This model therefore predicts that everyone will catch Covid eventually.

> **Did you know?**
> Both the University of Oxford and Imperial College London used stochastic models to model Covid cases and hospitalizations.

We can summarize all the bullet points and diagrams as one matrix, which will save both time and ink. This is called the transition matrix, as it shows all the transition rates for one period (and although this Covid model is using days, the discrete period could be weeks, months, or years). The way to read a transition matrix is that you travel from the states of the left to the states to the right (that is, the column headers). So, the rate in the ith row, jth column denotes the rate to go from state i to state j. For our process, the transition matrix is (Fig. 17.2):

Note that the transition matrix shows the expected rate of transitions, not the number of individuals that will move.

Now, we can start to use our model to make predictions. We might be most interested in estimating how many people will be infectious after one day, and how many will die in that time. Let's say we have a population of p

	S	I	D	R
S	$1-c$	c	0	0
I	0	$1-m-r$	m	r
D	0	0	1	0
R	0	0	0	1

Fig. 17.2 The transition matrix of the basic 4-state model. The rate to move from the Infectious state (I) to the Deceased state *D* is *m*

people,[6] who at time 0 (the beginning of the model, before any transitions have occurred) are all susceptible to Covid. During the first day, some of those people will move to the infectious state. How many will move? Well, in stochastic models there is some randomness[7] in exactly how many will move, but we would expect that $p \times c$ people will move to become infectious. Remember, in a stochastic model, each person in a state has a probability of moving, not an absolute directive to do so.

Suppose the chance of moving to a different state was 50%, and each person in a state flipped a coin to see if they'd move states today. Perhaps, due to the randomness of the coin flip, no one would move at all. So, when we talk about numbers of people in a state, we mean the expectation of the number of people in a state, in the same way that we would expect 5/10 coin flips to be tails, but the actual result could be 8/10. This is how the model is stochastic even though we have specified parameters: we have only specified the expected transition rate.

If we have pc people expected to become infectious, that leaves $p - pc$ people still susceptible. To give an estimate of the number of people who die of Covid during the first day, we need to estimate the number of people moving to the deceased state. On the first day, no one could yet move into the dead or immune states, because no one started the day in the Infectious state, and in our simplified process you have to go through the Infectious state to get to the Deceased or Immune states. This model therefore assumes that people can only die with Covid. Such assumptions should be stated clearly from the start. In this case, it simplifies the model considerably.

[6] Where p is a sufficiently large number that we don't have to worry about fractions of people in any of our calculations.

[7] In real life this isn't necessarily randomness, but variability caused by things such as how many infected contacts you have and your susceptibility to disease, and therefore not truly random. However, this variability can be treated as if it were random. We can account for some of this variability by modeling more susceptible states to represent different levels of susceptibility to the disease (as we will see in the next section) and creating a c parameter for each. However, stochastic models assume that all individuals within a state may jump to the next state at random, at a rate given by the parameter (and that the population within a state is homogenous).

We expect that after one day we would have $p - pc$ in the Susceptible state, and pc people Infectious (and no one dead or immune).

Using algebra instead of numbers in our calculations allows an easy assessment of effect of the different parameters on the model outcome. For example, if $p = 10$ and $c = 0.1$. At the end of the first day we would expect only one person infected. Our situation under these parameters would be quite different from the situation where $c = 0.5$, and we therefore expected 5 people to be infectious. And this is only for 10 people in a population. If our population is very large (say, 50 million), even slightly different c values will have huge consequences for millions of people. For example, the difference between $c = 0.15$ and $c = 0.17$ after just one day in a population of 50 million is one million more people infected! Returning to $p = 10$, we can visualize the difference in outcomes if c changes from 0.1 to 0.5 below (Fig. 17.3).

Our model allows us to predict further into the future by considering movement for the second day too. Remember, individuals in our model have a chance to move, but there isn't a strict quota on how many people must move in each period. Because we are now starting from an estimated position (the estimate of the number of people in each state at the end of day one), the estimated position at the end of day two will contain even more uncertainty than the previous day, due to the randomness in the stochastic model.

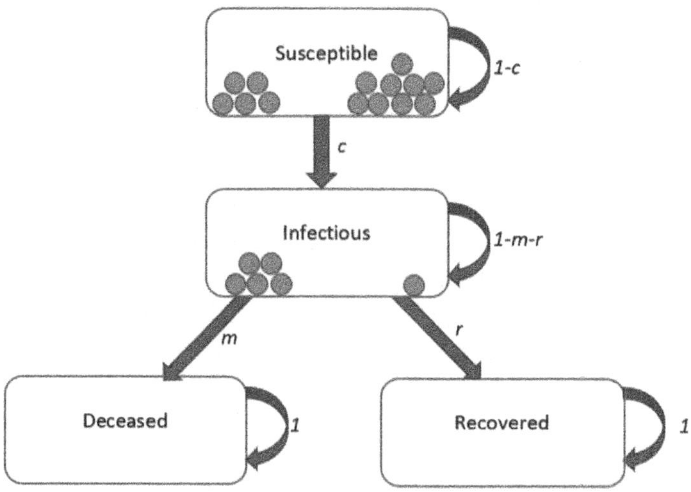

Fig. 17.3 Changing the transition rates (here, parameter c) gives different outcomes. With $p = 10$ and $c = 0.1$, the orange dots on the right show the expected outcome at the end of one day. If $p = 10$ but $c = 0.5$, the green dots on the left show the expected result

However, we can calculate the expected number of people in each state after two days as follows:

- We start with p Susceptible people at the beginning of our model run, and pc people are expected to move during day 1. Therefore, we expect $p - pc$ start in the Susceptible state after one day. We still expect people to move at a rate of c, so expect $(p - pc) - c(p - pc)$ to remain after two days. We could write this as $p(1 - c)^2$.
- Let's skip ahead to the end states: after two days, we could expect some people to have died, as the $p \times c$ people who were infectious after one day could now move to either death or recover. They die at a rate of m, so we expect mpc deaths after two days.
- Using a similar logic, we expect rpc people have recovered after two days.
- The Infectious state needs to take account of those people who have moved out of the state, as well as those who came in. By adding the new arrivals and subtracting those who move on, we get $pc + c(p - pc) - mpc - rpc$. Algebra fans might write this as $pc(1 - m - r) + p(1 - c)c$.

On and on we could go, for three days, four days...... until no more people can move in the system, or until we run out of enthusiasm. However, keeping track of the numbers in each state and all their possible movements to get from one state to another can spiral out of control very quickly in a larger model.

Example 17.1

People who are recovered at the end of day 4 took one of the following routes:

- Becoming infectious on day 1, recovering on day 2
- Becoming infectious on day 1, recovering on day 3
- Becoming infectious on day 1, recovering on day 4
- Becoming infectious on day 2, recovering on day 3
- Becoming infectious on day 2, recovering on day 4
- Becoming infectious on day 3, recovering on day 4

Exercise 17.1

How many routes are there to still being Susceptible at time 5? How might you express the number of people still Susceptible at time 5?

Using the transition matrix removes the requirement to keep track of all the possibilities. Previously the one-step transition matrix was defined, i.e., a matrix that defines all the possible movements in one period.

$$\text{Our transition matrix, } T = \begin{bmatrix} 1-c & c & 0\ 0 \\ 0 & 1-m-r & m\ r \\ 0 & 0 & 1\ 0 \\ 0 & 0 & 0\ 1 \end{bmatrix}$$

With $c = 0.1$, $m = 0.01$, and $r = 0.1$

$$T = \begin{bmatrix} 0.9 & 0.1 & 0 & 0 \\ 0 & 0.89 & 0.01 & 0.1 \\ 0 & 0 & 1 & 0 \\ 0 & 0 & 0 & 1 \end{bmatrix}$$

By multiplying the transition matrix by itself, we can calculate the two-step transition matrix. Now, we don't have to do all the calculations for the end of day one just to get estimates for the end of day two. However, a two-step transition matrix means you don't get the estimated results for the end of day one, so if more granular model results are needed then stick with the one-step matrix.

For our 4-state Covid model, parameterized with $c = 0.1, m = 0.01, r = 0.1$, the two-step transition matrix can be calculated as:

$$T \times T = \begin{bmatrix} 0.9 & 0.1 & 0 & 0 \\ 0 & 0.89 & 0.01 & 0.1 \\ 0 & 0 & 1 & 0 \\ 0 & 0 & 0 & 1 \end{bmatrix} \times \begin{bmatrix} 0.9 & 0.1 & 0 & 0 \\ 0 & 0.89 & 0.01 & 0.1 \\ 0 & 0 & 1 & 0 \\ 0 & 0 & 0 & 1 \end{bmatrix}$$

$$= \begin{bmatrix} 0.81 & 0.179 & 0.001 & 0.01 \\ 0 & 0.7921 & 0.0189 & 0.189 \\ 0 & 0 & 1 & 0 \\ 0 & 0 & 0 & 1 \end{bmatrix}$$

The great advantage with stochastic models like this is that if you can draw the system, it is easy to create a transition matrix. And if you have the transition matrix, it is easy to do the relevant calculations. We'll see some extensions to this simple model in the next section. The other benefit is that it can be easier to collect data on the rate of change rather than the number of people in each state in real life. For example, the rate of infection might be taken from public health statistics. If there were 150,000 new infections a day in a population of 60,000,000, that is a daily infection rate of 0.25%. This data could form our estimate for parameter c. With stochastic models we can quickly give estimates for:

- The number of people infectious at any one time
- The number of deaths attributable to the disease
- The level of immunity in the population
- The reproduction number, often denoted R. This is the average number of people that one infectious person will pass the disease onto. If R is less than 1, eventually the disease will fizzle out. However, if it sneaks above 1, the disease will see exponential growth in the population.

Everybody Loves R...

If you listened to the news at all in 2020 or 2021, you probably heard of the R number (and you probably learned more about the Greek alphabet too). R is the reproduction number, which tells you how many people each one infected people passes the virus on to, on average. If the R number is 1, then on average each ill person infects one other person. If it is 1.5, each person infects a person and a half (on average, of course!), then those 1.5 people infect another 1.5 each and so on, and cases will grow exponentially. An R number less than 1 means that the virus is in decline.

The reason this was constantly in the news was that R changed with time and with people's behavior. If people had lots of close contacts, they were likely to infect more people. It also changes depending on the variant. The Omicron variant was more infectious than the Delta variant so it gave rise to a higher R number. So, R is a measure both of people's behavior and the disease, thereby giving an intuitive way of predicting how strong future interventions might need to be. An R number of 1.5 means that we need to reduce disease transmission by at least a third (to get it to or below 1) to stop the epidemic spiraling out of control. In the initial stages of the Covid pandemic when many countries were tracing contacts of infected people, R was easily calculated. However, like most modeling in a crisis, it's not ideal.

Firstly, R is really hard to measure when you aren't counting contacts. R does not have a natural timescale: one infected person could infect two people immediately and then no one else, or one person now and one person in 10 days' time—which means that it cannot easily be used to predict the number of cases in a day, week or month. Instead, R relates to generations of the disease; one infectious person creates two cases of the next generation. Because it takes some time for the cases in the new generation to be counted, we can only estimate R for the previous generation, which may have been two weeks ago (or more).

Secondly, R is an average transmission over all the cases in a generation. Therefore, it doesn't distinguish between severe and asymptomatic cases. A variant that has an R of 1.2 might be more severe and therefore of more concern than a variant with R of 1.8. However, R can be calculated for different ages, regions, behaviors and other factors, if sufficient data is available. But it doesn't account for 'super-spreaders'—the name given to the 20% of people who seemed to cause 80% of Covid cases. Reducing the contacts of this group therefore had a more pronounced effect than changing the behavior of a different group.

R also doesn't give any information about how many infected people there are. If R is 5 but there are only two infectious people, you still might be able to get the disease under control. If R is 1.2 but there are 2 million infectious people, that might cause a problem for hospitals (as well as businesses, schools, and food supply chains).

All of this uncertainty means that R is usually given as a range. If the bottom of the range is greater than 1, the number of cases will keep growing—and then we get to talk about the R number (and hopefully other metrics) some more!

All of these metrics can be used to decide public health policy. A benefit of this kind of modeling is that the numbers of people in each state at every time period can be found. Also, the number of people in each state can be readily updated with actual data once it is available, without further modification of the calculations.

Example 17.2

A 4-state Covid model is created (let's not worry what the parameters were for this example). It predicts that after five days, there will be 61.5 million susceptible, with 2 million infectious, 1 million recovered, and 50,000 deceased.

After large-scale public testing initiative, it is found that in fact, 5 million people are infectious and 2 million recovered. The number of Covid deaths is 55,000 according to official figures. The actual figures after five days (in thousands) are:

$$(S_5, I_5, D_5, R_5) = (57945, 5000, 2000, 55)$$

To find the estimate for day 6, instead of running the model from day 1 again, we use the actual values from day 5 as our starting population, and our transition matrix, T. Day 6 estimates are therefore calculated as:

$$(S_5, I_5, D_5, R_5) \times \text{Transition Matrix} = (57945, 5000, 2000, 55) \times \begin{bmatrix} 0.9 & 0.1 & 0 & 0 \\ 0 & 0.89 & 0.01 & 0.1 \\ 0 & 0 & 1 & 0 \\ 0 & 0 & 0 & 1 \end{bmatrix}$$

$$= (52150.5, 10244.5, 2050, 555)$$

So, it is quick and easy to give updated estimates based on real data using this method, which is very important consideration in a crisis.

Stochastic models are also used in pensions or life assurance products. In these models, death by any cause is allowed and the states can include working, retired, in ill-health (as sometimes life assurance pays a benefit in this case), and deceased. Or the model could be set up to consider joint lives, where people move from the state of married to the states of widowed, then to deceased. There really are so many variations here, but the next section will consider just a few.

17.3 Re-modeling (and Re-modeling and Re-modeling) Coronavirus Spread

The 4-state Covid model introduced in the previous section is a huge simplification of the real situation. As time goes on, research might suggest that other factors affect value of the model parameters, particularly the rate of catching Covid and the mortality associated with it. The ways in which a simple stochastic model can be adjusted are almost endless, but some caution is required. Complex models may be more predictive, but they are also more difficult to explain; they require more parameters which require estimation (leading to more potential for uncertainty); and they can take longer to build and run. Another potential problem is that complexity is mistaken for accuracy; a model can be complicated but entirely incorrect and a simple model can be a correct approximation. Adding complexity to the model should only be done when the addition is thought to have a material impact on the results; when the inclusion of the additional factor makes a significant[8] difference to the results.

Below are some changes to the basic 4-state Covid model that will incorporate factors that have since been found to be significant to Covid transmission and severity.

- Vaccination

As the primary purpose of vaccination is to prevent individuals being susceptible to a disease, let's assume they are 100% effective in that purpose. Now, individuals can move out of the Susceptible state without passing through the Infectious state, into the Recovered state (which should now be named 'Has Antibodies' for clarity). In order to make this transition, a new parameter, v, is created which is the rate at which susceptible people become vaccinated and therefore are no longer susceptible. Then, people move from the Susceptible state to the Has Antibodies state at a rate of v, bypassing the Infectious stage entirely.

- Waning Immunity

Waning immunity means that people with antibodies (either through vaccination or contracting the disease) can become susceptible again. If we assume that the effect totally wears off and does so instantly—that someone with previous immunity instantly can join the leagues of the susceptible again—a transition from the Has Antibodies state back to the Susceptible state can be added. A numerical estimate of the rate of waning immunity parameter is also required and perhaps can be taken from pharmaceutical studies.

- Seasonality

People catch and are affected more severely by viruses in colder months. This too can be incorporated into a stochastic model. If we wanted to keep it simple, we could just change the transition rates of certain time periods, for example letting $c = 0.3$ and $m = 0.05$ for time periods 0 to 5, but further changing the transition matrix so that $c = 0.33$ and $m = 0.055$ for time

[8] This can be either be statistically significant or significant as per general parlance.

periods 6 to 12. This can be interpreted to mean that seasonality increases both the rate of catching Covid and the mortality rate by 10%.

In the real modeling performed by the University of Warwick,[9] it was assumed that the effect of seasonality did not appear as a sudden jump in parameters, but was a smooth curve. Assuming that the impact of seasonality was actually a sine curve, peaking at +10% and troughing at −10%, modelers created a smooth seasonal effect on the transmission parameter, avoiding awkward step changes.

- Asymptomatic Covid

A turning point in the pandemic was certainly the widespread availability of testing. This meant that for the first time, people could determine whether they had Covid in a non-hospital setting and therefore better information was available for both individuals and government advisors. Testing also revealed asymptomatic cases, where the person tested positive for Covid but had no symptoms. The discovery of asymptomatic Covid meant that previous estimates of cases were materially understated. However, asymptomatic cases were also less likely to die from the disease. To include this in the model, a new state is required: the Infectious (Asymptomatic) state. People either move from Susceptible to Infectious (Asymptomatic) or Susceptible to Infectious (Symptomatic). We therefore need to estimate three new parameters: the rate of moving from Susceptible to Asymptomatic, the rate of dying if asymptomatic, and the rate of recovering if asymptomatic. Now, asymptomatic and symptomatic cases can have different mortality rate, and the model now predicts the number of people in those two states separately.[10]

- Length of Infection

One simplification with stochastic models is that the model is memoryless. It doesn't remember any of the previous time periods and instead the only relevant information is the current number of people in each state and the relevant transition matrix. This makes calculations simple, but it means that people could become infectious and in the next time period, recover. This doesn't fit with our understanding of Covid, as people are usually infectious

[9] This University of Warwick Stochastic Modeling paper is much more advanced than this chapter, but the details on the final page show that the researchers have created stochastic models for each local authority population with only three unknown parameters!

[10] Splitting the Infectious state into Asymptomatic and Symptomatic Covid can be seen in the diagrams of a real-life modeling team A Two-Phase Stochastic Dynamic Model for COVID-19 Mid-Term Policy Recommendations in Greece.

for at least three days. To allow for this, we need to add two more states: Infectious Day 1 and Infectious Day 2. The state previously known as Infectious is renamed "Infectious 3+ days". People move from being Susceptible to Infectious Day 1. Then, some die (so we need a new parameter to estimate this mortality rate), but all others move onto Infectious Day 2. Here too, some die (again, a new mortality parameter), but all others move onto Infectious 3+ days. In the Infectious 3+ days, people can die or recover as in the 4-state model, although we should check if our parameter estimates for mortality and recovery are still reasonable. From this model, we can investigate the impact of changing the isolation guidelines. For example, if the isolation period is cut from two days to one day, how many people are released from isolation? Using this in conjunction with an economic model, we can estimate the impact of the length of isolation periods on the economy.

- Infection rate based on Infectious people

One thing we know about contagious diseases, including Covid, is that transmission occurs during contact with an infectious person. To make our model more realistic, the rate of catching Covid should be related to the number of people who have Covid. This makes the calculations a little more complex but will make a material difference to our model results, allowing cases to grow exponentially just as they did in 2020. In the basic 4-state model, the rate of catching Covid was c, estimated as a constant. Now, let's relax that assumption and let c be a function of the number of people in the Infectious state at the most recent previous time, that is, time $t - 1$, which we can call I_{t-1}. Now define a function that links the transmission rate at a given time with the number of cases at some previous time.

A parameter of $c_t = N \times \frac{I_{t-1}}{p} \times 0.33$ could be interpreted to mean that if a person has N contacts per day (which are assumed to be independent from each other), those contacts are infected at the rate of the general population, and contact with an infected person will mean an individual has a 1 in 3 chance of catching Covid. The model predicts both the exponential effect of contagious diseases (that is, the more cases in the population, the faster it will grow) and the effect of number of contacts on the spread of Covid. This could be used to model the effects of lockdown on the spread of disease.

We could also do the same for mortality. If we assume that mortality is constant unless hospitals are overwhelmed with cases, we could write a function that increases mortality if cases are above a certain point, such as $m = \max\left[0.05, \left(1 + \frac{I_t}{p} - l\right) \times 0.05\right]$. Here, if the percentage of people currently infected is above a certain limit l, which could be interpreted as the maximum safe capacity of hospitals, the rate of mortality will increase. The maximum function here means that our mortality parameter is either 5%, or higher if the number of infectious people exceeds a certain limit.

Stochastic models, of whatever complexity, can be used to test different scenarios. For example, if we assume mask-wearing reduces the transmission parameter by a certain percentage, we can fairly easily rerun the model to investigate potential outcomes. We could split out a new Infectious state to represent older people, who are more likely to die from Covid, or schoolchildren, who are less likely to die (similar to how Asymptomatic Covid can be dealt with). With these new states the effect of targeted interventions could be investigated, such as vaccinating older people, or allowing schools to remain open while the rest of the country is locked down.

At all levels of complexity, stochastic models contain an element of randomness. Therefore, although the model prediction is the expected number of people in each state at every time period, that is only the statistical expectation of the result. In reality, stochastic models are run hundreds or thousands of times, each time allowing the transitions to be an expression of a random variable. To demonstrate how this works to give a final result, let's revert to the original 4-state model, with parameters $(c, m, r) = (0.3, 0.05, 0.25)$, and starting with an entirely susceptible population of 100 people.

In the first time period, although we expect 30 people ($pc = 100 \times 0.3 = 30$) to become infected on the first day, each potential infection is actually an independent expression of a random variable. Each person independently does or does not get Covid with a probability of 30% and 70%, respectively.

Exercise 17.2

If you have a 10-sided die or a random number generator, model whether the first ten people do or do not catch Covid if the probability of infection is 30%.

This is actually a binomial distribution with 100 trials and a success probability of 0.3 (although I define 'success' very loosely here!), the outcome of which gives you a number of successes. Of course, the number of successes will be between 0 and 100, echoing the adage that you can't succeed if you don't try. So, to speed up the calculation within our stochastic model, instead of generating 100 random variables and then counting the number of people who do or don't move, it is quicker to generate one binomial random variable which gives the number of people who moved. If we started with 100 people in the Susceptible state, the model creates one expression of a binomial random variable (trials = 100, success probability = 0.3) which, in one step, gives the number of people who become infectious at the first time period. Suppose this number is 26.

Immediately jumping on this outcome as a result is not good modeling practice as if this were rerun, it is unlikely that the outcome will be exactly the same. So, the model is run 100 times. In this trial (and some others) 26 people became infectious after one day, but there may be runs where 30, 31, 35, or even 50 people will be infectious. By allowing the randomness in our parameters to express itself, a range of results are generated and can be plotted.

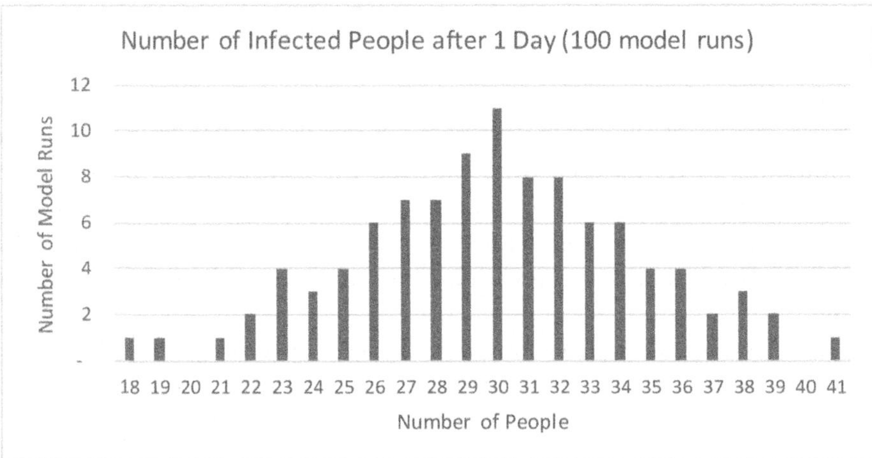

The more runs completed, the closer this plot will tend toward the statistical distribution underlying the parameter. Even with only 100 runs (1000 or 10,000 would be more common for actuarial work) the graph already seems to suggest that 30 is the most common number of infections. However, 50% of the runs suggest that between 28 and 33 people would be infected, and in 90% of the runs it was between 24 and 38 people. To put some meaning to that, our model is 90% sure that the number infected will be in the range of 24–38. Of course, we can also calculate other confidence intervals or ranges like this.

However, in a pandemic we are predominantly concerned with the worst-case scenario in order to plan and prepare, so the analyst might wish to focus more on the upper end of the results. In 100 runs, our model gave more than 40 infectious people 1% of the time, and more than 36 just over 10% of the time. Actuaries often decide on a level to be the 'realistic worst-case' scenario. This might be the scenario which is only exceeded 5% of the time (or 1%

of the time, if more caution is required). In this model, the realistic worst-case scenario exceeded only 5% of the time would be 39.[11] Planning for 39 infectious people would mean that we would have sufficient preparations 95% of the time, according to this model.

This section has provided a bit more detail on extending a Covid model to be more realistic and, of course, the same principles can apply to other modeling purposes too. But while predictive models are interesting, what did the pandemic mean in reality for actuaries?

17.4 Covid Corollaries: The Effect of the Pandemic on Insurance and Pensions

The pandemic has changed many aspects of our lives, whether that is around attitude to risk or the day-to-day logistics of getting to work. Some might continue to hoard toilet paper and flour for many years, and some are keen to jump on the next flight to Mexico and get back to normality. Some people are doing both. As we all start to unravel and deal with the effects of the pandemic on ourselves personally, there have been considerable effects for the professional lives of actuaries.

Firstly, as alluded to throughout this chapter, many actuaries have helped to shape governments' Covid responses. As these models were widely reported in mainstream media, some actuaries working in other fields may be drawn to public service roles. However, actuaries in the private sector have faced professional challenges too, in investments, insurance, and pensions.

Between 28th February 2020 and 20th March 2020, the FTSE 100 dropped by 30%. This sudden and unexpected volatility led investment actuaries to quickly re-model assets and liabilities. Notably, some sectors that have previously been seen as safe bets, such as airlines, suddenly dropped in outlook as their operations stopped completely. The changing Covid travel regulation also gave rise to an uncertain future outlook for these companies. Post-pandemic, economic uncertainty remains high, especially as so many governments took on record levels of peace-time debt to finance Covid-related economic relief programs. Inflation risk is the other big uncertainty for investment actuaries, and the high levels of inflation seen in many countries at the end of 2021 and throughout 2022 created significant uncertainty that requires mitigation. This inflation is somewhat a hangover from the

[11] There are 3 runs where there are 39 infectious people or above, but 6 runs where there are 38 people. Interpolation doesn't make much sense in this model, as it is not possible to infect a fractional person! Therefore, 39 is the realistic worst-case scenario exceeding 5%.

pandemic. During 2020 demand was stifled, so in 2021 demand surged suddenly, creating a rise in prices that may change erratically. In addition, some manufacturing nations such as China have continued to have lockdowns well into 2022, affecting factory output. The immediate challenge for governments is to control the level of inflation without creating stagnation or even a recession. The challenge for investment actuaries is to quantify the effects of inflation risk or the risk of recession on company investments, taking into account likely government interventions. The aftermath of the pandemic has its own risks too.

For **general insurance** companies, the pandemic led to unexpectedly high claims costs on some lines of business due to a sharp increase in business interruption and travel claims. Business interruption claims in particular have been controversial, as many UK businesses were only covered for business interruption in the event of property damage. Those businesses are therefore ineligible to claim—a great disappointment to many small business owners who were forced to close by government regulation. However, other businesses did have cover for non-property damage business interruption, particularly those who had coverage in case of prevention of access or disease. These clauses pay business interruption claims in the case that customers are unable to get to the business (for example, due to road closure, nearby property damage making the area unsafe, or even for contagious diseases such as measles that prompt a local closure to contain the spread). Some insurers had explicitly excluded coronaviruses from coverage after the SARS outbreak, but other insurers said that a national lockdown was not envisaged at the point of underwriting, and prevention of access or disease was only intended to cover short-term interruption claims (measles usually only leads to small-scale closures of up to two weeks). However, in January 2021, the Financial Conduct Authority ruled that many insurers without specific exclusions were in fact liable and would have to pay these claims, leading to some very large payouts. The intricacies of contract law are beyond the scope of this book, but this ruling cost insurers billions of pounds.

Although business and travel policies have experienced higher claims costs, motor insurers have seen a decrease in the frequency of claims as many drivers reduced their mileage significantly in the pandemic. However, this creates a difficult situation for actuaries: to what extent will this trend continue? Should 2020 and 2021 be included in pricing data or reserving claims triangles and if not, how should actuaries deal with this lack of recent data? Is pre-Covid data still relevant? If not, then what can be used? There are many questions that actuaries will have to tackle in the short term, not helped by the fact that it's not yet clear what the new normal will be.

However, the pandemic has brought some opportunities for general insurers too. Analysts at Swiss Re estimate that by the end of 2022, insurance premiums will be at a historic high as people have been reminded of the need for insurance in a risky world, and as inflation climbs to post-millennium highs. Some segments are booming. For example, due to a generation of lockdown puppies, pet insurance is seeing high growth. Travel insurance is rebounding quickly, as customers are keen to take a foreign holiday again (although Covid is largely being excluded from these policies by insurers as a precaution). Even in the face of high inflation, general insurance is largely a necessary product and relatively inelastic as people buy it because they require it, rather than out of any particular desire. As such, the post-pandemic growth looks steady, even good, but high inflation and the cost of living crisis remains a challenge for insurers' profitability.

Conversely, health insurance is well-known in the industry to be subject to elastic demand. It is very price sensitive, with the number of policies sold correlating strongly to the price of cover and the general economic situation. While post-pandemic high inflation should therefore mean a bleak picture for health insurers, there has actually been a slight uptick in optional health-care coverage. In the UK, which has a public healthcare system, individuals who are able to afford it are buying private healthcare policies as a solution to long National Health Service waiting lists, which have been exacerbated by the closure of hospitals to all but emergency patients and staff absence during the pandemic. While sales may be increasing, there is also a potential for a huge number of unreported health claims as people put off seeing a healthcare professional during Covid, and therefore will receive a delayed diagnosis. This will likely result in higher costs, as later diagnoses require more costly treatments. However, in the medium term the outlook is uncertain. Will the general population stick to the daily walking or exercise habits started in the pandemic or will working from home create more sedentary and less healthy people? Will high inflation mean that potential customers are unable to afford their healthcare premiums? Will governments increase taxes on insurance products to recoup money spent during the pandemic (remember, tax on insurance products is often called a 'stealth tax', as they raise money for the government but are collected by the insurer, so customers may not realize that they are paying tax during this transaction)? This would further increase the price of health insurance and therefore deter customers. There are lots of questions and not enough understanding yet of what the new normal is.

Pensions and life assurance companies have been frantically calculating and re-calculating the effect of Covid throughout the pandemic. They spend a huge amount of time modeling when people might die, so an event like this that led to a high number of excess deaths (that is, deaths over and above usual causes and numbers) has a direct impact on their results. As countries continue to move past their Covid peaks, it is not obvious what the short-term mortality experience will be. Will it be higher than pre-Covid, as people suffer with either Covid or other illnesses that they have not been exposed to for a while? Will it be the same as it was pre-pandemic? Or will it actually be lower than pre-pandemic levels? A lower mortality rate could occur if Covid mostly brought forward the deaths of people who would have otherwise died post-pandemic, in a phenomenon chillingly called harvesting. Early views of 2022 experience shows that mortality is returning to pre-pandemic levels, but pensions and life assurance companies are still considering a range of levels of post-pandemic mortality (plus the effects of another wave of Covid). In the medium to long term, these companies face uncertainties around the effect of long Covid, future variants or spread, and lockdowns. These effects could be in the form of an increase in mental health conditions, organ problems due to long Covid, or permanent respiratory tissue damage due to Covid. In the medium to long term, the effect of prolonged loneliness experienced in lockdowns may even increase the development of conditions such as dementia. Life assurance companies who provide ill-health benefits might be particularly interested in not just life expectancy, but of the quality of life in those years remaining.

The effect on the pensions industry doesn't just include the effect on the generations close to retirement. For those born between 1965 and 1980 (affectionately known as Generation X), lockdown has meant that more people in this age bracket have struggled to contribute to their **Defined Contribution** pension, leading to concerns that they will not have adequate pension provision when they reach retirement age. In a study by the International Longevity Centre, this is particularly acute for self-employed workers, 39% of whom said that they saved less because of Covid. This again raises the question of intergenerational fairness, and no doubt this will be replicated for Generation Y (born 1980–1995, un-affectionately known as Millennials) and Generation Z (1995–2010, known as Zoomers), who at the time of writing look likely to spend the early part of their career in a post-pandemic period of high inflation, therefore affecting their ability to save early in their career when the compounding effect of interest means their savings are worth more. Actuaries are trying to quantify this effect and making recommendations[12]

[12] Primary recommendations include a change in tax treatment for pension savings and auto-enrolment for self-employed workers.

to governments for public policy that incentivizes pension saving in a bid to stem rising generational inequality.

The pandemic has increased public awareness of risk and statistical modeling, and likewise increased interest and literacy in statistics. The dissection of statistical models by the mainstream media has led to complicated explanations about why models didn't predict what happened and the extent to which politics should be led by data. It created a nation of armchair epidemiologists and statisticians, like the ones I overheard in the supermarket. I sincerely hope some join the ranks of the professionals; the actuarial community needs more critical thinkers, more problem-solvers, and more people who care deeply about getting things right.

18

Careers in Actuarial Science

Chapter Objectives

- Understand common entry requirements to actuarial roles, and alternatives to actuarial roles
- Consider different routes for sitting professional exams
- Understand the importance of professional development for actuaries and examples of useful development paths
- Understand how demand for actuarial roles is changing globally and within industries

Find out what you like doing best and get someone to pay you for doing it.
—Katherine Whitehorn, journalist

Barbie has held a variety of roles, including sign language teacher, yogi, Chancellor of Germany, a US Air Force Thunderbirds pilot, Canadian Mountie, UN ambassador, bee-keeper, and multi-discipline Olympic medal-winner. But she has never been an actuary. While Barbie inspires children to dream of pursuing a multitude of important and interesting careers, the first time most people learn of actuaries' existence is when they receive a statement about their pension. Hopefully, this book has revealed the mysterious art of actuarial science. You might even have found the previous chapters interesting, and want to learn more about becoming an actuary yourself. As

© The Author(s), under exclusive license to Springer Nature Switzerland AG 2022
C. Townsend, *A Risky Business*,
https://doi.org/10.1007/978-3-031-11673-5_18

Barbie has so far neglected this task, this chapter will answer some common questions about careers in actuarial work. If this is of no interest to you, then please feel free to skip over this chapter, or read it to affirm your own choice.

18.1 What Do You Need to Be an Actuary?

The main component of actuarial science is quantifying risk, so it should be no surprise that strong numerical skills are needed to succeed as an actuary. You really need to enjoy and be good at working with statistics, as it will form the majority of your day-to-day work, at least in the early years of your career. You don't need to be a math fanatic; if you're not a fan of the theoretical and abstract areas such as topology, that's not a problem. But if you love solving logic puzzles or working out optimal solutions to real-life problems, then actuarial work might interest you. Actuaries need to be resourceful when tasked to solve problems with very little data and they need to be curious about data and problems. They also need to be good communicators and have a strong sense of objectivity. Actuaries, perhaps unlike pure mathematicians, also need to have a good deal of common sense.

The usual route to becoming an actuary to complete a bachelor's degree or equivalent (or higher!) in a numerically based subject; mathematics, statistics, operational research, computer science, economics, or even physics and chemistry degrees will be sufficient. Some colleges and universities even offer degrees in actuarial science. While these courses are obviously very relevant to actuarial work and often give you lots of exemptions to professional exams (more on those later), it is a very specialized degree which might limit your non-actuarial options. Having a more general degree not only gives you more options if you decide not to become an actuary, but it might also give you a bit more range in your skillset. For example, someone with a Natural Sciences degree might be well-suited to model natural catastrophe or climate change risk, or someone with a computer science degree might particularly excel as a pricing actuary working with big data. Even if a person with a Natural Sciences degree doesn't end up modeling their specialism, their experience of trying to model complex phenomena can be applied to almost any statistical modeling problem.

Once you have a bachelor's degree, the next step is to get a job as an actuarial analyst. Depending on your location and the jobs market in that area, you might be able to go straight from your degree to an actuarial job with no specific actuarial skills demonstrated—this is the most common route in the UK. In the USA and Canada, most employers look for a candidate that not only has a numerical degree but also has passed some of the early actuarial professional exams. Wherever you are in the world, you can improve your chances of landing a role by working on some of your other technical skills. Having advanced skills in spreadsheet packages such as Microsoft Excel and basic coding skills (again, Microsoft VBA is ubiquitous) will prove to employers that you will be able to get stuck in with work straight away. Free coding languages such as Python or R are also relatively widely used, but companies might pay for a data software such as SAS or SQL. Therefore, having a specific coding language on your CV might appeal to some employers, but might add little value for others, depending on their specific setup. Though non-essential, a familiarity and ability to learn about data and programming languages will stand you in good stead.

While landing a role as an actuarial analyst is largely based on your technical skills, working as an actuary requires additional skills. Actuaries have traditionally had a reputation of being bright but anti-social. However, now computers can take the heavy lifting out of calculations, actuarial skillsets have had to expand to stay relevant. A key part of actuarial work is to communicate results, often to non-actuaries. For example, actuaries will write reports on the reserves at a pension fund which needs to be understood by the members of the public. They need to explain the assumptions and sensitivities of a model to the board of directors, or work with earth scientists to elicit the required inputs for their catastrophe models. As such, actuaries need strong communication skills; not just being able to present and write fluently and accurately, but having soft skills that build trust and understanding.

Through their code of conduct, actuaries commit to acting in the public good. The type of person who likes exploiting loopholes and pushing the boundaries of acceptable conduct is probably not going to be well-suited to actuarial work. The two most common reasons inexperienced actuaries face disciplinary action are not keeping up to date with continuous professional development requirements and cheating in exams. The punishment is usually a fine for a first offence for the former, and for the latter, being barred from sitting any further exams for some years. However, dishonest or illegal activity in your personal life can also bring disciplinary action. The Institute and Faculty of Actuaries' 2020 report states that one individual received sanctions for using a student discount card (to give a discount for rail journeys) when they were not eligible to do so. Actuaries need a strong moral compass, and the integrity to act on their convictions. Being too easily influenced is not ideal for someone in a position of such authority and power, as actuaries often are. Being objective is a valuable skill as it is not for actuaries to tell their client what they want to hear, but to acquire and interpret the data as accurately as possible.

Finally, actuaries are very determined people. The volley of exams, usually undertaken while working in your first job, can be too much for a lot of people. Most actuaries have missed parties or events while studying for an exam. It takes technical ability to pass exams, but organization, determination and persistence to even get to the exam hall. If you want to be an actuary, or anything else that is not straightforward or easy, you have to keep going and apply yourself over several years. It is thousands of hours of work, which culminates in just a few short letters after your name.[1]

18.2 Professional Qualifications

To become a qualified actuary, you must pass a series of professional exams. These vary by region, depending on the institute you become a student member of. In this section I will focus on the exam paths in North America. These are prescribed by the Society of Actuaries (SOA) and the Casualty Actuarial Society (CAS). This section also includes exams by the UK-based Institute and Faculty of Actuaries (IFoA). The IFoA syllabus aligns closely to the International Actuarial Association. Do check your local institute's website for updates, more detail about the exam syllabi, past papers, and more information.

Most institutes have two levels of being an actuary; associate and fellow. Usually when we refer to qualified actuaries, we mean fellows, as they have passed all the exams for their specialism and are able to take on reserved roles.

As you can see from the table below, there are quite a few exams to take! In addition to these, there is also mandatory online content and courses, as well as evidence of experience in an actuarial workplace. Once all the general skills have been demonstrated, you can apply to become an associate of your institute. Alternatively, you can carry straight onto the specialist tracks to become a fellow (Table 18.1).

18.2.1 IFoA Exams—The Detail

- The Core Statistics and Core Mathematics exams comprise of two papers each; one written paper and one computer assessment using R. The Core

[1] Fellows of CAS, SOA and the IFoA may use the letters FCAS, FSA and FIA after their name respectively.

Table 18.1 Summarizing the exams for the IFoA, SOA, and CAS

	UK	Canada & USA	
	IFoA	SOA	CAS
General Technical Skills	Core Statistics 1	Exam P	
	Core Statistics 2	Exam FM	
	Core Mathematics 1	Exam FAM	Exam 3F
	Core Mathematics 2	Exam ALTAM or ASTAM	MAS-I
	Core Business 1	Exam SRM	MAS-II
	Core Business 2	Exam PA	Exam 5
			Exam 6
General Practical Skills	Core Business 3		
	Core Practice 1		
	Core Practice 2		
	Core Practice 3		
Specialist Skills	Specialist Principles Exam	Specialist Technical Exam	Exam 7
	Specialist Principles Exam	Specialist Technical Exam	Exam 8
	Specialist Advanced Exam	Specialist Advanced Exam	Exam 9

Mathematics papers cover topics such as annuities and life-dependent annuities, as well as option pricing and modeling asset returns. The Core Statistics papers cover statistical distributions, population modeling using stochastic models, and traditional **reserving** methods such as Bornhuetter-Ferguson and chain ladder.

- Core Business 1 and 2 are written papers covering economics-type content such as supply and demand and understanding balance sheets. Core Business 3 is an online assessment.
- Core Practice 1 is a big exam with lots of content. It consists of two papers, one of which is a case study situation. Both test your holistic knowledge of actuarial work and the industries in which actuaries commonly work.
- Core Practice 2 also has two papers, but it is an online exam testing your spreadsheet modeling and documentation skills.
- Core Practice 3 is a single online test which assesses your skill in communicating technical concepts to non-technical audiences.
- The Specialist Principles and Specialist Advanced exams are dependent on your specialism. For example, pensions actuaries will likely sit Specialist Principles 2 and Specialist Principles 4 (about Life Insurance and Pensions respectively), then Specialist Advanced 4. Actuaries in **general insurance** usually take Specialist Principles 7 (reserving and capital modeling),

Specialist Principles 8 (pricing), and Specialist Advanced 3. You don't have to take the exam that matches your current role, but experience in the field definitely helps secure a pass!

18.2.2 SOA and CAS Exams—The Detail

SOA and CAS share the first two exams, Probability, and Financial Mathematics. This is really helpful if you haven't quite decided which area interests you most. If you are interested in life, health, and pensions, SOA is the institute for you. For general insurance, CAS is the relevant institute. Note that there is a separate institute for Canadian actuaries (rather alluringly called the CIA, although as you'd imagine in this context it stands for the Canadian Institute of Actuaries), but they don't administer their own exams. To become a member of the CIA, reach fellow or associate level in SOA or CAS and the qualification can be transferred with minimal paperwork and no further exams. The other difference between SOA and CAS is that SOA exams are named as a flurry of acronyms, whereas CAS exams are numbered, but neither naming method provides much of a clue as to the content.

After the first two exams, SOA then requires four further exams to reach associateship. Then, you specialize and sit a further three exams related to that specialism. The student will have already usually have chosen to work in life and health rather than general insurance by this point (as otherwise they would have likely joined CAS). The general exams for SOA are:

- Exam FAM (Fundamentals of Actuarial Mathematics), which is followed by either the ALTAM or ASTAM (Advanced Long/Short Term Actuarial Mathematics) exam—the candidate can choose.
- Exam SRM (Statistics for Risk Modeling), which is a hefty 3.5-hour multiple choice exam, consisting of 35 questions on time series, linear models and clustering.
- Exam PA (Predictive Analytics), which is a project-based scenario exam. Lasting over five hours, candidates use Excel, R, and Word to investigate and document a business problem.
- Finally, to become a SOA fellow, you then choose one of six specialism tracks, including Individual Life and Annuities, Group and Health or Corporate Finance. Each track has its own final three written exams and some online modules.

After Probability and Financial Mathematics, CAS has eight further exams:

- Exam 3F (Financial Economics), which covers option pricing and portfolio theory.
- MAS-I and MAS-II (Modern Actuarial Statistics I and II) which covers topics such as statistical distributions, credibility theory, and linear models.
- Exam 5 has the catchy subtitle 'Basic Techniques for Rate Making and Estimating Claim Liabilities', a more detailed look at the pricing techniques from Chapters 8 and 9. It's a four-hour exam, and as well as calculation-based questions, there are also some essay questions to test the candidate's skill at communicating these concepts and explaining the pricing process.
- Exam 6 is location specific, so there are different versions of Exam 6 for USA candidates, Canadians, and those in Taipei. For those sitting the exam in any other location there is a generic international version. The content is about the regulation applicable in the area, including capital requirements and financial reporting. There is also a section on government-administered insurance programs, such as the US flood scheme.
- Exam 7 is focused on reserving and, to give the full subtitle, the 'Estimation of Policy Liabilities, Insurance Company Valuation and Enterprise Risk Management'.
- Exam 8 is also known as Advanced Ratemaking, so builds on pricing techniques covered in Exam 5. Some of the techniques in this exam will be familiar to you—**Generalized Linear Models** are covered here, as well as Increased Limit Factors and pricing **reinsurance**—and this four-hour exam also includes some essay-style questions about catastrophe models and reinsurance.
- Exam 9, subtitled 'Financial Risk and Rate of Return', is an economics-heavy paper that covers portfolio theory, options and securities pricing, and assessing profitability, among other topics.
- Note that the CAS exams doesn't contain any tracks to choose from but Exams 7, 8, and 9 are only for those aiming for fellowship, and so contain very specialist general insurance skills.

18.3 Exam FAQs

I've covered some of this content before, do I still have to do the exam?

If your university or college has an agreement with the relevant institute, you are able to apply for an exemption from the exam. This usually involves submitting a transcript of your grades for each college course. If your college doesn't have such an agreement, you can sometimes apply directly to your institute, providing evidence of where your college courses met every single clause in the syllabus (this can be from multiple courses).

Which order should I take the exams in?

There isn't a rigidly prescribed order, but the later exams do assume a knowledge of the earlier ones. So, sitting the earlier exams first is safe bet, and it will give you time to build your practical experience of the industry in preparation for the later specialist exams. If you haven't yet decided between SOA and CAS, take Probability and Financial Mathematics as your first two papers (the two that are common between the institutes) to give yourself time to secure a role as an actuarial analyst and then decide.

After that, decide based on your current skills. If you did a lot of calculus or economics at college, starting with CB1, CB2 or Exam 3 will use your prior knowledge and calculus skills. SOA's SRM and PA exams naturally complement each other as they both use R, so sitting those close together will keep those coding skills fresh in your mind.

Also consider how much time you have to study and therefore how many exams you can realistically take in one sitting. CP2 and CP3 exams are testing your practical skills so there isn't much technical content to remember, and so these exams can easily be sat together, or pair one of these practical exams with a technical one. For example, sitting CP2 and SP2 in one sitting might be easier to manage than CB2 and SP2 together.

How long will it take me to pass?

To pass all the exams to fellowship level will probably take you five years. Some people do it in less, particularly if they have exemptions, but these are generally exceptions. Lots of people also take a break from exams: if you are feeling burnt out from three or four years of continually balancing work and studying, taking just one sitting off can really improve your motivation again.

You can work as an actuarial analyst until you qualify, so passing one sitting quicker won't dramatically improve your career path.

What is the pass mark?

Sit the exam and see! No, really, the pass marks for actuarial exams are set after the paper has been sat. In order to maintain standards, if the paper was slightly easier in a sitting, the pass mark is set a little higher than normal. This is determined by the distribution of marks: lots of people doing well is taken to mean it was an easier paper. However, you might be more interested in the pass rate. For Exam P, the average pass rate since 2008 is 42%. It reached a giddy height of 65% in March 2020, but has come back down and in November 2021 it was 44%. The average pass rate for CS2 is 39%.

With pass rates like these, you need to accept that you might fail some modules first time round. You also need to develop effective study techniques!

How do I pass?

Passing an actuarial exam starts the day after the previous sitting. On the first day, order your new study materials. As soon as they arrive, create a plan. You'll need to make a quick note of how many chapters, assignments, end of unit question sets and past exam papers there are. Then, check when the exam is. Now starting with the week before the exam, assign every task to a week. The week before the exam you might want to do 6 past exam papers. The week before that, another 6. Repeat until you have assigned all the tasks to a week on your calendar. Covering the course material and doing at least 10 past papers or exam-style papers is the bare minimum. Doing 15 exam-style papers feels like a good effort and 20 is setting you up for exam success.

Past exam papers are the best form of studying because they will get you used to how the material will appear in an exam rather than a textbook. It'll also help you see how the marks are allocated and what does and doesn't score. But they are particularly good for making you think. They are a really active form of studying which gets your brain thinking. Reading course notes (or reading exam solutions) isn't nearly so helpful as answering questions as you won't retain the information as completely.

Find out what study techniques works for you. I make detailed notes while reading the course material, which not only prevents me from skim-reading but also gives me a handy summary of the chapter. I also love flashcards. You could try making up your own exam questions, or record yourself reading the summary notes to create an audiobook. Paid tuition can also be very effective and your employer might be willing to pay for tutorials or study support from a specific company—if this is offered, do accept and make the most of the sessions.

How can I work and study effectively?

Coming up with a detailed plan as above is a very important step. But while creating this, check your work diary. When are your busy periods at work? When are your holidays? If so, mark these on your plan as well. Plan less study for busy times, and more when you think there will be some flex. However, even in busy times you will likely need to study in order to cover what you need. Think critically about your day: when do you work best? If you work best in the morning, try getting to your desk an hour earlier than you need and working from there first (put headphones on so you aren't disturbed). Or try doing a little bit at lunch, or reading through notes on your commute in. After work is the obvious choice and will give you a bigger chunk of study time, but allow yourself time to relax too.

Actuarial work is predominantly screen-based so finding ways to study that aren't just writing at a desk will help break it up a bit. Flashcards and audiobooks of notes are great for squeezing in some revision in a short time. As with all learning, don't count the hours, count the outcomes. Instead of saying 'I will study for an hour', commit to studying Chapter 7 (or a portion of it) and making thorough notes, or commit to completing a past paper question and then understanding the answer.

But work and study aren't necessarily in competition. Find ways to study through work, such as asking colleagues to explain concepts to you (both theory and practical experience will be useful eventually in the exams) or working in a different area for a short period. If you have colleagues sitting the same exam, you might find group study helpful (or you might not!). Getting someone else to mark your script after a mock exam or assignment can be so useful, both in terms of areas to study but also your exam technique. You could swap scripts with another analyst, or a senior analyst might be willing to do you a favor.

When I pass, then what?

Most employers will offer salary increases after every passed exam. The amount depends on the size of the exam and your own employer: the first exams usually only attract a small raise of $1,000 or so, whereas the later exams could bump you up $5,000, particularly if you are then qualified. It is also worth watching out for any office traditions for qualifiers. Bringing a case of champagne into the office, picking up the bar tab for the whole actuarial department's drinks and buying a specific cake from a famous London bakery are all traditions that I have been a beneficiary of.

18.4 Professional Development

In this book, I've argued that risks in society are changing and with it, actuarial work. Even in the past five years, there has been an increasing awareness and quantification of the risks related to climate change, decarbonization **transition risk**, and biodiversity risk. Actuarial exams are not meant to be a comprehensive list of all possible situations. Instead, they provide actuaries with the framework to approach and solve problems that they will face throughout their careers. Does that mean that actuaries flying blind when it comes to new risks?

Like many professions, actuaries are required to complete continuous professional development (abbreviated to CPD). This is usually an annual requirement of a specified number of hours of learning or training related to your work. Lots of activities can be included, but here are some ideas:

- Industry or institute conferences or webinars.
- Internal company training on actuarial skills. This could be training on a new tool, or a presentation on how an industry issue or regulation will affect actuaries in your line of work. Many actuarial departments run regular learning sessions throughout the year to help employees cover their CPD requirements and develop their skills.
- Being a member of a working party. These are groups of actuaries, organized by the institute, to research a specific issue. There are working parties for all kinds of topics such as intergenerational fairness, mortality, Third-Party Injury, and asbestos. As new issues emerge, working parties are key to understanding current practice, setting industry best practice, and suggesting guidelines or regulations for the industry.
- Professional skills training. This aims to increase awareness and experience of the actuaries' code of conduct. Training might include discussions, role-playing, and 'What would you do if…' type scenarios.
- Learning to code. While you may have covered the basics of R or Python during college or actuarial exams, if you're not using these skills regularly

then you can easily forget them. Learning a coding language may help you automate processes and make your work more efficient, as well as introduce you to new techniques. Try to find a course that includes both teaching and practice.[2]

If you enjoy formal learning and have a particular area of interest, there are further professional courses run by actuarial institutes. Each institute has its own offering, but here are some examples from the institutes in the UK and USA/Canada:

- Data Science: both the IFoA and SOA offer certificates in data science, designed to complement actuarial exams. The certificates also cover some machine learning and predictive analytics topics. The IFoA course is a 10-week program (approximately 80 hours of work in total), and SOA's predictive analytics course takes 5 months with a recommended time commitment of 180 hours .
- Enterprise Risk is another big area of post-qualification study. The Chartered Enterprise Risk Actuary (CERA) credential is popular with those working, or interested in working, in actuarial positions focused on finance and investment. Being a CERA is usually achieved by sitting a specific specialist skills exam set by your institute, but CERA is a globally recognized qualification.
- International Financial Reporting for Insurers certificate, by SOA. This covers the theory and practical application of new international reporting standards, supporting members through this regulatory transition.
- SOA's newest certificate is called 'Ethical & Responsible Use of Data & Predictive Models'. If you find the ethical dilemmas of using vast quantities of data interesting, this certificate teaches you to spot bias in data and systems and quantify the risk associated with biased analytics.
- The IFoA's newest course is on Climate Risk and Sustainability. In eight weeks, members are guided through the techniques and best practice of quantifying climate risk, including transition and biodiversity risks.

Actuarial institutes are there to support their members, not just through qualification, but throughout their careers. Or into new careers; learning and accreditations gained through a professional body are highly respected and can help actuaries venture into a new role or field of work. Many of the certificates above have been introduced in the past five years. This shows that there

[2] I highly recommend DataCamp as an effective way to learn Python, R, or SQL.

is a real emphasis on preparing actuaries to respond to current and future challenges.

18.5 The International Market for Actuarial Science

There are about 80,000 qualified actuaries practicing worldwide. In contrast, there are 2 million accountants.[3] But actuaries are far from an endangered species. Their numbers have been rising steadily for several years now (according to the International Association of Actuaries, there were 68,000 qualified actuaries in 2019) and are expected to continue rising. In 2005, about 75% of the world's actuaries were based in the UK, USA, or Canada. As more countries have recognized the importance and skills offered by actuarial science, more countries now offer actuarial science courses and are growing their own actuaries. China and India have seen particularly strong growth in the number of actuaries, and data from the UK-based Institute of Actuaries show that their Indian student members surpass those living in the UK.

Demand for actuarial work is growing strongly too. In India, it is estimated that the demand for experienced, qualified actuaries is about three times the supply, although over time as the influx of students start to qualify, this will level out. More generally, we have seen throughout this book that actuarial skills are relevant to many industries and problems, including some outside of the traditional stomping ground of insurance, pensions, and investments.

Although getting your first actuarial role can be difficult, qualified actuaries are rarely short of a job offer. In some areas, such as **Defined Benefit** pension schemes, demand is waning as these schemes become less popular. But equally, other fields spring up, such as a growing importance of climate risk—and with it, climate actuary roles. Actuaries in developing countries may be employed in government roles, helping global insurers reach the local market, or working to set up microinsurance schemes. The growing global middle class also feeds demand for actuaries; this group is more likely to be thinking about pension provision and health insurance, and will start to require property insurance for their items. Even within traditional areas, actuaries are taking on a wider variety of roles. No longer confined to the reserving or pricing desk, insurers may have a claims actuary (an actuary who quantifies risk due to claim surges and claims processes), or an actuary leading a data science team.

[3] According to the International Federation of Accountants.

Actuaries are highly qualified professions with an extensive transferable skill set, and there is high demand for their abilities. As such, actuarial work is well paid. Exact salaries will depend hugely based on the region, the company (or public employer), area of work, seniority of the role, experience of the individual, and much more. It has already been mentioned that passing exams often come with pay rises or bonuses to move actuarial analysts up the pay scale. However, exact figures collected by official agencies usually don't give this level of detail and only give an average for the profession.

The US Bureau of Labor Statistics estimates the 2020 median salary was $111,000, although those in general insurance companies earned slightly less on average ($100,000). It also estimates that about 5,000 additional actuarial roles will be needed by 2029 to meet demand. An inexact survey of job adverts suggests that unqualified actuaries should expect to be earning between $45,000–$75,000.

In the UK, a graduate actuarial analyst could expect a salary of £30,000-£40,000, depending on exam passes/exemptions. With a couple of years of experience and a bit of exam success, actuarial analysts or newly qualified actuaries can expect to increase this to £55,000. The average salary for a qualified actuarial manager was £75,000, and senior managers can earn £100,000+ .

In Australia, new analysts can expect to earn between $55–60,000 AUD, with the median actuarial salary across all levels being $129,000 in 2021. The top 25% of actuarial earners receive $171,000 AUD or more.

As in many professions, there is a difference in actuarial salaries by gender. In 2013, the IFoA found that about a quarter of qualified actuaries were female, although noted that their student membership was over 35% female. In 2015, it found that female actuaries leave the profession 13 years earlier on average than men, and so there is less representation at senior levels which affects career-wide averages split by gender. This pattern is reflected around the world. The average Australian male actuary is five years older than the average female actuary, and earns over $55,000 more. This largely reflects that female actuaries are younger, more junior and are more likely to work part-time than male colleagues; when allowing for part-time working, the gender pay gap is 7%.[4] This gap gets worse the more experienced the actuary is. Among those who have been qualified over 15 years, the gender pay gap is 29%. Viewed optimistically, this Australian data could indicate that the gender pay gap is narrowing over time.

[4] 2016–17 Tax Stats released | Australian Taxation Office (ato.gov.au). https://www.ato.gov.au/Media-centre/Media-releases/2016-17-Tax-Stats-released/

Actuarial roles tend to earn less than investment or financial jobs and, as we saw earlier, actuaries in this sphere earn more than general insurance actuaries. But the salary, although generous, is not the only draw of actuarial science as a career. Compared to the financial sector, hours in general insurance and pensions companies are generally more regular. There are some notable exceptions; during reserving quarters, reserving or reporting actuaries will likely need to put in many more additional hours to meet deadlines—working until 8 pm or even 10 pm during a reserving month is not uncommon. Similarly, actuaries who work with catastrophe risk may need to put in more hours at short notice to give updated views on catastrophe claim costs as they emerge. However, actuarial work tends to give a good work-life balance and, to address the disparity in career longevity between genders, institutes worldwide are encouraging companies to promote flexible or part-time contracts. Post-pandemic, actuarial roles are largely homeworking and recruiters are suggesting that this change is being welcomed by both employees and employers, and so actuarial roles will continue to be a blend of office and homeworking for the foreseeable future.

18.6 A Day in the Life of an Actuarial Analyst

On telling people I'm an actuary, two questions tend to follow. First is usually 'What is that? Is it a pensions thing?' (or a blank stare, which I assume is the same question). Hopefully, this book gives the answer.[5] The second, only slightly impertinent, question is 'What do you *do* all day?'.

As shown in this book, actuarial work can be hugely varied. From analyzing investment risk in New York to encouraging microinsurance in Sub-Saharan Africa, the day-to-day work of actuaries is not easy to generalize. Because of this, in this section I will avoid generalizing completely and instead, look at two days based on my own experience, representative of my time as a reserving analyst and a pricing actuarial analyst in a large general insurance company respectively.

18.6.1 A Day in the Life of a Pricing Analyst

8.30 am: Log on, check emails and set priorities for the day. I can see that one of the data engineers has provided an updated dataset to

[5] Although rather mischievously, I am also known to say 'I do some modelling…' and momentarily enjoy the confused look on their faces before explaining.

model on, and that an **underwriter** has asked a question about trade groups—I'll set up a meeting to answer their questions, but it looks like they are busy until next week.

9 am: Team meeting to check what we're working on. As I give my priorities for the day, a senior analyst says they have some code that would make my task easier so I'll follow up on that.

9.30 am: Check the updated dataset for errors or issues. It seems fine, so as soon as I have the code from the senior analyst, I can use this for analysis.

10.30 am: Meeting with the team that deploys the new price structures. They want to know what factors are in the models so they can start to code it. We also talk about what key performance indicators they'll need and more generally about the oddities in the dataset.

11.30 am: The senior analyst in the team goes through their code with me. I can see that this will speed up my task, but also that it needs some amendments. I need to add a section of code to create a graph that isn't included and point it toward my new data.

1.30 pm: After lunch, I start making these changes and updates to the code and annoyingly it doesn't work. I spend some time trying to work out the issue. Eventually, I ask the senior analyst who looks over it and finds a syntax error in my data import. I feel a bit annoyed at myself for not spotting it sooner, and I make a cup of tea while it runs.

3 pm: With the new dataset available and checked, I start on a Fire Frequency risk model. I notice an interesting trend: some very old and rural vehicles which appear to be incredibly flammable. Presumably these are tractors, but I'll do a proper investigation tomorrow.

5 pm: I think I've finished on the Fire Frequency model, so I document what's been done and send it across to the senior analyst for review. I finish up the day by writing a list for tomorrow. I also book that meeting in the underwriter's calendar next week—hopefully they won't cancel again.

18.6.2 A Day in the Life of a Reserving Analyst

8.30 am: It's the second week of a reserving month, so it's all systems go. I glance at my emails but there's nothing that can't wait until tomorrow. Or maybe next week. I get straight to it, and start looking at claims triangles. We usually use the chain ladder method for older years, so I'm trying to select development factors. I default to using the average of the most recent eight periods, excluding the highest and lowest, but I need to check that this method is suitable for each period. Where it isn't, I select something else and jot down my rationale—I know my reviewer will want to hear my thought process.

10 am: One particular year has got a very odd development pattern. Looking at the claim movement in the period, I suspect this might be related to some large claims (the movements are roughly multiples of £250,000 which seems suspiciously round to me). I was planning to look at the large claims tomorrow, but I bring this forward because I want to take it into consideration for my development factor choices.

11 am: I can see three large claims have had a big movement this quarter. From my pre-reserving meeting with the portfolio managers, I was expecting two movements. The third one is seemingly newly reported £2 million claim. I find the claim number, and give the portfolio manager a call to check.

11.30 am: Phew, false alarm: the claim was erroneously attributed into this portfolio, but actually it relates to another portfolio. I also pop over to my colleague's desk to let them know they are missing a £2 m claim, taking a coffee for them to soften the blow. They've almost finished their selections for that portfolio, so this will mean redoing a lot of the work. I briefly commiserate with them before getting back to my own reserving.

12 pm: I don't have time to rerun all my data again, so exclude this year from my development factor calculation. Once I've finished and checked my selections, I let my reviewer know.

1 pm: I start on my projections for the most recent year. It is too undeveloped to use a chain ladder method, so we have a spreadsheet to do an Expected Ultimate Loss method, with a few adjustments. I start updating the file links and updating the information, adding in detailed notes on the large losses (including the £2 m that isn't mine).

2.30 pm: Meeting with my reviewer for the other portfolio I did the reserving projections for. She only had a couple of changes to make to the development factors, but now I have to make sure all the files are updated and the new number is coming through. I check this twice: these spreadsheets are a key output and will be shown to the Chief Actuary, so I want to make sure my numbers are correct.

4 pm: I start on the development factor selections for another portfolio. This is one is smaller so the claim movements are more volatile. However, it's a smaller portfolio and hasn't had any large claims that I need to worry about, so I can blast through fairly quickly, check it, then let my reviewer know.

5.30 pm: The portfolio with large claims will hit the reinsurance threshold so I need to reduce the reserves according. The reinsurance team have just sent the data with the exact amounts of reinsurance recoveries, so I check that and incorporate that into my spreadsheet.

6.30 pm: The office is still full of analysts trying to complete their work. Tomorrow lots of us have review meetings, so there's a level of stress but also a sense of camaraderie. I eat another bar of chocolate, put some upbeat music on, and power on for another hour. The busy reserving period finishes in a few days, so I'm looking forward to an easier week next week.

18.7 Final Remarks on Actuarial Careers

Actuarial skills are highly in demand worldwide. With such diversity in actuarial roles, there is enough scope to imagine and reimagine your career many times over to suit your interests, passions, and curiosity. But there are also many interesting technical non-actuarial roles related to insurance, pensions, investments, banking, population modeling, or epidemiology.

If you do decide to become an actuary, I wish you a varied and interesting career. Do let me know how you get on. May your data be clean, your estimates accurate, your commutation functions short and your assumptions stated. I wish you the best of luck.

19

Looking Forward: The Future of Actuarial Work

Life can only be understood backwards; but it must be lived forwards.
— Søren Kierkegaard

This book aimed to give you an introduction into actuarial techniques and industries. Whether you read it cover to cover or cherry-picked chapters, I hope you found something that piqued your interest.

I hope that readers come away with an understanding of the context from which actuarial work has emerged, the broad principles that underpin this, and the application and relevance of this in today's market and wider society. Whether in respect of the growing volume and utilization of data, the somewhat morbid role of life expectancy calculations in pensions and **annuity** products, the theory and application of modeling in insurance pricing, or the criticality of effective **reserving** and capital modeling in ensuring the ongoing resilience of financial systems; actuaries combine a robust set of technical skills with an ability to communicate effectively on topics relating to risk and uncertainty, and in doing so support an array of services essential to our society.

While earlier chapters focused predominantly on traditional actuarial skills and roles, later chapters have explored the application of these skills and aimed to highlight the current and future areas where actuaries can add value. As we face large scale challenges such as catastrophes, climate change, and **latent claims** which have significant and widespread impact; actuaries can help to quantify these risks. Armed with that information, they can help

© The Author(s), under exclusive license to Springer Nature Switzerland AG 2022
C. Townsend, *A Risky Business*,
https://doi.org/10.1007/978-3-031-11673-5_19

governments and society to manage and mitigate them. Actuaries can create risk transfer systems that increase social justice and incentivize considerate and collaborative behavior.

I started writing this book during lockdown in 2021. Chapter 17 is a recognition of all the work that goes into epidemic modeling, both during a pandemic and in the aftermath. It is both an acknowledgment of the flaws of statistical modeling and a celebration of the existence of those who repeatedly put their models under public scrutiny in order to advise governments and the public of the risks of inaction.

If any of the above intrigued you, I hope you are inspired to find out more. Finding your local actuarial institute online is a good start, as they create resources on hot topics and current events that are accessible and interesting. Perhaps you have simply enjoyed looking at the world through a lens of risk quantification and mitigation, or, as in Chapter 18, you might want to find out more about becoming an actuary yourself. Society will always need technically capable people who can understand and communicate risk and opportunity, and who work in the public interest above all else.

19.1 It's The End of The World as We Know it—And We'll Be Fine

Actuarial science has a long, rich, and global history. From the canny ship-owners from Lombardy, sixteenth-century Dutch life insurers, the first employer-sponsored pension scheme from American Express, to Otto von Bismarck's universal pension idea, people throughout history have craved a way to protect themselves from life's uncertainties. Insurance has played a part in many of history's dramas: the 'Glorious Revolution', the Great Fire of London, the San Francisco earthquake, the sinking of the Titanic, and more recently, the largest oil spill in history, Hurricane Katrina, 9/11 and the Covid-19 pandemic. Actuarial science has not only coped with the change in society but enabled it, by insuring businesses, oil rigs, and motor car drivers.

Against this backdrop, actuarial science is ideally placed to help society through the turbulent times of both the present and the future. When it comes to appreciation in the media, actuaries have generally been overlooked in favor of lawyers, medics, and politicians.[1] However, like a technical crew at a theatre production, their unobtrusive yet reassuring presence is the sign of a good job done. Most people don't worry daily about the availability of

[1] I base this assertion largely on the fact that there has never been a TV drama about actuaries.

affordable insurance or the investment risk of their pension. Actuaries silently take this mental load, add in thousands of other worries, and almost magically, return something that can be planned for. That most people don't even notice their work speaks to their success.

And yet, the risks facing society today are larger and more complex than ever seen previously. Firstly, there are the global risks: everyone faces risk because of anthropogenic climate change. That might be a direct physical risk, in the case of Pacific Islander populations or those in hurricane-prone areas, or it might be the risk of the change to a low-carbon economy. There is a risk that our investments are suddenly worthless, whether that is share and shares in some companies, or physical assets like oil rigs or petrol cars. Then, there is the risk of an increased level of zoonotic diseases, the consequences of living longer, the risk of ill-health or even premature death from pollutants, including plastic. Can we share risk retrospectively, and if so, how can Indigenous populations be remunerated for the risks to their health and livelihoods that have been imposed upon them? How can these risks be fairly transferred back to the creators of the risk (the major polluters), and would doing so give the best outcome for the public?

Secondly, there are the risks that our society has created. Over the past fifty years, most countries have shifted more toward individuals bearing their own risk, rather than risk being aggregated. That has led to poorer outcomes for the public, particularly the most vulnerable. How can that be addressed, acknowledging that businesses and governments no longer want to accept the aggregate risk either? Collective Defined Contribution pension schemes may be one answer to this, but the same questions are relevant to healthcare too. Who pays? And who profits? Our society also needs to tackle intergenerational inequality. The assumption that the next generation will be better off—economically or in terms of social mobility, healthcare and longevity—appears to have broken down in the past twenty years. How can this effect be quantified, and how can risk be redistributed so that young people can succeed as past generations did? As the demographics change and our society ages, how can social care and state pensions be fairly paid for? Can the risks of doing so unfairly be quantified? Can private insurance be the answer—or is that a risk in itself to state provision?

Finally, there are so risks arising from technological advance, as all generations have seen. For our great-grandparents it was the risk of changing from horses to motor cars. For us, it is the risk and opportunities of increased data: automated systems leave us vulnerable to data loss, cyber-attack (which can also cause physical harm), and systematic bias from machine systems. How can the idea of pooled risk work in a world with so much possible stratification of risk? How can we quantify the risks of bias in automated processes? How can we mitigate cyber risk when it is encouraged by states as a weapon?

This book has discussed how the challenges of society are reflected in actuarial work. I hope I have conveyed how actuaries can make a positive impact in a variety of ways. Actuarial work is surprisingly varied: although most actuaries start in traditional insurance and pensions industries, there are lots of other options to use actuarial skills to make a difference. That could be microinsurance in sub-Saharan Africa, expanding health insurance coverage in developing countries, or it could be advising the government about issues around social care, demographic change, or statistical models of epidemiology. Or there is potential to work in a non-traditional role in a traditional industry: increasingly insurers are appointing actuaries in data roles to quantify bias, or under the newly created title of 'climate actuary' to mitigate **transition risk**. I argue that these new roles are new applications of the same task: quantifying risk, then communicating the results.

While passing the initial exams feels all-consuming, there are many opportunities to learn—which is just as well, as there are so many aspects of actuarial science to learn about! The rise of Collective Defined Contribution schemes, biodiversity metrics, new financial instruments to spread risk: there is vast innovation in the profession to meet the challenges of our modern, interconnected world. It is anything but a static role. More data and more analytics capabilities to build more complicated models may make the task of quantifying risk easier, but it makes the communication of results more difficult. Perhaps we can look forward to more insuring more robots, or using them as part of claims assessments to give better **FNOL** estimates. Automating data preparation, or even the modeling itself, will be an essential part of the future, but we will still need actuaries to cast a careful eye over the output of automation. Being an actuary requires integrity and a good deal of common sense, two attributes that computers lack. The role evolves to use computers more, but the skillset of an actuary can't be replicated by a machine.

Throughout this book I argue that actuarial work is crucial to a modern, civilized society. That is even more true when there is so much uncertainty in the world that we have little control over. We can't control the events—hurricanes, floods, cyber-attacks, pandemics—but we can control the financial loss arising from them. Sharing risk allows us to take more risks. And we take so many risks daily: owning a house, driving a car, running a business, looking after our health, planning for the future. We crave security in a world of uncertainty and we can't just ignore the dangers. Actuaries don't remove risk, but by sharing and **pooling** risk, they bring more certainty to the world. Life is a risk worth sharing.

And so, actuarial science prepares for the future in the same way it has always done. Over a cup of coffee, two parties make a promise: if things go wrong, I'll help you rebuild, recover and plan for your future. And what could be more civilized than that?

Appendices

Appendix A: Table of Compound Interest, for 3% and 5% Effective Interest Per Annum

	3%		5%			
n	$(1+i)^n$	a_n	$(1+i)^n$	a_n		$i^{(12)}$
1	1.03000	0.9709	1.05000	0.9524	3%	0.029595
2	1.06090	1.9135	1.10250	1.8594	5%	0.048889
3	1.09273	2.8286	1.15763	2.7232		
4	1.12551	3.7171	1.21551	3.5460		
5	1.15927	4.5797	1.27628	4.3295		
6	1.19405	5.4172	1.34010	5.0757		
7	1.22987	6.2303	1.40710	5.7864		
8	1.26677	7.0197	1.47746	6.4632		
9	1.30477	7.7861	1.55133	7.1078		
10	1.34392	8.5302	1.62889	7.7217		
11	1.38423	9.2526	1.71034	8.3064		
12	1.42576	9.9540	1.79586	8.8633		
13	1.46853	10.6350	1.88565	9.3936		
14	1.51259	11.2961	1.97993	9.8986		
15	1.55797	11.9379	2.07893	10.3797		
16	1.60471	12.5611	2.18287	10.8378		
17	1.65285	13.1661	2.29202	11.2741		
18	1.70243	13.7535	2.40662	11.6896		
19	1.75351	14.3238	2.52695	12.0853		
20	1.80611	14.8775	2.65330	12.4622		

(continued)

© The Editor(s) (if applicable) and The Author(s), under exclusive license to Springer Nature Switzerland AG 2022
C. Townsend, *A Risky Business*,
https://doi.org/10.1007/978-3-031-11673-5

(continued)

	3%		5%		
n	$(1+i)^n$	a_n	$(1+i)^n$	a_n	$i^{(12)}$
21	1.86029	15.4150	2.78596	12.8212	
22	1.91610	15.9369	2.92526	13.1630	
23	1.97359	16.4436	3.07152	13.4886	
24	2.03279	16.9355	3.22510	13.7986	
25	2.09378	17.4131	3.38635	14.0939	

Appendix B: Life Tables for UK Male and Female Lives

National life tables: UK—Office for National Statistics (ons.gov.uk) (https://www.ons.gov.uk/peoplepopulationandcommunity/birthsdeathsandmarriages/lifeexpectancies/datasets/nationallifetablesunitedkingdomreferencetables).

Age	Males				Females			
x	q_x	p_x	T_x	K_x	q_x	p_x	T_x	K_x
0	0.004267	0.995733	79.37	79.00	0.003536	0.996464	83.06	83.00
1	0.000243	0.999757	78.71	78.00	0.000213	0.999787	82.35	82.00
2	0.000132	0.999868	77.72	77.00	0.000127	0.999873	81.37	81.00
3	0.000101	0.999899	76.73	76.00	0.000098	0.999902	80.38	80.00
4	0.000097	0.999903	75.74	75.00	0.000068	0.999932	79.39	79.00
5	0.000085	0.999915	74.75	74.00	0.000086	0.999914	78.39	78.00
6	0.000088	0.999912	73.76	73.00	0.000082	0.999918	77.40	77.00
7	0.000069	0.999931	72.76	72.00	0.000062	0.999938	76.41	76.00
8	0.000067	0.999933	71.77	71.00	0.000064	0.999936	75.41	75.00
9	0.000059	0.999941	70.77	70.00	0.000053	0.999947	74.42	74.00
10	0.000074	0.999926	69.78	69.00	0.000064	0.999936	73.42	73.00
11	0.000085	0.999915	68.78	68.00	0.000066	0.999934	72.42	72.00
12	0.000105	0.999895	67.79	67.00	0.000060	0.999940	71.43	71.00
13	0.000127	0.999873	66.79	66.00	0.000080	0.999920	70.43	70.00
14	0.000121	0.999879	65.80	65.00	0.000101	0.999899	69.44	69.00
15	0.000174	0.999826	64.81	64.00	0.000112	0.999888	68.45	68.00
16	0.000227	0.999773	63.82	63.00	0.000147	0.999853	67.45	67.00
17	0.000316	0.999684	62.84	62.00	0.000159	0.999841	66.46	66.00
18	0.000400	0.999600	61.86	61.00	0.000227	0.999773	65.47	65.00
19	0.000448	0.999552	60.88	60.00	0.000200	0.999800	64.49	64.00
20	0.000508	0.999492	59.91	59.00	0.000190	0.999810	63.50	63.00
21	0.000513	0.999487	58.94	58.00	0.000211	0.999789	62.51	62.00
22	0.000504	0.999496	57.97	57.00	0.000229	0.999771	61.53	61.00

(continued)

(continued)

Age	Males				Females			
x	q_x	p_x	T_x	K_x	q_x	p_x	T_x	K_x
23	0.000505	0.999495	57.00	57.00	0.000222	0.999778	60.54	60.00
24	0.000552	0.999448	56.03	56.00	0.000222	0.999778	59.55	59.00
25	0.000598	0.999402	55.06	55.00	0.000254	0.999746	58.57	58.00
26	0.000581	0.999419	54.09	54.00	0.000262	0.999738	57.58	57.00
27	0.000621	0.999379	53.12	53.00	0.000292	0.999708	56.60	56.00
28	0.000695	0.999305	52.15	52.00	0.000314	0.999686	55.61	55.00
29	0.000732	0.999268	51.19	51.00	0.000320	0.999680	54.63	54.00
30	0.000771	0.999229	50.23	50.00	0.000370	0.999630	53.65	53.00
31	0.000835	0.999165	49.26	49.00	0.000394	0.999606	52.67	52.00
32	0.000835	0.999165	48.30	48.00	0.000453	0.999547	51.69	51.00
33	0.000929	0.999071	47.34	47.00	0.000486	0.999514	50.71	50.00
34	0.000957	0.999043	46.39	46.00	0.000556	0.999444	49.74	49.00
35	0.001073	0.998927	45.43	45.00	0.000582	0.999418	48.76	48.00
36	0.001146	0.998854	44.48	44.00	0.000645	0.999355	47.79	47.00
37	0.001309	0.998691	43.53	43.00	0.000784	0.999216	46.82	46.00
38	0.001261	0.998739	42.59	42.00	0.000732	0.999268	45.86	45.00
39	0.001399	0.998601	41.64	41.00	0.000852	0.999148	44.89	44.00
40	0.001529	0.998471	40.70	40.00	0.000902	0.999098	43.93	43.00
41	0.001670	0.998330	39.76	39.00	0.000986	0.999014	42.97	42.00
42	0.001818	0.998182	38.83	38.00	0.001073	0.998927	42.01	42.00
43	0.002002	0.997998	37.90	37.00	0.001171	0.998829	41.05	41.00
44	0.002093	0.997907	36.97	36.00	0.001316	0.998684	40.10	40.00
45	0.002347	0.997653	36.05	36.00	0.001425	0.998575	39.15	39.00
46	0.002488	0.997512	35.13	35.00	0.001564	0.998436	38.21	38.00
47	0.002696	0.997304	34.22	34.00	0.001695	0.998305	37.27	37.00
48	0.002852	0.997148	33.31	33.00	0.001840	0.998160	36.33	36.00
49	0.003189	0.996811	32.40	32.00	0.001943	0.998057	35.40	35.00
50	0.003379	0.996621	31.50	31.00	0.002169	0.997831	34.47	34.00
51	0.003606	0.996394	30.61	30.00	0.002358	0.997642	33.54	33.00
52	0.003907	0.996093	29.72	29.00	0.002557	0.997443	32.62	32.00
53	0.004125	0.995875	28.83	28.00	0.002697	0.997303	31.70	31.00
54	0.004478	0.995522	27.95	27.00	0.002914	0.997086	30.78	30.00
55	0.004760	0.995240	27.07	27.00	0.003194	0.996806	29.87	29.00
56	0.005389	0.994611	26.20	26.00	0.003542	0.996458	28.97	28.00
57	0.005856	0.994144	25.34	25.00	0.003816	0.996184	28.07	28.00
58	0.006394	0.993606	24.49	24.00	0.004230	0.995770	27.17	27.00
59	0.006930	0.993070	23.64	23.00	0.004616	0.995384	26.29	26.00
60	0.007595	0.992405	22.80	22.00	0.005054	0.994946	25.41	25.00
61	0.008313	0.991687	21.97	21.00	0.005492	0.994508	24.53	24.00
62	0.009223	0.990777	21.15	21.00	0.006264	0.993736	23.67	23.00
63	0.010178	0.989822	20.34	20.00	0.006714	0.993286	22.81	22.00
64	0.010947	0.989053	19.55	19.00	0.007264	0.992736	21.96	21.00

(continued)

(continued)

Age	Males				Females			
x	q_x	p_x	T_x	K_x	q_x	p_x	T_x	K_x
65	0.012025	0.987975	18.76	18.00	0.007988	0.992012	21.12	21.00
66	0.013330	0.986670	17.98	17.00	0.008573	0.991427	20.29	20.00
67	0.014437	0.985563	17.22	17.00	0.009380	0.990620	19.46	19.00
68	0.015740	0.984260	16.46	16.00	0.010330	0.989670	18.64	18.00
69	0.017289	0.982711	15.72	15.00	0.011296	0.988704	17.83	17.00
70	0.018286	0.981714	14.99	14.00	0.012437	0.987563	17.02	17.00
71	0.020276	0.979724	14.26	14.00	0.013337	0.986663	16.23	16.00
72	0.022326	0.977674	13.54	13.00	0.015241	0.984759	15.44	15.00
73	0.025500	0.974500	12.84	12.00	0.017347	0.982653	14.68	14.00
74	0.028123	0.971877	12.16	12.00	0.019167	0.980833	13.93	13.00
75	0.031402	0.968598	11.50	11.00	0.021437	0.978563	13.19	13.00
76	0.035115	0.964885	10.85	10.00	0.024217	0.975783	12.47	12.00
77	0.038838	0.961162	10.23	10.00	0.027308	0.972692	11.76	11.00
78	0.043521	0.956479	9.63	9.00	0.030961	0.969039	11.08	11.00
79	0.048099	0.951901	9.04	9.00	0.034481	0.965519	10.42	10.00
80	0.053982	0.946018	8.47	8.00	0.038463	0.961537	9.77	9.00
81	0.060071	0.939929	7.93	7.00	0.043634	0.956366	9.14	9.00
82	0.066512	0.933488	7.40	7.00	0.048958	0.951042	8.54	8.00
83	0.075396	0.924604	6.89	6.00	0.056269	0.943731	7.95	7.00
84	0.084755	0.915245	6.41	6.00	0.063935	0.936065	7.39	7.00
85	0.094664	0.905336	5.96	5.00	0.072462	0.927538	6.87	6.00
86	0.106853	0.893147	5.53	5.00	0.083094	0.916906	6.36	6.00
87	0.118585	0.881415	5.14	5.00	0.093462	0.906538	5.89	5.00
88	0.133359	0.866641	4.76	4.00	0.106434	0.893566	5.45	5.00
89	0.149851	0.850149	4.41	4.00	0.118942	0.881058	5.04	5.00
90	0.159530	0.840470	4.10	4.00	0.134387	0.865613	4.65	4.00
91	0.179055	0.820945	3.79	3.00	0.150755	0.849245	4.30	4.00
92	0.196950	0.803050	3.51	3.00	0.167079	0.832921	3.97	3.00
93	0.215044	0.784956	3.24	3.00	0.184344	0.815656	3.67	3.00
94	0.238086	0.761914	2.99	2.00	0.204467	0.795533	3.38	3.00
95	0.261012	0.738988	2.77	2.00	0.228210	0.771790	3.12	3.00
96	0.286714	0.713286	2.58	2.00	0.250765	0.749235	2.90	2.00
97	0.304113	0.695887	2.41	2.00	0.267058	0.732942	2.70	2.00
98	0.325892	0.674108	2.25	2.00	0.291260	0.708740	2.51	2.00
99	0.369540	0.630460	2.09	2.00	0.309526	0.690474	2.33	2.00
100	0.384386	0.615614	2.02	2.00	0.343363	0.656637	2.15	2.00

Appendix C: Net Present Value of Life-Dependent Annuities for 3% and 5% Interest Rates

Actuarial Tables I Internal Revenue Service (irs.gov) (https://www.irs.gov/retirement-plans/actuarial-tables).

Interest at 3.0%				Interest at 5.0%			
Age	**Annuity**	**Age**	**Annuity**	**Age**	**Annuity**	**Age**	**Annuity**
0	29.2854	**55**	16.8667	**0**	19.1799	**55**	13.3935
1	29.3716	**56**	16.4852	**1**	19.2763	**56**	13.1536
2	29.2677	**57**	16.1007	**2**	19.2501	**57**	12.9089
3	29.1556	**58**	15.7139	**3**	19.2192	**58**	12.66
4	29.0377	**59**	15.3243	**4**	19.1851	**59**	12.4064
5	28.9151	**60**	14.9317	**5**	19.1485	**60**	12.1477
6	28.7879	**61**	14.5365	**6**	19.1095	**61**	11.8844
7	28.6568	**62**	14.1398	**7**	19.0685	**62**	11.617
8	28.5212	**63**	13.742	**8**	19.0251	**63**	11.3459
9	28.3809	**64**	13.3432	**9**	18.9791	**64**	11.0711
10	28.2361	**65**	12.9434	**10**	18.9306	**65**	10.7925
11	28.0866	**66**	12.5391	**11**	18.8795	**66**	10.5073
12	27.9327	**67**	12.1317	**12**	18.8258	**67**	10.2162
13	27.7752	**68**	11.7225	**13**	18.7702	**68**	9.9204
14	27.6153	**69**	11.3126	**14**	18.7133	**69**	9.6208
15	27.4536	**70**	10.9031	**15**	18.6558	**70**	9.318
16	27.2907	**71**	10.4937	**16**	18.5978	**71**	9.0119
17	27.1259	**72**	10.0854	**17**	18.5391	**72**	8.7032
18	26.9589	**73**	9.6795	**18**	18.4793	**73**	8.393
19	26.7889	**74**	9.2777	**19**	18.418	**74**	8.0829
20	26.6148	**75**	8.8816	**20**	18.3543	**75**	7.7742
21	26.437	**76**	8.492	**21**	18.2886	**76**	7.4677
22	26.2551	**77**	8.1094	**22**	18.2205	**77**	7.1639
23	26.0684	**78**	7.7341	**23**	18.1495	**78**	6.8631
24	25.8759	**79**	7.3666	**24**	18.0749	**79**	6.566
25	25.6769	**80**	7.008	**25**	17.9962	**80**	6.2736
26	25.4713	**81**	6.6584	**26**	17.9131	**81**	5.9861
27	25.2588	**82**	6.3184	**27**	17.8253	**82**	5.7042
28	25.0395	**83**	5.9885	**28**	17.733	**83**	5.4285
29	24.8139	**84**	5.6691	**29**	17.6363	**84**	5.1594
30	24.5825	**85**	5.3605	**30**	17.5354	**85**	4.8976
31	24.3448	**86**	5.0631	**31**	17.4302	**86**	4.6433
32	24.1011	**87**	4.7772	**32**	17.3205	**87**	4.3973
33	23.8511	**88**	4.5027	**33**	17.2061	**88**	4.1594
34	23.5956	**89**	4.24	**34**	17.0875	**89**	3.9302
35	23.3341	**90**	3.9892	**35**	16.9643	**90**	3.7101
36	23.0666	**91**	3.7502	**36**	16.8364	**91**	3.4989

(continued)

(continued)

Interest at 3.0%				Interest at 5.0%			
Age	Annuity	Age	Annuity	Age	Annuity	Age	Annuity
37	22.7931	**92**	3.5228	**37**	16.7038	**92**	3.2969
38	22.5135	**93**	3.3075	**38**	16.5661	**93**	3.1046
39	22.2281	**94**	3.1034	**39**	16.4236	**94**	2.9213
40	21.937	**95**	2.9104	**40**	16.2762	**95**	2.7471
41	21.6401	**96**	2.729	**41**	16.1238	**96**	2.5825
42	21.3377	**97**	2.5583	**42**	15.9665	**97**	2.427
43	21.0289	**98**	2.3977	**43**	15.8035	**98**	2.28
44	20.7145	**99**	2.2465	**44**	15.6353	**99**	2.1411
45	20.3943	**100**	2.106	**45**	15.4617	**100**	2.0116
46	20.0681	**101**	1.9729	**46**	15.2823	**101**	1.8884
47	19.7364	**102**	1.8499	**47**	15.0975	**102**	1.7743
48	19.3989	**103**	1.7281	**48**	14.9069	**103**	1.6609
49	19.0555	**104**	1.6161	**49**	14.7103	**104**	1.5566
50	18.7057	**105**	1.5056	**50**	14.5071	**105**	1.4537
51	18.3496	**106**	1.3681	**51**	14.2974	**106**	1.3247
52	17.9872	**107**	1.2125	**52**	14.081	**107**	1.1784
53	17.6188	**108**	0.9638	**53**	13.858	**108**	0.941
54	17.2451	**109**	0.4854	**54**	13.6286	**109**	0.4762

Glossary

Accumulated Value: the future value of a payment today, which accounts for the interest that would be earned on the amount.

Accumulation: an accumulation of risk is present if multiple policies have the potential to be affected by a single event. This is usually based on geographic factors, but can be by product or other risk factors.

Annual Venture: the activity of a Lloyd's syndicate relating to one calendar year. All policies written as part of an annual venture are considered a cohort.

Annuity: a stream of cashflows paid at regular intervals (usually yearly or monthly).

Annuity factor: denoted a_n is the net present value of an annuity of \$1 paid for n years, at a given interest rate. The annuity factor can also be given for a life-dependent annuity, denoted a_x, where x is the current age of the annuity recipient.

Anti-selection: when an insurer or pensions provider attracts customers that are riskier than average. This is usually due to lax underwriting restrictions or mis-pricing.

Best Estimate: an estimate that is neither prudent or conservative, i.e., it is the best guess at what the estimate will truly be.

Burning Cost: the cost of claims divided by the exposure during a certain period.

Cashflow: a monetary transaction. A positive cashflow is money received, a negative cashflow is money outgoing.

Claim ratio or Loss Ratio (LR): the cost of claims divided by gross earned premiums. See also Ultimate Loss Ratio.

Claims-made basis: where cover is provided for any claims that are reported during the policy period.

© The Editor(s) (if applicable) and The Author(s), under exclusive license to Springer Nature Switzerland AG 2022
C. Townsend, *A Risky Business*,
https://doi.org/10.1007/978-3-031-11673-5

Combined Operating Ratio (COR): the sum of the loss ratio and expense ratio. A COR lower than 100% suggests an insurer is profitable, higher than 100% and it is likely unprofitable. COR does not take into account investment returns.

Defined Benefit (DB): a type of pension where the benefits are specified upon joining the scheme. Benefits are usually a function of salary and years of service.

Defined Contribution (DC): a type of pension when the benefits are directed related to the contributions made. As such, the benefits are unknown until retirement.

Earned Premium: the premium for risk covered in that period. It can be gross or net of reinsurance.

ESG: Environmental, Social, and Governance (Sustainability).

Excess: the amount of a claim that the policyholder must pay before the insurer pays anything toward the claim. For example, with an excess of $500, a $750 claim would cost the policyholder $500 and the insurer $250.

Excess of loss (XL): a type of reinsurance where the reinsurer covers any losses above a set threshold, usually up to an upper limit.

Expected Maximum Loss (EML): the biggest loss that is reasonably expected to occur from one event. It is smaller than the Possible Maximum Loss, which considers a wider range of events.

Expense ratio: the cost of expenses divided by gross written premiums. Used to calculate the Combined Operating Ratio.

First Notification Of Loss (FNOL): the first point at which the insurer is made aware of the claim. The FNOL team will collect information about the loss event and give an initial claim estimate.

(Claim) Frequency: how often a claim type occurs. It is calculated as the number of claims divided by the exposure measure, over a given time period (this is usually one year or the period of the policy).

General insurance: insurance that isn't life, health, or pensions. It is also known as Property and Casualty.

Generalized Linear Model (GLM): a type of statistical model, popular because it is a more flexible version of linear models.

Gross Written Premium (GWP): see Written premium.

Indemnity: a core principle of insurance where after a loss, a policyholder is restored to the same position as they were before the loss.

Insurable interest: if a person would incur financial loss due to a loss event, they have an insurable interest in the risk.

Insurtech: technology that makes insurance more efficient.

Latent claim: a claim due to a risk that the insurer didn't know about at the time of underwriting the policy. Latent claims can also be defined as claims that are reported decades after the loss event.

Law of Large Numbers: a statistical law that states that the larger the sample size, the closer the sample mean will get to the population average.

Lloyd's Chain of Security: the sequence of funds that Lloyd's of London has in order to ensure adequate capital is available in the case of large losses. It comprises of syndicate levels fund, Member's Funds at Lloyd's, then the Lloyd's Central Fund, and they will be exhausted in that order.

Lloyd's Realistic Disaster Scenarios: a set of events (some historic, some hypothetical) that must be tested in catastrophe modeling for all Lloyd's of London syndicates. The set aims to be representative of the risks that syndicates are exposed to, and is reviewed annually for omissions.

Long-tailed: insurance where the average time between the loss event and claim settlement is long. This is subjective, but more than 7 years is usually considered long-tailed.

Loss event: an event where an insured loss occurs.

Losses-occurring basis: where cover is provided for any and all valid loss events that occur in the policy period.

Members' Funds at Lloyd's: a fund held by Lloyd's of London that is attributable to a single member. This is used in the case that syndicate funds are insufficient. Members must top up their Funds at Lloyd's during a process called 'coming into line', where their fund is assessed against the amount of risk that they are undertaking.

Moral hazard: when a policyholder acts in a different way because they are covered by insurance. This usually manifests in an insured policyholder taking more risks or acting less carefully than they would otherwise.

Net Present Value (NPV): the value today of a future payment, or stream of future payments, allowing for the compounding effect of interest that would accrue on a payment made today.

Original loss: the loss to the policyholder if they didn't have insurance.

Peril: an event that can cause a loss. Examples include fire, storm, public liability, etc.

Periodic Payment Order (PPO): a court order mandating that a claim must be paid not as a lump sum, but as a regular stream of payments.

Perpetuity: an annuity paid indefinitely.

Pooling: holding a group of similar risks in order to reduce volatility and capital requirements for claims costs. This works on a similar basis to the Law of Large Numbers.

Quota share: a type of reinsurance where risks are shared in a given proportion. This proportion is the same for all risks covered by the reinsurance.

Rating factor: a factor that is used in pricing or underwriting. Ideally, these would be the same as the risk factors but in practice this might not be possible.

Reinsurance: insurance for insurers. An insurer buys reinsurance to transfer or share risk.

Reinsurance To Close (RITC): A system that allows Lloyd's syndicates to share profits with members after three years, even if some claims are not settled. The subsequent year's Annual Venture assumes responsibility for the open claims and any claims yet to be reported. For this, they are paid a premium.

Reserving: the process of projecting the ultimate claim cost for a cohort of policies, and using this figure to set the level of the reserves, the amount of money set aside to pay claims.

Reversionary annuity: a type of life-dependent annuity. On the death of the policyholder, the annuity reverts to being based on the life of another nominated individual. This is usually the policyholder's spouse. The annuity paid to the spouse is usually lower than the annuity paid to the policyholder. Also known colloquially as a widow's pension.

Risk factor: a factor that affects the level of risk.

(Claim) Severity: the cost of a claim or the average cost of a type of claim. It is calculated as the total cost of the claim divided by the number of claims that cost relates to.

Short-tailed: insurance where the average time between the loss event and claim settlement is short. This is subjective, but less than 3–5 years is usually considered short-tailed.

Solvency Ratio: the ratio of the free assets (i.e., the assets less the amount that is set aside to pay claims) to written premiums, net of reinsurance. The higher the solvency ratio, the more resilient a company will be to catastrophes.

Syndicate: a group of Lloyd's members who insure risks as a collective.

Transition risk: the risk created while transitioning from one system to another.

Ultimate Loss Ratio (ULR): the cost of ultimate claims divided by gross earned premiums. This is the output of reserving exercises.

Underinsured: when a policyholder has insurance, but only for cover or a sum insured less than what they would ideally transfer. They may be knowingly underinsured, for example because they can't afford the higher premium that more coverage would cost, or it can be unknowingly underinsured, for example because they have underestimated their requirements or mis-understood the terms of the insurance.

Underwriter: someone who assesses a risk and decides if it should be written by an insurer. They also set the terms and conditions of cover, any exclusions and, for large heterogenous risks, set the price.

Written premium: the premium relating to policies written in that period. It can be gross or net of reinsurance. Gross Written Premiums are frequently used and abbreviated to GWP.

Index

© The Editor(s) (if applicable) and The Author(s), under exclusive license to Springer Nature Switzerland AG 2022
C. Townsend, *A Risky Business*,
https://doi.org/10.1007/978-3-031-11673-5

The manufacturer's authorised representative in the EU is Springer Nature Customer Service Centre GmbH, Europaplatz 3, 69115 Heidelberg, Germany. If you have any concerns regarding our products, please contact ProductSafety@springernature.com

Printed and bound by CPI Group (UK) Ltd, Croydon, CR0 4YY
30/09/2026
02225216-0020